IFAC

International Federation of Automatic Control

MODEL BASED PROCESS CONTROL

Other IFAC Publications

AUTOMATICA

the journal of IFAC, the International Federation of Automatic Control

Editor-in-Chief: G. S. Axelby, 211 Coronet Drive, North Linthicum, Maryland 21090, USA

Published bi-monthly

IFAC PROCEEDINGS SERIES

General Editor: Janos Gertler, Department of Electrical and Computer Engineering, George Mason University, Fairfax, Virginia, USA

MODEL BASED PROCESS CONTROL

Proceedings of the IFAC Workshop
Atlanta, Georgia, USA, 13–14 June, 1988

Edited by

T. J. McAVOY

Department of Chemical and Nuclear Engineering
University of Maryland, USA

Y. ARKUN

School of Chemical Engineering,
Georgia Institute of Technology, USA

and

E. ZAFIRIOU

Department of Chemical and Nuclear Engineering and
Systems Research Center, University of Maryland, USA

Published for the

INTERNATIONAL FEDERATION OF AUTOMATIC CONTROL

by

PERGAMON PRESS

OXFORD · NEW YORK · BEIJING · FRANKFURT
SÃO PAULO · SYDNEY · TOKYO · TORONTO

U.K.	Pergamon Press plc, Headington Hill Hall, Oxford OX3 0BW, England
U.S.A.	Pergamon Press, Inc., Maxwell House, Fairview Park, Elmsford, New York 10523, U.S.A.
PEOPLE'S REPUBLIC OF CHINA	Pergamon Press, Room 4037, Qianmen Hotel, Beijing, People's Republic of China
FEDERAL REPUBLIC OF GERMANY	Pergamon Press GmbH, Hammerweg 6, D-6242 Kronberg, Federal Republic of Germany
BRAZIL	Pergamon Editora Ltda, Rua Eça de Queiros, 346, CEP 04011, Paraiso, São Paulo, Brazil
AUSTRALIA	Pergamon Press Australia Pty Ltd., P.O. Box 544, Potts Point, N.S.W. 2011, Australia
JAPAN	Pergamon Press, 5th Floor, Matsuoka Central Building, 1-7-1 Nishishinjuku, Shinjuku-ku, Tokyo 160, Japan
CANADA	Pergamon Press Canada Ltd., Suite No. 271, 253 College Street, Toronto, Ontario, Canada M5T 1R5

First edition 1989

British Library Cataloguing in Publication Data

Model based process control: proceedings of
the IFAC workshop, Atlanta Georgia, USA, 13–14 June 1988.
1. Process control
I. McAvoy, Thomas J. (Thomas John), *1940–*
II. Arkun, Y III. Zafiriou, E.
IV. International Federation of Automatic Control
670.42'7

ISBN: 9780080357355

These proceedings were reproduced by means of the photo-offset process using the manuscripts supplied by the authors of the different papers. The manuscripts have been typed using different typewriters and typefaces. The lay-out, figures and tables of some papers did not agree completely with the standard requirements: consequently the reproduction does not display complete uniformity. To ensure rapid publication this discrepancy could not be changed: nor could the English be checked completely. Therefore, the readers are asked to excuse any deficiencies of this publication which may be due to the above mentioned reasons.

The Editors

Transferred to digital print 2009

Printed and bound in Great Britain by CPI Antony Rowe, Chippenham and Eastbourne

IFAC WORKSHOP ON MODEL BASED PROCESS CONTROL

Sponsored by
(IFAC) — International Federation of Automatic Control
Working Group on Process Control

Organized by
American Automatic Control Council

International Programme Committee (IPC)
T. McAvoy (Chairman), USA
J. Balchen, The Netherlands
D. Bonvin, Switzerland
D. W. Clarke, UK
T. Edgar, USA
E. D. Gilles, FRG
H. Koivo, Finland
M. Kummel, Denmark
M. Morari, USA
A. J. Morris, Canada
G. Schmidt, FRG
D. Seborg, USA
T. Takamatsu, Japan

National Organizing Committee (NOC)
Y. Arkun (Chairman)
A. Palazoglu
J. Schork
E. Zafiriou

FOREWORD

This volume contains the proceedings of the IFAC sponsored Workshop on Model Based Process Control. The Workshop was also sponsored by the American Automatic Control Council and IFAC's Working Group on Process Control. The Workshop was held in Atlanta, Georgia on June 13 and 14, 1988. A total of 99 people from 18 countries attended the Workshop. Of the attendees 39 came from industry. This large industrial attendance indicates the strong interest of practicioners in model based control techniques. The Workshop format consisted of an invited tutorial and four invited case studies. All of the case studies deal with actual industrial applications. A key message that comes through from these papers is that industry has bought into the model based technology. Multivariable, constrained model based control is being applied effectively today. The case studies were followed by 15 papers that were selected from 31 contributed papers. The authors of the selected papers came from 8 different countries. Thus, the Workshop provided an international forum to discuss the area of model based process control.

Several important research topics and needs emerged from the Workshop. These included process identification, state space approaches to the prediction of the effect of disturbances, control of nonsquare plants, real-time detection of and recovery from abnormal operating conditions due to ill-conditioning and feasibility with respect to constraints, the effect of hard constraints on the stability of the model predictive methods, model predictive control of distributed parameter systems, nonlinear model predictive control and robustness considerations. Academic and industrial case studies demonstrated the need for further research in these areas.

T. J. McAvoy

Y. Arkun

E. Zafiriou

CONTENTS

TUTORIAL

INVITED INDUSTRIAL CASE STUDIES

CONTRIBUTED PAPERS

MODEL PREDICTIVE CONTROL: THEORY AND PRACTICE

Manfred Morari*, Carlos E. Garcia and David M. Prett****

**Chemical Engineering, 206–41, California Institute of Technology, Pasadena, California 91125, USA*
***Shell Development Company, P.O. Box 1380, Houston, Texas 77001, USA*

ABSTRACT

We refer to Model Predictive Control (MPC) as that family of controllers in which there is a direct use of an explicit and separately identifiable model. Control design methods based on the MPC concept have found wide acceptance in industrial applications and have been studied by academia. The reason for such popularity is the ability of MPC designs to yield high performance control systems capable of operating without expert intervention for long periods of time. In this paper the issues of importance that any control system should address are stated. MPC techniques are then reviewed in the light of these issues in order to point out their advantages in design and implementation.

A number of design techniques emanating from MPC, namely Dynamic Matrix Control, Model Algorithmic Control, Inferential Control and Internal Model Control, are put in perspective with respect to each other and the relation to more traditional methods like Linear Quadratic Control is examined. The flexible constraint handling capabilities of MPC are shown to be a significant advantage in the context of the overall operating objectives of the process industries and the $1-, 2-$, and ∞ norm formulations of the performance objective are discussed. The application of MPC to nonlinear systems is not covered for brevity. Finally, it is explained that though MPC is not inherently more or less robust than classical feedback, it can be adjusted more easily for robustness.

INTRODUCTION

The petro-chemical industry is characterized as having very dynamic and unpredictable marketplace conditions. For instance, in the course of the last 15 years we have witnessed an enormous variation in crude and product prices. It is generally accepted that the most effective way to generate the most profit out of our plants while responding to marketplace variations with minimal capital investment is provided by the integration of all aspects of automation of the decision making process. These are:

- *Measurements:* The gathering and monitoring of process measurements via instrumentation.
- *Control:* The manipulation of process degrees of freedom for the satisfaction of operating criteria. This typically involves two layers of implementation: the single loop control which is performed via analog controllers or rapid sampling digital controllers; and the control performed using realtime computers with relatively large CPU capabilities.
- *Optimization:* The manipulation of process degrees of freedom for the satisfaction of plant economic objectives. It is usually implemented at a rate such that the controlled plant is assumed to be at steady-state. Therefore, the distinction between control and optimization is primarily a difference in implementation frequencies.
- *Logistics:* The allocation of raw materials and scheduling of operating plants for the maximization of profits and the realization of the company's program.

Each one of these automation layers plays a unique and complementary role in allowing a company to react rapidly to changes. Therefore, one layer cannot be effective without the others. In addition, the effectiveness of the whole approach is only possible when all manufacturing plants are integrated into the system.

Although maintaining a stable operation of the process was possibly the only objective of control systems in the past, this integration imposes more demanding requirements. In the petro-chemical industries control systems need to satisfy one or more of the following practical performance criteria:

- *Economic:* These can be associated with either maintaining process variables at the targets dictated by the optimization phase or dynamically minimizing an operating cost function.
- *Safety and Environmental:* Some process variables must not violate specified bounds for reasons of personnel or equipment safety, or because of environmental regulations.
- *Equipment:* The control system must not drive the process outside the physical limitations of the equipment.
- *Product Quality:* Consumer specifications on products must be satisfied.
- *Human Preference:* There exist excessive levels of variable oscillations or jaggedness that the operator will not tolerate. There can also be preferred modes of operation.

In addition, the implementation of such integrated systems is forcing our processes to operate over an ever wider range of conditions. As a result, we can state the control problem that any control system must solve as follows:

On-line update the manipulated variables to satisfy multiple, changing performance criteria in the face of changing plant characteristics.

The whole spectrum of process control methodologies in use today is faced with the solution of this problem. The difference between these methodologies lies in the particular assumptions and compromises made in the mathematical formulation of performance criteria and in the selection of a process representation. These are made primarily to simplify the mathematical problem so that its solution fits the existing hardware capabilities. The natural mathematical representation of many of these criteria is in the form of dynamic objective functions to be minimized and of dynamic inequality constraints. The usual mathematical representation for the process is a dynamic model with its associated uncertainties. The importance of uncertainties is increasingly being recognized by control theoreticians and thus are being included explicitly in the formulation of controllers. However, one of the most crucial compromises made in control is to ignore constraints in the formulation of the problem. As we explain below these simplifications can deny the control system of its achievable performance. It is a fact that in practice the operating point of a plant that satisfies the overall economic goals of the process will lie at the intersection of constraints (Arkun, 1978; Prett and Gillette, 1979). Therefore, in order to be successful, any control system must anticipate constraint violations and correct for them in a systematic way: violations must not be allowed while keeping the operation close to these constraints. The usual practice in process control is to ignore the constraint issue at the design stage and then "handle" it in an *ad-hoc* way during the implementation. Since each petro-chemical process (or unit) is unique we cannot exploit the population factor as is done in other industries (e.g., aerospace). That is, we cannot afford extreme expenses in designing an ad-hoc control system that we know will not work in another process and therefore its cost cannot be spread over a large number of applications. Due to the increase in the number of applications of this type (resulting from the need to achieve integration), this implies an enormous burden both in the design and maintenance costs of these loops. In our experience, these costs more than offset the profitability of any ad-hoc control system.

In conclusion, economics demand that control systems must be designed with no ad-hoc fixups and transparent specification of performance criteria such as constraints. Our experience has demonstrated that *Model Predictive Control* (MPC) techniques provide the only *methodology* to handle constraints in a systematic way during the design and implementation of the controller. Moreover, in its most general form MPC is not restricted in terms of the model, objective function and/or constraint functionality. For these reasons, it is the only methodology that currently can reflect most directly the many performance criteria of relevance to the process industries and is capable of utilizing any available process model. This is the primary reason for the success of these techniques in numerous applications in the chemical process industries.

In this paper the MPC methodology is reviewed and compared with other seemingly identical techniques. We particularly emphasize the unconstrained version of MPC since it is only in this form that it is possible to compare it with other schemes. Then the several existing forms of constrained MPC are reviewed, concluding with the nonlinear MPC approaches. Although the issue of model uncertainties in MPC techniques is not dealt with in this paper, some comments on robustness of MPC are included.

HISTORICAL BACKGROUND

The current interest of the processing industry in Model Predictive Control (MPC) can be traced back to a set of papers which appeared in the late 1970's. In 1978 Richalet et al. described successful applications of "Model Predictive Heuristic Control" and in 1979 engineers from Shell (Cutler & Ramaker, 1979; Prett & Gillette, 1979) outlined "Dynamic Matrix Control" (DMC) and report applications to a fluid catalytic cracker. In both algorithms an *explicit* dynamic model of the plant is used to predict the effect of future actions of the manipulated variables on the output. (Thus the name "Model Predictive Control"). The future moves of the manipulated variables are determined by optimization with the objective of minimizing the predicted error subject to operating constraints. The optimization is repeated at each sampling time based on updated information (measurements) from the plant.

Thus, in the context of MPC the control problem including the relative importance of the different objectives, the constraints, etc. is formulated as a dynamic optimization problem. While this by itself is hardly a new idea, it constitutes one of the first examples of large scale dynamic optimization applied routinely in real time in the process industries.

The MPC concept has a long history. The connections between the closely related minimum time optimal control problem and Linear Programming were recognized first by Zadeh & Whalen (1962). Propoi (1963) proposed the moving horizon approach which is at the core of all MPC algorithms. It became known as "Open Loop Optimal Feedback". The extensive work on this problem during the seventies was reviewed in the thesis by Gutman (1982). The connection between this work and MPC was discovered by Chang & Seborg (1983).

Since the rediscovery of MPC in 1978 and 1979, its popularity in the Chemical Process Industries has increased steadily. Mehra et al. (1982) review a number of applications including a superheater, a steam generator, a wind tunnel, a utility boiler connected to a distillation column and a glass furnace. Shell has applied MPC to many systems, among them a fluid catalytic cracking unit (Prett and Gillette, 1979) and a highly nonlinear batch reactor (Garcia, 1984). Matsko (1985) summarizes several successful implementations in the pulp and paper industries.

Several companies (Bailey, DMC, Profimatics, Setpoint) offer MPC software. Cutler & Hawkins (1987) report a complex industrial application to a hydrocracker reactor involving seven independent variables (five manipulated, two disturbance) and four dependent (controlled) variables including a number of constraints. Martin et al. (1986) cites seven completed applications and ten under design. They include: fluid catalytic cracker – including regenerator loading, reactor severity and differential pressure controls; hydrocracker (or hydrotreater) bed outlet temperature control and weight average bed temperature profile control; hydrocracker recycle surge drum level control; reformer weight average inlet temperature profile control; analyzer loop control. The latter has been described in more detail by Caldwell & Martin (1987). Setpoint (Grosdidier, 1987) has applied the MPC technology to: fixed and ebulating bed hydrocrackers; fluid catalytic crackers; distillation columns; absorber/stripper bottom C_2 composition control and other chemical and petroleum refining operations.

In academia MPC has been applied under controlled conditions to a simple mixing tank and a heat exchanger (Arkun et al., 1986) as well as a coupled distillation column system for the separation of a ternary mixture (Levien, 1985; Levien and Morari, 1986). Parish and Brosilow (1985) compare MPC with conventional control schemes on a heat-exchanger and an industrial autoclave.

Most applications reported above are multivariable and involve constraints. It is exactly these types of problems which motivated the development of the MPC control techniques. Largely independently a second branch of MPC emerged, whose main objective is *adaptive* control. Peterka's predictive controller (1984), Ydstie's extended-horizon design (1984) and EPSAC developed by DeKeyser et al. (1982, 1985) are in this category as well as Clarke's generalized predictive control algorithm (1987a,b). These developments are essentially limited to SISO systems with extension to the MIMO case conceptually straightforward but very involved when the details are considered. The constrained case is not considered in any detail in these papers. Because of the different underlying philosophy these algorithms are outside the focus of this paper. Nevertheless some cross references will be useful at times because these algorithms were largely developed for the nonadaptive case with the adaptation added in an *ad hoc* manner based on recursive least squares (or similar) parameter estimates. The stability and robustness of the *adaptive* scheme was generally not analyzed.

MODELS

All derivations in this paper will be carried out for general MIMO systems. Occasionally, in the interest of providing special insight SISO systems are going to be discussed separately. The idea of MPC is not limited to a particular system description, but the computation and implementation depend on the model representation. Depending on the context we will readily switch between state space, transfer matrix and convolution type models. We will assume the system to be described in state space by

$$x(k) = Ax(k-1) + Bu(k-1) \tag{1}$$

$$y(k) = Cx(k) \tag{2}$$

For zero-initial conditions the equivalent transfer matrix representation is

$$y(z) = P(z)u(z) \tag{3}$$

where

$$P(z) \triangleq C(zI - A)^{-1}B \tag{4}$$

Because most chemical engineering processes are open-loop stable our discussion will be limited to stable systems. The extension of the presented results to unstable systems is described elsewhere (for example, Morari et al., 1988). When A is stable the inverse in (4) can be expanded into a Neuman series

$$P(z) = \sum_{i=0}^{\infty} CA^i Bz^{-i-1} \tag{5}$$

$$\triangleq \sum_{i=1}^{\infty} \bar{H}_i z^{-i} \tag{6}$$

where $\bar{H}_i$ are the impulse response coefficients, whose magnitude vanishes as $i \to \infty$. Thus, in the time domain we have the truncated *impulse response* model

$$y(k) = \sum_{i=1}^{n} \bar{H}_i u(k-i) \tag{7}$$

and with the definitions

$$\bar{H}_i = H_i - H_{i-1} \tag{8}$$

$$H_i = \sum_{j=1}^{i} \bar{H}_j \tag{9}$$

the truncated *step response* model

$$y(k) = \sum_{i=1}^{n} H_i \Delta u(k-1) \tag{10}$$

where

$$\Delta u(k) = u(k) - u(k-1) \tag{11}$$

and H_i are the step response coefficients. Depending on the time delay structure of the system the leading step response coefficient matrices may be zero or have zero elements.

MPC ALGORITHM FORMULATIONS

The name "Model Predictive Control" arises from the manner in which the control law is computed (Fig. 1). At the present time k the behavior of the process over a horizon p is considered. Using a model the process response to changes in the manipulated variable is predicted. The moves of the manipulated variables are selected such that the predicted response has certain desirable characteristics. Only the first computed change in the manipulated variable is implemented. At time $k+1$ the computation is repeated with the horizon moved by one time interval.

We will demonstrate how Dynamic Matrix Control (DMC) and Model Algorithmic Control (MAC) are derived. All other MPC algorithms which have been proposed are very similar.

Dynamic Matrix Control

The manipulated variables are selected to minimize a quadratic objective.

$$\min_{\Delta u(k)\ldots\Delta u(k+m-1)} \sum_{\ell=1}^{p} \|\hat{y}(k+\ell|k) - r(k+\ell)\|^2_{\Gamma_\ell}$$

$$+\|\Delta u(k+\ell-1)\|^2_{B_\ell} \tag{12}$$

$$\hat{y}(k+\ell|k) = \sum_{i=1}^{\ell} H_i \Delta u(k+\ell-i) + \sum_{i=\ell+1}^{n} H_i \Delta u(k+\ell-i)$$

$$+\hat{d}(k+\ell|k) \tag{13}$$

$$\hat{d}(k+\ell|k) = \hat{d}(k|k) = y_m(k) - \sum_{i=1}^{n} H_i \Delta u(k-i) \tag{14}$$

$$\sum_{\ell=1}^{p} C^j_{y_\ell} \hat{y}(k+\ell|k) + C^j_{y_\ell} u(k+\ell-1) + c^j \le 0; \qquad j = 1, n_c \tag{15}$$

$\hat{y}(k+\ell\|k)$	=	predicted value of y at time $k+\ell$ based on information available at time k
$\hat{d}(k+\ell\|k)$	=	predicted value of additive disturbances at process output at time $k+\ell$ based on information available at time k
$y_m(k)$	=	measurement of y at time k
$\Delta u(k+\ell)$	=	$u(k+\ell)-u(k+\ell-1)$
$H_i, i=1$, n	=	model step response matrix coefficient
n	=	truncation order
n_c	=	number of constraints
p	=	horizon length (in general $p >> n$)
m	=	number of manipulated variable moves in the future ($\Delta u(k+\ell)=0$ $\forall \ell \geq m; m < p$)
$\|\|x\|\|_Q^2$	=	$x^T Q x$
Γ_ℓ, B_ℓ	=	weighting matrices
$C_{y\ell}^j, C_{u\ell}^j, c^j$	=	constant matrices

The prediction of the output (13) involves three terms on the RHS: The first term includes the present and all future moves of the manipulated variables which are to be determined as to solve (12). The second term includes only past values of the manipulated variables and is completely known at time k. The third term is the predicted disturbance $\hat{d}$ which is obtained from (7). $\hat{d}(k+\ell|k)$ is assumed constant for all future times ($\ell \geq 0$). At time k it is estimated as the difference between the measured output $y(k)$ and the output predicted from the model. In block diagram notation (14) corresponds to a model $\tilde{P}$ in parallel with the plant P (Fig. 2A) with the resulting feedback signal equal to $\hat{d}(k|k)$. Equations (12)–(15) define a Quadratic Program which is solved on-line at every time step. This "controller" is represented by block Q in Fig. 2A.

Though computationally more involved than standard linear time invariant algorithms, the flexible constraint handling capabilities of MPC are very attractive for practical applications: A stuck valve can be simply specified by the operator on the console as an additional constraint for the optimization program. The algorithm will automatically adjust the actions of all the other manipulated variables to compensate for this failure situation as well as possible. In an unexpected emergency which a traditional fixed- logic scheme might find difficult to cope with, MPC will keep the process operating safely away from all constraints or allow the operator to shut it down in a smooth manner.

Model Algorithmic Control

MAC is distinctive from DMC in three aspects.

1. Instead of the step response model involving Δu, an impulse response model involving u is employed. If the input u is penalized in the quadratic objective, then the controller does not remove offset. This can be corrected by a static offset compensator (Garcia and Morari, 1982). If the input u is not penalized then extremely awkward procedures are necessary to treat nonminimum phase systems (Mehra and Rouhani, 1980).
2. The number of input moves m is not used for tuning ($m=p$).
3. The disturbance estimate (14) is filtered. Let $y(k)$ be the measurement and $\hat{y}(k)$ the model prediction. Then the disturbance estimate is defined recursively as

$$\hat{d}(k+\ell|k) = \alpha\hat{d}(k+\ell-1|k)+(1-\alpha)(y_m(k)-\hat{y}(k)) \quad (16)$$

with $\hat{d}(k|k) = 0, 0 \leq \alpha < 1$. Equation (16) adds a first order exponential filter with adjustable parameter α in the feedback path (Fig. 2A). For $r=0$ this is equivalent to augmenting Q. α is a much more direct and convenient tuning parameter than the weights, horizon length, etc. in the general MPC formulation. α is directly related to closed loop speed of response, bandwidth and robustness, but does not affect nominal ($P = \tilde{P}(z)$) stability. Garcia & Morari (1982, 1985a,b) have analyzed the effect of this filter in detail.

ANALYSIS

The MPC formulation (5)-(8) looks reasonable and attractive, and has been used extensively in industrial applications. However, a complete and general analysis of its properties (stability, robustness and performance) is not possible with the currently available tools. In general, the resulting control law is time varying and cannot be expressed in closed form. We would like to compare MPC with other design techniques and discuss alternate formulations and extensions. For this purpose we will concentrate first on the unconstrained case because only here a rigorous analysis is possible. The we discuss the different ways by which constraints can be handled and their implications. Finally we will review some extensions to nonlinear systems.

UNCONSTRAINED MPC

Without constraints (12)-(14) is a standard linear least squares problem which can be solved explicitly quite easily. With the moving horizon assumption a linear time invariant controller is found. Garcia and Morari (1985b) have shown how to obtain the controller transfer function from the linear least squares solution.

Structure

Garcia and Morari (1982) were the first to show that the structure depicted in Figs. 2A and C is inherent in all MPC and other control schemes. It will be referred to as Internal Model Control (IMC) structure in this paper. Here P is the plant, $\tilde{P}$ a model of the plant and Q (Q_1 and Q_2) the controller(s). y is the measured output, r the reference signal (setpoint), u the manipulated variable and d the effect of the unmeasured disturbances on the output. The total MPC system which has to be implemented consists of the model $\tilde{P}$ and the controller(s) Q (Q_1 and Q_2) and is indicated by the shaded box in Fig. 2. In this section and throughout most of the paper we will assume that the model $\tilde{P}$ is a perfect description of the plant ($P=\tilde{P}$). We will retain the super tilde ($\tilde{}$), however, to emphasize the distinction between the real plant P and the model $\tilde{P}$ which is a part of the control system. The IMC structures in Figs. 2A and C have largely the same characteristics. Initially we will concentrate our analysis on Fig. 2A. Subsequently the advantages of employing two controller blocks Q_1 and Q_2 as in Fig. 2C will be addressed.

The following three facts are among the reasons why MPC is attractive.

Fact 1: The IMC structure in Fig. 2A and the classic control structure in Fig. 2B are equivalent in the sense that any pair of external inputs $\{r, d\}$ will give rise to the same internal signals $\{u, y\}$ if and only if Q and C are related by

$$Q = C(I + \tilde{P}C)^{-1} \quad (17)$$

$$C = Q(I - \tilde{P}Q)^{-1} \quad (18)$$

Fact 2: If $P = \tilde{P}$ then the relation between any input and output in Fig. 2A is *affine** in the controller Q. In particular

$$y = PQ(r - d) + d \quad (19)$$

$$e = y - r = (I - PQ)(d - r) \quad (20)$$

Fact 3: If P is stable, then the MPC system in Fig. 2A is internally stable if and only if the classic control system with C defined by (18) is internally stable. In particular when $P = \tilde{P}$ the MPC system is stable if and only if Q is stable.

These facts have the following important implications:

Because of *Fact 1* the performance of unconstrained MPC is not inherently better than that of classic control as one might be led to believe from the literature. Indeed, for any MPC there is an equivalent classic controller with identical performance.

Q can be considered an alternate parametrization of the classic feedback controller C, albeit one with very attractive properties: The set of all controllers C which gives rise to closed loop stable systems is essentially impossible to characterize. On the contrary the set of all controllers Q with the same property is simply the set of all stable $Q's$ (*Fact 3*). Furthermore all important transfer functions (e.g., (19) and (20)) are affine in Q but nonlinear functions of C (*Fact 2*). From a mathematical point of view it is much simpler to optimize an affine function of Q by searching over all stable $Q's$ than it is to optimize a nonlinear function of C subject to the complicated constraint of closed loop stability. From an engineering viewpoint it is attractive to adjust a controller Q which is *directly* related to a setpoint and disturbance response (19) and (20) and where (in the absence of model uncertainty) closed loop stability is automatically guaranteed as long as Q is stable. On the other hand even when C is a simple PID controller it is usually not at all obvious how closed loop performance is affected by the three adjustable parameters and for what parameter values the closed loop system is stable. As apparent from (19) Q plays the role of a *feedforward* controller. The design of feedforward controllers is generally much simpler than that of feedback controllers.

The main limitation of MPC in Fig. 2A is apparent from (20): Both the disturbances d and the reference signals r affect the error e through the same transfer matrix $(I - PQ)$. If r and d have different dynamic characteristics it is clearly impossible to select Q simultaneously for good setpoint tracking and disturbance rejection. For the "Two-Degree-of-Freedom- Structure" in Fig. 2C and the equivalent classic structure in Fig. 2D (for $P = \tilde{P}$) we find

$$e = (I - \tilde{P}Q_1)d - (I - \tilde{P}Q_2)r \quad (21)$$

Here in the absence of model error the two controller blocks make it possible to design independently for good disturbance response and setpoint following. The equivalent classic feedback controller is described by

$$u = Q_1(I - \tilde{P}Q_1)^{-1}Q_1^{-1}Q_2 r - Q_1(I - \tilde{P}Q_1)^{-1}y \quad (22)$$

$$\triangleq C_1 r - C_2 y$$

An excellent historical review of the origins of the structure in Fig. 2A which has as its special characteristic a model in parallel with the plant is provided by Frank (1974). It appears to have been discovered by several people simultaneously in the late fifties. Newton, Gould and Kaiser (1957) use the structure to transform the closed loop system into an open loop one so that the results of Wiener can be applied to find the H_2–optimal controller Q. When the Smith Predictor (Smith, 1957) is written in the form shown in Fig. 3 where $\tilde{P}^*$ is the SISO process model without time delay it can be noticed that its structure also contains a process model in parallel with the plant. Independently Zirwas (1958) and Giloi (1959) suggested the predictor structure for the control of systems with time delay. Horowitz (1963) introduced a similar structure and called it "model feedback".

Frank (1974) first realized the general power of this structure, fully exploits it and extends the work by Newton et al. (1957) to handle persistent disturbances and setpoints. Youla et al. (1976) extend the convenient "Q–parametrization" of the controller C to handle unstable plants. In 1981 Zames ushers in the era of H_∞–control utilizing for his developments the Q–parametrization. At present it is used in all robust controller design methodologies.

Unaware of all these developments the process industries both in France (Richalet et al., 1978) and in the U.S. (Cutler and Ramaker, 1979; Prett and Gillette, 1979) exploit the advantages of the parallel model/plant arrangement, Brosilow (1979) utilizes the Smith Predictor parametrization to develop a robust design procedure and Garcia and Morari (1982, 1985 a,b) unify all these concepts and refer to the structure in Fig. 2A as Internal Model Control (IMC) because the process *model* is explicitly an *internal* part of the controller.

The two-degree-of-freedom structure is usually attributed to Horowitz (1963). It has been analyzed by many people since (e.g., Vidyasagar, 1985).

Tuning Guidelines

As we will analyze in more detail below the problem (12)-(14) is very closely related to the standard Linear Quadratic Optimal Control problem, for which a wealth of powerful theoretical results is available. In particular, exact conditions on the tuning parameters are known which yield a stabilizing feedback control law. Because of the finite horizon in (12) most conditions which guarantee that (12)-(14) will lead to a stabilizing controller are only sufficient. Thus, at this time the tuning of MPC has to proceed largely by trial and error with these sufficient conditions as guidelines.

For simplicity, the theorems below (Garcia & Morari, 1982) are formulated for SISO systems without delays. Equivalent results for MIMO systems have been derived (Gar-

* The relation between x and y is called *affine* when $y = A + Bx$.

cia & Morari, 1985b) but are somewhat more complicated because specific nonsingularity conditions have to be imposed.

Theorem 1: For $\Gamma_\ell \neq 0, B_\ell = 0$, selecting $m = p \leq n$ yields the model inverse control $Q(z) = z^{-1}P(z)^{-1}$.

This output deadbeat control law is only stable if all zeros of $P(z)$ are inside the unit circle. Even when it is stable it is generally very aggressive and can lead to intersample rippling if $P(z)$ has zeros close to (-1,0). This control law can be acceptable if the sampling time is relatively large.

Theorem 2: There exists a finite $B^* > 0$ such that for $B_\ell \geq B^*(\ell = 1, \ldots, m)$ the control law is stable for all $m \geq 1, p \geq 1$ and $\Gamma_\ell > 0$.

This theorem implies that penalizing control action can stabilize the system regardless of the other parameter choices.

Theorem 3: Assume $\Gamma_\ell = 1, B_\ell = 0$. Then for sufficiently small m and sufficiently large $p > n + m - 1$ the closed loop system is stable.

Thus the horizon length p (relative to the number of manipulated variable moves m) plays a similar role as the input penalty parameter B_ℓ.

Several other more specific results are available. Reid et al. (1979) derived the weighting matrices to yield a state deadbeat control law. Systems with a monotone discrete step response are analyzed by Garcia & Morari (1982). See also the review by Clarke & Mohtadi (1987) for further stability conditions. It can also be easily shown that the control law (12)-(14) is of Type 1, i.e., no offset for step-like inputs (Garcia & Morari, 1982).

Linear Quadratic Control (LQC) Law Computation

The objective of this section is to define a specific LQC problem and to relate it to the MPC problem (12)-(14). Let the process be described by

$$x(k+1) = Ax(k) + Bu(k) + \bar{w}_1(k) \tag{23}$$

$$d(k+1) = d(k) + \bar{w}_2(k) \tag{24}$$

$$y(k) = Cx(k) + d(k) + \bar{w}_3(k) \tag{25}$$

where $\bar{w}(k) = [\bar{w}_1(k), \bar{w}_2(k), \bar{w}_2(k)]^T$ forms a sequence of zero mean uncorrelated (in time) vector stochastic variables with the variance matrix $\bar{V}$. The disturbance d is a Wiener process. It can be pictured as a sequence of random steps whose amplitude is described by a normal distribution, and the time of occurrence follows a Poisson distribution (Chang, 1961, p. 56). The deterministic equivalent of (24) is $d = constant$, as was assumed for MPC. It is only meaningful to define an optimal control problem for the system (23)-(25) if it is stabilizable and detectable. $\left\{\begin{bmatrix} A & 0 \\ 0 & I \end{bmatrix}, \begin{bmatrix} B \\ 0 \end{bmatrix}\right\}$ is clearly not stabilizable. By differencing (23)-(25) become

$$\begin{bmatrix} \Delta x(k+1) \\ \bar{y}(k+1) \end{bmatrix} = \begin{bmatrix} A & 0 \\ CA & I \end{bmatrix} \begin{bmatrix} \Delta x(k) \\ \bar{y}(k) \end{bmatrix} + \begin{bmatrix} B \\ CB \end{bmatrix} \Delta u(k) + \begin{bmatrix} w_1(k) \\ w_2(k) \end{bmatrix} \tag{26}$$

$$y(k+1) = \bar{y}(k+1) + w_3(k+1) \tag{27}$$

where $w(k) = [w_1(k), w_2(k), w_3(k)]^T$ is again a sequence of zero mean vector stochastic variables whose correlation function can be derived from $\bar{V}$. $\bar{y}$ is an additional state. If we assume that $\bar{w}_1 = 0$ and $\bar{w}_3 = 0$ then $w(k)$ is again uncorrelated and the variance matrix of $\bar{w}(k)$ and $w(k)$ becomes

$$\bar{V} = V = \begin{bmatrix} 0 & 0 & 0 \\ 0 & V_0 & 0 \\ 0 & 0 & 0 \end{bmatrix} \tag{28}$$

We will also assume that the covariance matrix of the state estimate $(\Delta x, \bar{y})^T$ at the initial point is

$$V' = \begin{bmatrix} 0 & 0 \\ 0 & V_1 \end{bmatrix} \tag{29}$$

Equations (28) and (29) imply that Δx is known perfectly all the time and that there is no measurement noise.

Controllability. The system (26)

$$\left\{\begin{bmatrix} A & 0 \\ CA & I \end{bmatrix}, \begin{bmatrix} B \\ CB \end{bmatrix}\right\}$$

is controllable if and only if

1. $\{A, B\}$ is controllable
2. $\begin{bmatrix} A-I & B \\ CA & CB \end{bmatrix}$ has full row rank

(The proof follows the arguments by Morari and Stephanopoulos, 1980). Condition 2 requires the number of manipulated variables to be at least as large as the number of controlled outputs. It can be interpreted further in the context of the steady state form of (26)

$$\Delta y = [-CA(A-I)^{-1}B + CB]\Delta u \tag{30}$$

If $dim\ y = dim\ u$ then according to Schurs formula (Gantmacher, 1959) condition 2 becomes

$$det \begin{bmatrix} A-I & B \\ CA & CB \end{bmatrix} = det(A-I) \cdot$$

$$det(-CA(A-I)^{-1}B + CB) \neq 0 \tag{31}$$

Comparing (30) and (31) we note that controllability of (26) requires the steady state gain matrix between Δu and Δy to be nonsingular which is clearly a very reasonable requirement. In summary, conditions 1 and 2 are satisfied in any practical situation.

Detectability. The system (26)

$$\left\{\begin{bmatrix} A & 0 \\ CA & I \end{bmatrix}, [0, I]\right\}$$

is detectable if and only if $\{A, CA\}$ is detectable. (This can be proved using the arguments by Morari and Stephanopoulos, 1980). Because A is stable, $\{A, CA\}$ is detectable.

Optimal Control Problem. Rewrite problem (12) for the system (26)-(29) in a form similar to the one used by Kwakernaak and Sivan (1972, p. 539).

$$\min_{\Delta u(k) \ldots \Delta u(k+p-1)} E\{\sum_{\ell=1}^{p} [y(k+\ell) - r(k+\ell)]^T R_3(\ell)\, [y(k+\ell) - r(k+\ell)] + \Delta u(k+\ell-1)^T R_2(\ell) \Delta u(k+\ell-1)\} \tag{32}$$

where $E\{\cdot\}$ is the expected value operator, $R_2 > 0$, $R_3 > 0$ and $R_2(\ell) = \infty$ for $\ell > m + 1$.
The unique optimal solution is the feedback law

$$\Delta u(k) = F_1(k)\Delta\hat{x}(k) + F_2(k)[\hat{y}(k) - r(k)] \tag{33}$$

where the formula for $F(k)$ can be found in Kwakernaak and Sivan (1972, p. 494) and $\Delta\hat{x}(k)$ and $\hat{y}(k)$ are the state estimates obtained from the optimal observer (Kwakernaak and Sivan, 1972, p. 530)

$$\Delta\hat{x}(k+1) = A\Delta\hat{x}(k) + B\Delta u(k) \tag{34}$$

$$\hat{y}(k+1) = CA\Delta\hat{x}(k) + y(k) + CB\Delta u(k) \tag{35}$$

Note that because of the specific noise assumptions the optimal observer is time invariant. According to (34) the states Δx are estimated in open loop fashion from the process model. (35) is identical with the corresponding part of (26) except that the measurement $y(k)$ appears instead of $\bar{y}(k)$: the estimate $\hat{y}(k+1)$ is the measurement $y(k)$ plus the effect of the manipulated variables as expressed through the model. In block diagram form the estimator (34) and (35) corresponds to a model in parallel with the plant as is characteristic of MPC.

Comparison of Unconstrained MPC and LQC

1. The control laws resulting from the two computational procedures are essentially equal. The only difference arises from the fact that the state space model and the truncated step response model are not identical (but can be made arbitrarily close to each other).
2. The MPC computation requires the solution of a (possibly large) linear least squares problem. LQC involves the solution of some recursive matrix relations. The order of the system which determines the dimension of the matrices involved is generally much less than the truncation order of the step response model.
3. For implementation it is desirable to use a time invariant control law. In the general LQC context time invariance is achieved by making the weights R_2 and R_3 constant and extending the horizon to infinity. Unconstrained MPC is time invariant because only the first control move obtained from the finite horizon problem is applied at each time step.
4. The time invariant LQC is tuned by adjusting the constant weighting matrices R_2 and R_3. MPC can be tuned by adjusting R_2 and R_3 but a more frequently used tuning parameter is the number of moves m. The horizon p is usually selected such that $p \geq n$. There appears to be consensus in the literature that it is easier to pick a reasonable m *a priori* than R_2 and R_3.
5. If the control problem is really of a stochastic nature, the implicit assumptions in MPC are quite restrictive (no measurement and state excitation noise) and might lead to performance which is inferior to what could be obtained by a full Linear Quadratic Gaussian (LQG) approach. If – as is usually the case – the covariance matrices are *de-facto* tuning parameters then the MPC performance should be no better or worse than LQG.
6. It is well known that the LQG controller leads to a closed loop stable system as the horizon is extended to infinity. A similar result can be proven for unconstrained MPC (Garcia and Morari, 1982, 1985b).
7. Both controllers reject sustained step like disturbances, i.e., include implicitly integral action.

Internal Model Control (IMC)

The motivation behind the development of IMC was to combine the advantage of the different unconstrained MPC schemes and to avoid their disadvantages: easy on-line tuning via adjustment of physically meaningful parameters and without any concern about closed loop stability; good performance without intersample rippling; ability to cope with inputs other than steps. IMC utilizes the controller structure shown in Fig. 2A (and C) and uses a filter for tuning in a similar manner as MAC. The design procedure which is described in detail by Morari et al. (1988) and Zafiriou & Morari (1986b) consists of two steps:

1. First, $\hat{Q}$ is designed to yield a good response (usually in the least square sense) without regard for constraints on the manipulated variables and robustness.
2. The IMC controller Q is found by augmenting $\hat{Q}$ with a lowpass filter $F(Q = \hat{Q}F)$ whose parameters are adjusted either off–line or on–line to reduce the action of the manipulated variable and to improve the robustness (the controller $\hat{Q}$ is "detuned").

The controller $\hat{Q}_0(z)$ which minimizes $\sum_{k=0}^{\infty} e_k^2$ for a disturbance input d is derived by Zafiriou and Morari (1986b) (also see Morari et al., 1988):

$$\hat{Q}_0(z) = z(P_M(z)d_M(z))^{-1}\{(P_A(z)^{-1}d_M(z)z^{-1}\}_* \tag{36}$$

where the plant P is factored into an allpass P_A and a minimum phase part P_M. We have

$$P = P_A \cdot P_M \tag{37}$$

$$P_A(z) = z^{-N} \prod_{j=1}^{n} \frac{(1-\bar{\varsigma}_j^{-1})(z-\varsigma_j)}{(1-\varsigma_j)(z-\bar{\varsigma}_j^{-1})} \tag{38}$$

where $\varsigma_j, j = 1, \ldots, n$, are the zeros of $P(z)$ outside the unit circle and the integer N is such that $n^N P(z)$ is semi–proper, i.e., its numerator and denominator polynomials have the same degree. The overbar denotes complex conjugate. Similarly d is factored into d_A and d_M with $z^{N_d}d(z)$ semi-proper. Finally, the operator $\{\cdot\}_*$ in (36) implies that after a partial fraction expansion only the strictly proper stable (including poles at $z = 1$) terms are retained.

The H_2–optimal controller $\hat{Q}_0(z)$, however, may exhibit intersample rippling caused by poles of $\hat{Q}_0(z)$ close to $(-1, 0)$. A detailed study of the advantages and disadvantages and the theoretical reasons behind them, led Zafiriou and Morari (1985a) to a simple method for obtaining the IMC-controller $\hat{Q}(z)$:

$$\hat{Q}(z) = \hat{Q}_0(z)q_-(z)B(z) \tag{39}$$

where $q_-(z)B(z)$ replaces all the poles of $\hat{Q}_0(z)$ with negative real part with poles at the origin. The introduction of poles at the origin incorporates into the design some of the advantages of a deadbeat type response while at the same time known problems of deadbeat controllers, like overshoot, are avoided. Let $\pi_j, j = 1, \ldots, \rho$ be the poles of $\hat{Q}_0(z)$ with negative real part. Then

$$q_-(z) = z^{-p} \prod_{j=1}^{p} \frac{z-\pi_j}{1-\pi_j} \quad (40)$$

$$B(z) = \sum_{j=0}^{m-1} b_j z^{-j} \quad (41)$$

where m is the "Type" of the input d and the coefficients $b_j, j = 0, \ldots, m-1$ are computed so that $\hat{Q}(z)$ produces no steady-state offset (Zafiriou and Morari, 1986). For offset-free tracking of steps and ramps we have

Type 1:

$$b_0 = 1 \quad (42)$$

Type 2:

$$b_0 = 1 - b_1, \quad b_1 = \sum_{j=1}^{p} \frac{\pi_j}{(1-\pi_j)} \quad (43)$$

Also the discrete filter $F(z)$ has to satisfy certain conditions for offset-free tracking. For steps and ramps we have:

Type 1:

$$F(1) = 1 \quad (44)$$

Type 2:

$$F'(1) = 0 \quad (45)$$

Commonly used filters are
Type 1:

$$F(z) = \frac{(1-\alpha)z}{z-\alpha} \quad (46)$$

Type 2:

$$F(z) = \frac{(1-\alpha)z}{z-\alpha}(\beta_0 + \beta_1 z^{-1} + \ldots + \beta_w z^{-w}) \quad (47)$$

with

$$\beta_0 = 1 - (\beta_1 + \ldots + \beta_w) \quad (48)$$

$$\beta_k = \frac{-6k\alpha}{(1-\alpha)w(w+1)(2w+1)}, \quad k = 1, \ldots, w \quad (49)$$

Any $w \geq 2$ will satisfy the no-offset property. However the higher the w, the closer (47) approximates (46).

Instead of the filter, a term penalizing input variations can be included in the quadratic objective function with very similar effect (Scali and Morari, 1988). However, while the filter time constant α has direct physical significance (it is directly related to the closed loop time constant/inverse bandwidth), the input action penalty term is rather artificial and the weight is difficult to select without trial and error using simulations. Also, any change in the penalty weight requires the optimal control problem to be resolved (linear least squares problem in DMC and MAC, Riccati equation or spectral factorization for LQC). A change in α, on the contrary, can be implemented directly without any computational effort. In their simplest form with filter (46) or (47) SISO IMC controllers are "one-knob controllers" with the knob corresponding to closed loop bandwidth.

Conceptually, the Two-Step IMC design procedure extends to MIMO systems in a straightforward manner. In the first step the LQ optimal controller can be obtained from a MIMO version of (36). For rational plants the necessary all-pass factorization can be performed with standard software (Enns, 1984). The resulting controller does not necessarily lead to a decoupled response, which might be desired. Alternatively, Zafiriou and Morari (1985b,1987) show how to design the controller $\hat{Q}$ such that the closed loop transfer function has a specified structure.

There is much freedom for designing the filter. In the simplest case F is diagonal with one tuning parameter (the closed loop time constant) for each output. Often this is sufficient for achieving satisfactory response and robustness characteristics. For ill-conditioned systems robust performance is sometimes difficult to obtain without off-diagonal adjustable filter elements. Zafiriou and Morari (1986a) have developed a gradient search procedure for the filter parameters in order to optimize robust performance. Analytic expressions for the gradients are derived. Though this method performed well on a few test examples it is potentially plagued by the nonconvexity and non-differentiability of the objective function. The development of a more reliable technique is the subject of current research.

Comparison of IMC with DMC and MAC

1. The structure inherent in all MPC schemes, referred to as IMC- structure, corresponds to a very convenient way of parametrizing all stabilizing controllers for an open loop stable plant. This makes the structure very useful for design: a stable MP controller implies a stable closed loop system and vice versa.
2. IMC reduces the number of adjustable parameters of DMC and MAC to a minimum and expands the role of the MAC filter. The controller can also be tailored to accommodate inputs different from steps in an optimal manner. For low order models the IMC design procedure generates PID controllers with one adjustable parameter.
3. The advantage of unconstrained MPC (with the possible exception of IMC) over other LQ techniques is still to be demonstrated with more than case studies. Contrary to other techniques, however, the MPC ideas can be extended smoothly to generate nonlinear time varying controllers for linear systems with constraints.

CONSTRAINED MPC

As emphasized in the introductory sections of this paper, constraints are always present in any real life process control situation. Their importance has increased because supervisory optimizing control schemes frequently push the operating point toward the intersection of constraints (Arkun, 1978; Prett and Gillette, 1979). Most of today's control implementations handle constraints through split range controllers, overrides and more general min- -max selectors with some logic. These schemes are difficult to design, debug, explain to the operating personnel and maintain. For an example of a mildly complicated scheme of this type the reader is referred to Bristol (1980). The main attraction of MPC is that the engineer/operator can enter the constraints in a direct manner and that the algorithm will automatically find the best solution satisfying *all* of them.

Structure

Recall that the IMC structure (Fig. 2A and C) is effectively open-loop when $P = \tilde{P}$ and that it is internally stable if and only if P and Q are stable. This result holds trivially even when P and Q are nonlinear or when the inputs to P are subject to saturation constraints as long as the *constrained* inputs are also fed to the model $\tilde{P}$ so that $P = \tilde{P}$ is preserved. While input saturation causes complex stability problems for the classic feedback structure (Fig. 2B and D), it has no effect whatsoever on the stability of MPC. If input saturation is unlikely during normal operation and one is mainly concerned about stability under emergency conditions, the controller should simply be designed for the unconstrained system as discussed earlier in the paper and then *implemented* in the form suggested in Fig. 2A and C.

Because of the quasi open-loop structure, however, the unconstrained controller Q is unaware of the input constraints and the *performance* can suffer badly when the inputs saturate. Qualitatively, the inputs should be held longer at the constraints than what the controller in Fig. 1A and C does, in order to compensate for the restricted control action. Specifically, Q should be designed with the constraints included explicitly in the MPC formulation. As discussed by Campo and Morari (1986) and summarized next there are a number of possibilities for defining an appropriate objective function. In general Q becomes nonlinear and time varying and no stability analysis procedure is available when the constraints are active.

2-Norm Objective Function

The objective function (12) employs the 2-norm spatially (for the output and input vectors at a particular time) and temporally (over the horizon length). Apparently, a QP of the type (12)-(15) has been included in MAC for quite a while (Mehra et al., 1982) but the first detailed descriptions of solution procedures in the open literature were provided in the context of DMC by Garcia and Morshedi (1984) and Ricker (1985).

The advantage of the 2-norm is that at least for the unconstrained case an explicit expression for the control law can be found and its stability, robustness and performance characteristics can be analyzed with standard tools. A disadvantage is that for MIMO systems with many future moves m and long horizon p storage requirements can be quite formidable.

1-Norm Objective Function

Instead of the 2-norm the spatial and temporal 1-norm can be used to express the objective

$$\min_{\Delta u(k)\ldots\Delta u(k+m-1)} \sum_{\ell=1}^{p}\sum_{i=1}^{r} w_{\ell i}|\hat{y}_i(k+\ell|k)$$

$$-r_i(k+\ell)| \quad w_{\ell i} \geq 0 \qquad (50)$$

With the additional inequality constraints the resulting Linear Program (LP) is computationally simpler than the QP discussed previously. The long tradition of LP in optimal control was reviewed earlier in this paper. A 1-norm version of DMC was introduced by Morshedi et al. (1985)/

A severe disadvantage of the 1-norm formulation is that even in the unconstrained case a simple fundamental analysis of stability and performance is not possible because an explicit closed form expression for the (nonlinear) control law does not exist. Brosilow and coworkers (Brosilow et al., 1984, Brosilow and Zhao, 1987) circumvent this problem in the following manner: They first design a controller for the unconstrained process by some standard procedure. Then they minimize the 1-norm of the error not between the process output and setpoint as suggested by (12) but between the predicted constrained output and the ideal unconstrained output. In this formulation a term for penalizing the control action is not necessary, which eliminates some tuning parameters. If the constraints are not active the optimal value for the objective function found by the LP is zero and the process output is equal to what would be obtained by the linear time invariant controller which was designed in the first step. Obviously when the constraints become active an analysis of the properties of the control algorithm becomes impossible. However, because the response is kept close to the response of the unconstrained system it is reasonable to expect that its properties (stability, robustness, etc.) remain preserved.

∞-Norm Objective Function

Campo and Morari (1986) adopt Brosilow's idea but use a different norm. The unconstrained controller is designed by the IMC procedure. Then the ∞-norm of the error between the predicted constrained and the ideal unconstrained process output is minimized

$$\min_{\Delta u(k)\ldots\Delta u(k+m-1)} \max_{\substack{\ell=1,p\\ i=1,r}} w_{\ell i}|\hat{y}_i(k+\ell|k) - \hat{r}_i(k+\ell)| \quad (51)$$

In practice, the ∞-norm is particularly meaningful when peak excursions from desired trajectories are to be avoided. The 1- and 2-norms tend to keep average deviations small but allow large peak deviations. Also, in general, the ∞-norm results in an LP which can be solved more efficiently than the 1-norm LP.

Summary

Because of our deep understanding of linear systems it is preferable to employ an algorithm which is linear and for which a closed form expression can be derived and analyzed as long as the constraints are not active. QDMC and the formulation by Brosilow have these characteristics. The experience with different norms in conjunction with the Brosilow formulation has been too limited to date to draw any definite conclusions.

ROBUSTNESS

By robustness we mean roughly that the quality of performance of the feedback system is preserved when the dynamic behavior of the real plant is different from that assumed in the model. The importance of the robustness problem has become recognized universally in the last decade and much progress has been made toward its systematic solution (at least for linear systems. Very little is known about nonlinear systems). A detailed review and discussion is beyond the scope of this paper. The reader is referred to Doyle and Morari (1986).

In its linear form, unconstrained MPC is equivalent to classic feedback and therefore all the developed robustness analysis procedures can be applied. Whether MPC is robust or not depends on how it is designed. MPC is not inherently more or less robust than classic feedback as has been falsely claimed (Mehra et al., 1982). Also, the large number of parameters in a step or impulse response model as opposed to the three or four parameters in a parsimonious transfer function model does not add any robustness. By that argument one could include some parameters in the model which have no effect on the input-output description at all and obtain even more robustness.

On the other hand there is no doubt that MPC can be adjusted more easily for robustness than classic feedback. For example, the filter in MAC and IMC has a very direct effect on robustness (Garcia and Morari, 1982, 1985). This fact might be responsible for the misconceptions regarding MPC robustness.

CONCLUSIONS

Process control is driven by the need to respond competitively to a rapidly changing marketplace. Contrary to aerospace control the dynamics of the underlying system are simple (overdamped, no high frequency resonances). The difficulties arise from model uncertainty and the requirement to satisfy dynamically a number of constraints on states and inputs, which define the economically optimal operating condition. MPC is an attractive tool for addressing these issues. Tuning for robustness is direct and the constraints are considered explicitly in the algorithm. The future outputs predicted by the internal model could be displayed to the operator to give him confidence in the algorithm and to allow transparent on-line tuning.
Despite the progress during the last decade there are quite a few unresolved research issues:

- Linear Systems: The tuning procedures necessary to achieve robust performance for ill-conditioned MIMO systems (for an example study see Skogestad and Morari, 1986) are very complex. We lack the physical understanding of the connections between uncertainty and performance to suggest simpler techniques.
- Constraints: The LP and QP algorithms which have been employed for MPC are basically "off-the-shelf" with minor modifications. For large-scale applications more efficient tailor made algorithms will be needed. In the constrained algorithm model uncertainty is neglected. Thus, there is no guarantee that the real process variables will satisfy the constraints when the model variables do. Also, when there are constraints on both the inputs and states it is possible that the LP/QP is unable to find a feasible solution when a disturbance pushes the process outside the usual operating region. It is unclear how to recover gracefully in these situations. Some ideas have been presented by Gutman (1982), Garcia and Morshedi (1982) and Brosilow and Zhao (1987).
- Nonlinear Systems: The surface has barely been scratched. All important questions like nominal stability/model uncertainty and constraint handling are unanswered. The computational requirements are expected to be very high and the power of special computer architectures (e.g., hypercube) will have to be utilized.

ACKNOWLEDGEMENT

Partial support from the National Science Foundation and the Department of Energy are gratefully acknowledged.

REFERENCES

Arkun, Y. (1978). *Design of Steady-State Optimizing Control Structures for Chemical Processes*, Ph.D. Thesis, Univ. of Minnesota.

Arkun, Y., J. Hollett, W. M. Canney and M. Morari (1986). "Experimental Study of Internal Model Control" *Ind. Eng. Chem. Process Des. Dev.*, **25**, 102-108.

Bristol, E.H. (Nov. 1980). "After DDC. Idiomatic Control". *Chem. Eng. Progress,* **76**, No. 11, 84-89.

Brosilow, C. B. (1979). "The Structure and Design of Smith Predictors from the Viewpoint of Inferential Control". *Proc. Joint Automatic Control Conf., Denver, CO.*

Brosilow, C. B., G. Q. Zhao and K.C. Rao (1984). "A Linear Programming Approach to Constrained Multivariable Control", *Proc. American Control Conf., San Diego, CA*, 667-674.

Brosilow, C. B. and G. Q. Zhao (1987). "A Linear Programming Approach to Constrained Multivariable Process Control" in *Advances in Control and Dynamic Systems*, **27**.

Caldwell, J. M. and G. D. Martin (May 19-21, 1987). "On-Line Analyzer Predictive Control", *Sixth Annual Control Expo Conference, Rosemont, IL.*

Campo, P. J. and M. Morari (1986). "∞-Norm Formulation of Model Predictive Control Problems". *Proc. American Control Conf., Seattle, WA*, 339-343.

Chang, S. S. L. (1961). *Synthesis of Optimum Control Systems*, McGraw-Hill Book Company, Inc., New York.

Chang, T. S. and D. E. Seborg (1983). "A Linear Programming Approach to Multivariable Feedback Control with Inequality Constraints". *Int. J. Control*, **37**, 583-597.

Clarke, D. W., C. Mohtadi and P. S. Tuffs (1987). "Generalized Predictive Control - Part I. The Basic Algorithm", *Automatica*, **23**, 137-148.

Clarke, D. W., C. Mohtadi and P. S. Tuffs (1987). "Generalized Predictive Control - Part II. Extensions and Interpretations", *Automatica*, **23**, 149-160.

Clarke, D. W. and C. Mohtadi (1987). "Properties of Generalized Predictive Control", *Prof. 10th IFAC World Congress, Munich, Germany*, **10**, 63-74.

Cutler, C. R. and B. L. Ramaker (1979). "Dynamic Matrix Control - A Computer Control Algorithm". *AIChE National Mtg.*, Houston, TX; also *Proc. Joint Automatic Control Conf.*, San Francisco, CA.

Cutler, C. R. and R. B. Hawkins (1987). "Constrained Multivariable Control of a Hydrocracker Reactor", *Proc. American Control Conf.*, Minneapolis, MN, 1014-1020.

DeKeyser, R. M. C. and A. R. van Cauwenberghe (1982). "Typical Application Possibilities for Self-Tuning Predictive Control", *IFAC Symposium on Identification and System Parameter Estimation, Washington, D.C.*, **2**, 1552-1557.

DeKeyser, R. M. C. and A. R. van Cauwenberghe (1985). "Extended Prediction Self-Adaptive Control", *IFAC Symposium on Identification and System Parameter Estimation, York, UK*, 1255-1260.

DeKeyser, R. M. C., Ph. G. A. VandeVelde and F. A. G. Dumortier (1985). "A Comparative Study of Self-Adaptive Long-Range Predictive Control Methods", *IFAC Symposium on Identification and System Parameter Estimation, York, UK*, 1317-1322.

Enns, D. (1984). *Model Reduction for Control System Design.* Ph.D. Thesis, Stanford University.

Frank, P. M. (1974). *Entwurf von Regelkreisen mit vorgeschriebenem Verhalten*, G. Braun, Karlsruhe.

Gantmacher, F. R. (1959). *The Theory of Matrices, Vol. I.*, Chelsea Publ Co., New York.

Garcia, C. E. (1982). *Studies in Optimizing and Regulatory Control of Chemical Processing Systems.* Ph.D. Thesis, Univ. of Wisconsin, Madison).

Garcia, C. E. and M. Morari (1982). "Internal Model Control 1., A Unifying Review and Some New Results". *Ind. Eng. ΩChem. Process Des. Dev.*, **21**, 308-323.

Garcia, C. E. and M. Morari (1985a). "Internal Model Control 2. Design Procedure for Multivariable Systems". *Ind. Eng. Chem. Process Des. Dev.*, **24**, 472-484.

Garcia C. E. and M. Morari (1985b). "Internal Model Control 3. Multivariable Control Law Computation and Tuning ΩGuidelines". *Ind. Eng. Chem. Process Des. Dev.*, **24**, 484-494.

Garcia, C. E. and A. M. Morshedi (1984). "Quadratic Programming Solution of Dynamic Matrix Control (QDMC)". *Proc. American Control Conf., San Diego, CA* also *Chem. Eng. Commun.*, **46**, 73-87 (1986).

Garcia, C. E. and D. E. Prett (1986). "Advances in Industrial Model - Predictive Control" in *Chemical Process Control - CPC III* (M. Morari and T. J. McAvoy, eds.) CACHE and Elsevier, Amsterdam, 249-294.

Giloi, W. (1959). *Zur Theorie und Verwirklichung einer Regelung für Laufzeitstrecken nach dem Prinzip der ergänzenden Rückführung*, Ph.D. Thesis, TH Stuttgart

Grosdidier, P. (1987). Setpoint, Inc., Houston, personal communication.

Gutman, P. O. (1982). *Controllers for Bilinear and Constrained Linear Systems*, Ph.D. Thesis, Lund Institute of Technology.

Horowitz, I. M. (1963). *Synthesis of Feedback Systems*, Academic Press, London.

Kwakernaak, H. and R. Sivan (1972). *Linear Optimal Control Systems.* Wiley- Interscience, New York.

Levien, K. L. (1985). *Studies in the Design and Control of Coupled Distillation Columns*, Ph.D. Thesis, Univ. of Wisconsin, Madison.

Levien, K. L. and M. Morari (1986). "Internal Model Control of Coupled Distillation Columns". *AIChE J.*.

Martin, G. D., J. M. Caldwell and T. E. Ayral (1986). "Predictive Control Applications for the Petroleum Refining Industry", *Japan Petroleum Institute -Petroleum Refining Conference, Tokyo, Japan.*

Matsko, T. N. (1985). "Internal Model Control for Chemical Recovery", *Chem. Eng. Progress*, **81**, No. 12, 46-51.

Mehra, R. K. and R. Rouhani (1980). "Theoretical Considerations on Model Algorithmic Control for Nonminimum Phase Systems." *Proc. Joint Automatic Control Conf., San Francisco, CA.*

Mehra, R. K., R. Rouhani, J. Eterno, J. Richalet and A. Rault (1982). "Model Algorithmic Control: Review and Recent Development." *Eng. Foundation Conference on Chemical Process Control II, Sea Island, GA*, 287-310.

Morari, M. and J. C. Doyle (1986). "A Unifying Framework for Control System Design Under Uncertainty and its Implications for Chemical Process Control" in *Chemical Process Control - CPC III* (M. Morari and T. J. McAvoy, eds.), CACHE and Elsevier, Amsterdam, 5-52.

Morari, M., E. Zafiriou and C. G. Economou (1989). *Robust Process Control*, Prentice Hall, Englewood Cliffs, NJ.

Morari, M. and G. Stephanopoulos (1980). "Minimizing Unobservability in Inferential Control Schemes". *Int. J. Contr.*, **31**, 367-377.

Morshedi, A. M., C. R. Cutler and T. A. Skrovanek (1985). "Optimal Solution of Dynamic Matrix Control with Linear Programming Techniques (LDMC)". *Proc. American Control Conf., Boston, MA*, 199-208.

Newton, G. C., L. A. Gould and J. F. Kaiser (1957). *Analytical Design of Feedback Controls.* Wiley, New York.

Parrish, J. R. and C. B. Brosilow (1985). "Inferential Control Applications". *Automatica*, **21**, 527-538.

Peterka, V. (1984). "Predictor-Based Self-Tuning Control", *Automatica*, **20**, 39-50.

Prett, D. M. and R. D. Gillette (1979). "Optimization and Constrained Multivariable Control of a Catalytic Cracking Unit". *AIChE National Mtg., Houston, TX*; also *Proc. Joint Automatic Control Conf., San Francisco, CA* (1980).

Propoi, A. I. (1963). "Use of LP Methods for Synthesizing Sampled–Data Automatic Systems". *Automation and Remote Control*, **24**, 837.

Reid, J. G., R. K. Mehra and E. Kirkwood (1979). "Robustness Properties of Output Predictive Dead-Beat Control: SISO Case", *Proc. IEEE Conf. on Dec. and Control, Fort Lauderdale, FL*, 307.

Richalet, J. A., A. Rault, J. L. Testud and J. Papon (1978). "Model Predictive Heuristic Control: Applications to an Industrial Process". *Automatica*, **14**, 413-428.

Ricker, N. L. (1985). "The Use of Quadratic Programming for Constrained Internal Model Control". *Ind. Eng. Chem. Process Des. Dev.*, **24**, 925-936.

Rivera, D. E., S. Skogestad and M. Morari (1986). "Internal Model Control. PID Controller Design". *Ind. Eng. Chem. Process. Des. Dev.*, **25**, 252-265.

Scali, C. and M. Morari (1987), in preparation.

Skogestad, S. and M. Morari (1986). "Control of Ill–Conditioned Plants: High Purity Purity Distillation", *AIChE Annual Mtg., Miami Beach, FL.*

Smith, O. J. M. (1957). "Closer Control of Loops with Dead Time". *Chem. Eng. Progress*, **53**, No. 5, 217-219.

Vidyasagar, M. (1985). *Control System Synthesis - A Factorization Approach*, MIT Press, Cambridge, MA.

Ydstie, B. E. (1984). "Extended Horizon Adaptive Control", *9th World Congress of the IFAC, Budapest, Hungary.*

Youla, D. C., J. J. Bongiorno and H. A. Jabr (1976). "Modern Wiener–Hopf Design of Optimal Controllers Part I: The Single–Input–Output Case". *IEEE Trans. Autom. Control*, **AC-21**, 3-13.

Youla, D. C., H. A. Jabr and J. J. Bongiorno (1976). "Modern Wiener–Hopf Design of Optimal Controllers Part II: The Multivariable Case". *IEEE Trans. Autom. Control*, **AC-21**, 319-338.

Zadeh, L. A. and B. H. Whalen (1962). "On Optimal Control and Linear Programming". *IRE Trans. Autom. Control*, **7**, No. 4, 45.

Zafiriou, E. and M. Morari (1985a). "Digital Controllers for SISO Systems. A Review and a New Algorithm". *Int. J. Control*, **42**, 855-876.
Zafiriou, E. and M. Morari (1985b). "Robust Digital Controller Design for Multivariable Systems". *AIChE National Mtg., Chicago, IL.*
Zafiriou E. and M. Morari (1986a). "Design of the IMC Filter by Using the Structured Singular Value Approach". *Proc. of American Control Conf., Seattle, WA*, 1-6.
Zafiriou, E. and M. Morari (1986b). "Design of Robust Controllers and Sampling Time Selection for SISO Systems". *Int. J. Control*, **44**, 711-735.
Zafiriou, E. and M. Morari (1987), in preparation.
Zames, G. (1981). "Feedback and Optimal Sensitivity: Model Reference Transformations, Multiplicative Seminorms, and Approximate Inverses", *IEEE Trans. Autom. Control*, **AC-26**, 301-320.
Zirwas, H. C. (1958). *Die ergänzende Rückführung als Mittel zur schnellen Regelung von Regelstrecken mit Laufzeit.* Ph.D. Thesis, TH Stuttgart.

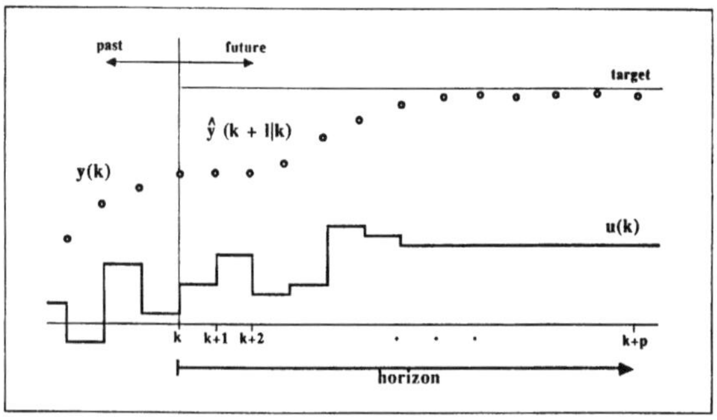

Fig. 1 The "moving horizon" approach of model-predictive control.

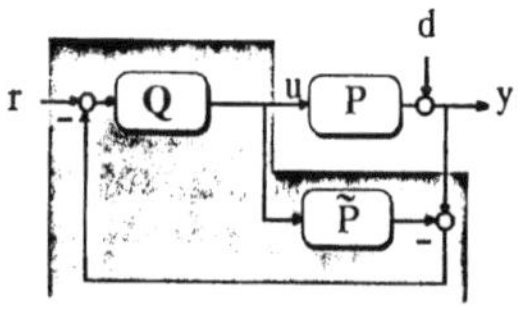

Fig. 2A IMC Structure

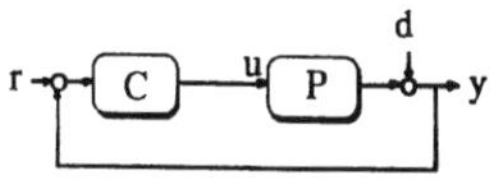

Fig. 2B Equivalent Classic Feedback Structure

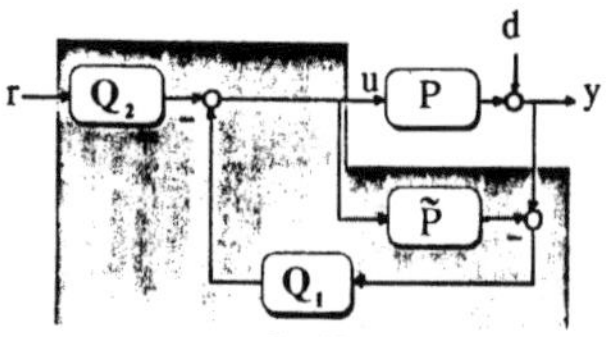

Fig. 2C
Two-Degree-of-Freedom IMC Structure

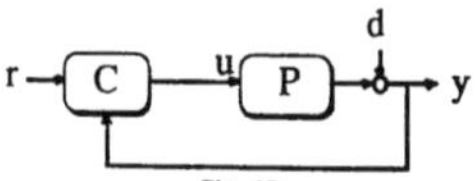

Fig. 2D
Equivalent Two-Degree-of-Freedom Classic Feedback Structure

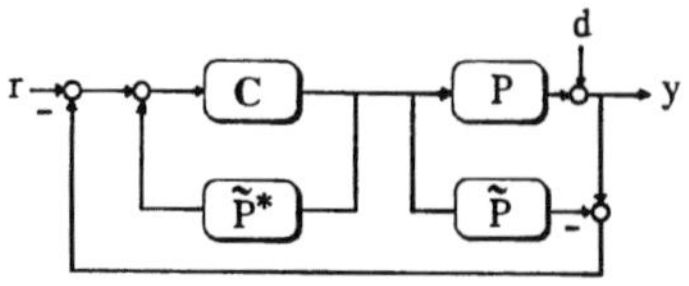

Fig. 3
Alternate representation of Smith Predictor Controller

CASE STUDIES OF MODEL-PREDICTIVE CONTROL IN PULP AND PAPER PRODUCTION

N. L. Ricker, T. Subrahmanian and T. Sim

Department of Chemical Engineering, BF-10, University of Washington, Seattle, WA 98195, USA

ABSTRACT

A case study of model-based predictive control of an evaporation process in a kraft pulp mill is presented. Two cases are considered. In the first, the process is represented by a linear model with two manipulated and two output variables. The control objectives are formulated as a quadratic programming (QP) problem, *i.e.*, minimization of a quadratic objective with linear inequality constraints on the manipulated variables. This approach is tested on the full-scale process and its performance is compared with that of PI control. The QP method meets the objectives of the plant operators and in general performs better than the PI controller. It was also found to be easier to tune once a model of the process had been determined.

In the second case the scope of the model is expanded to include additional up- and down-stream process elements. The control objective is reformulated to include constraints on both manipulated and output variables. Simulation is used to demonstrate the potential of this broader approach to the problem.

INTRODUCTION

Industrial processes must operate within constraints. Examples of common constraints in chemical processes include the limited range of flow-control valves, the capacities of storage tanks and other vessels, and specifications on acceptable product quality. Control strategies that include special provisions for the problems caused by constraints are a relatively new area of academic research (Ricker, 1985; Chang and Seborg, 1983; Brosilow, Zhao and Rao, 1984). Industry, on the other hand, has been forced to deal with such problems routinely and publications on sophisticated approaches began to appear around 1978 (Richalet and others, 1978; Prett and Gillette, 1980; Cutler, Haydel and Morshedi, 1983; Mehra and others, 1982). Although many of these describe applications to full-scale processes, important details of the control strategy and the application are often omitted for proprietary reasons and there is seldom an attempt to compare alternative methods. This provided the motivation for the present work.

THE EVAPORATION PROCESS

The basic evaporation process has been described in previous papers (Ricker and Sewell, 1984; Sewell, 1984). It is located in a kraft pulp mill owned by the Weyerhaeuser Company. As shown in Fig. 1, it is a six-effect, mid-feed design with a combination of co-current and counter-current flow patterns. The process objective is to evaporate water from the incoming *weak liquor* so that the *strong liquor* contains a specified amount of water, the remainder being non-volatile *solids*. The weak and product liquors are typically 17 and 55 % solids by weight, respectively. The heat source is steam with a nominal absolute pressure of about 300 kPa.

In addition to the the elements shown in Fig. 1 there are storage tanks from which the weak liquor is taken (hereafter called the *feed tank*) and to which the strong liquor is sent (the *product tank*). The feed tank capacity is 2544 m^3, equivalent to an 8-hour supply of weak liquor to the evaporators under normal operating conditions. The product tank capacity is 616 m^3 (again equivalent to about 8 hours).

The feed and product tanks act as buffers between the evaporator and the up- and down-stream processes. The up-stream processes are the *pulp-washing* and *pre-evaporation* steps in the kraft recovery cycle, which are subject to frequent temporary upsets as well as longer-term variations in production rates. The down-stream processes are the *concentrator* (a different type of evaporator), which removes additional water, and the *recovery boiler*. The control of the concentrator/recovery-boiler combination involves critical safety and efficiency problems. The plant operators believe that it is important to keep the concentration and flowrate of the liquor leaving the product tank as constant as possible.

Instrumentation

Measurements central to the control of the evaporator include: 1) the flowrate of the steam -- shown as F.101 in Fig. 1, 2) flowrates of the liquors entering and leaving the two tanks, 3) liquid levels in the feed and product tanks, and 4) the concentration of strong liquor -- inferred from a refractive index signal shown as D.149 in Fig. 1. The instrumentation also includes a refractometer for the weak liquor which we had hoped to use for feed-forward control of disturbances in weak liquor concentration. It was, however, "out of commission" for the duration of the experiments described here.

The long-standing control policy for the evaporator was essentially open-loop. Steam and liquor flowrates were controlled by feedback (*via* PID controllers implemented on a modern process-control computer system). The plant operators occasionally made step changes in the setpoints of these flow controllers in order to keep the tank levels and the strong liquor concentration near their nominal target values.

A PDP 11/34 was also available for research projects. Since most of our software development had been done on VAX systems we implemented our control algorithms on the 11/34. Weyerhaeuser provided hardware and software that allowed the 11/34 to sample the measurements listed above and to make changes in the *setpoints* for the steam and weak liquor flowrates.

To minimize aliasing problems (caused by 60 Hz noise) we had to install an analog pre-filter on the signal from the strong liquor refractometer. This eliminated frequency content above 0.1 Hz. Other signals contained very little high-frequency noise. All signals were sampled at 0.5 Hz and then filtered digitally to remove frequencies above the Nyquist frequency of the control algorithm (see *Results* for controller sampling rates).

Constraints

There were upper and lower bounds on both the steam and weak liquor flowrates. In the case of the weak liquor, these were a matter of *operating policy, i.e.,* the perceived limits of reliable operation of the evaporator. As long as the liquor pumps were operating normally, these bounds were well within those imposed by valve saturation.

In the case of the steam flowrate, however, the bounds *were* imposed by valve saturation. The process is often operated near the maximum steam flowrate. At any time the maximum possible steam flowrate depends on the valve characteristics (which may be assumed constant) and the pressure difference between the steam supply and the saturation pressure in the evaporator (both of which can vary significantly). *Thus the maximum possible steam flowrate varies in an unpredictable manner.* Occasionally the setpoint for the steam flowrate controller is above the maximum, in which case the valve saturates.

In our experiments the 11/34 computer could not measure the steam valve position or the output from the flow controller. Thus although the actual steam flowrate could be measured and the valve position could be noted on the operators' display, there was no reliable (automatic) way to sense that the steam valve was saturated. The effects of this are discussed in more detail in the *Results* section.

In the case of the output variables, the tank levels are, of course, constrained by the known capacities. As will be discussed later, it is good policy to define level constraints inside these limits in order to provide a safety margin. Constraints on the product liquor solids concentration are a matter of operating policy, as for the weak liquor flowrate.

Process model used for plant tests

Sewell derived a non-linear, lumped-parameter (18-state), dynamic model of the evaporator based on mass and energy balance equations (Sewell, 1984). He determined its adjustable parameters by running plant tests and showed that the model was representative under normal operating conditions. Sim then used discrete 5th-order transfer functions to approximate the input-output behavior of the non-linear model in the region of the steady state. Sim's transfer functions are given in Table 1 for a sampling time of 1 minute; they were used as the internal model for QP IMC in the plant tests. A much simpler approach would have been direct identification of a lower-order linear internal model from the plant data, but we wanted the non-linear model for simulations studies.

In Table 1, the output variables are defined as

$$y_i(t) = \frac{y_{i\,m}(t)}{y_{i\,ss}} - 1 \qquad i=1,2 \tag{1}$$

where

$y_i(t)$ is the ith normalized output variable at time t
$y_{i,m}(t)$ is the un-normalized value of the ith output signal at time t after all signal conditioning (*e.g.,* filtering)
$y_{i,ss}$ is the nominal steady-state value of $y_{i,m}(t)$

and the index i is defined such that

$y_1(t)$ is the normalized solids concentration in the strong (product) liquor
$y_2(t)$ is the normalized throughput

Similarly, the manipulated variables are defined as

$$u_j(t) = \frac{u_{j\,m}(t)}{u_{j\,ss}} - 1 \qquad j=1,2 \tag{2}$$

where the index j is defined such that

$u_1(t)$ is the normalized steam flowrate
$u_2(t)$ is the normalized weak liquor (WL) flowrate

A key point is the definition of $y_2(t)$, the throughput. As shown in Table 1, we used

$$y_2(t) = u_2(t) \tag{3}$$

i.e., it was defined to be equal to the WL flowrate. The advantage of this is the resulting partially decoupled internal model structure. The relative gain of the transfer function matrix is 1 at all frequencies, indicating that the closed-loop system will have relatively good robustness properties (Skogestad and Morari, 1987). A definition involving the *product* flowrate would have resulted in an internal model with full interaction (since both manipulated variables affect the product flowrate) and a greater likelihood of robustness problems. This is especially true because Sewell found that the product flowrate response was one of the more difficult to model accurately.

Process model for evaporation plus storage tanks

For the simulation studies of the process *including* the storage tanks the process model was as shown in Table 2. This lower-order linear model is based on that used previously by Ricker (1985). Its dynamics are similar but not exactly the same as those in the model of Table 1. We treated the flowrate entering the feed tank as a disturbance and also assumed that our control system could not regulate the flowrate leaving the product tank (since that would be under control of the concentrator/recovery-boiler system). We also assumed the presence of averaging level controllers on the two tanks -- in our case defined as proportional controllers with fixed gain. These were needed to make the open-loop system asymptotically stable, which simplifies the QP IMC design. The structure of the system is shown in Fig. 2. The controllers were tuned so that the tank level responded with a time constant of 30 minutes, on the same order as the dominant time constant of the evaporator process. The input and output variables are thus:

m_1 normalized level *setpoint* for the feed tank
m_2 normalized steam flowrate

y_1 normalized level in feed tank
y_2 normalized product concentration
y_3 normalized level in product tank

We could also have defined the problem as a 2-input, 2-output problem, omitting y_3. We retained it to see whether it would be possible to use the evaporator to coordinate between the demands of both the upstream and downsteam processes.

THEORETICAL BACKGROUND

Control algorithm for plant tests

Due to space limitations the QP IMC algorithm used in the plant tests can only be summarized here. Readers needing more details should consult Ricker (1985).

1. Discrete-time form of the internal model

As discussed in many texts, if we assume that all manipulated and output variables are sampled every *h* time units, and manipulated variables are held constant between the sampling instants, the input-output relationships given in Table 1 can be written in the equivalent discrete pulse-response form:

$$\mathbf{y}_{*k} = \mathbf{G}_* \ \mathbf{m}_{*k} + \mathbf{b}_{*k} \tag{4}$$

where the subscript *k* denotes a particular sampling instant, *i.e.,* $t = k\,h$, (usually the present) and

$\mathbf{y}_{*k}$ is a prediction of the *n* output variables over a *prediction horizon* of *p* future sampling intervals. The prediction is made at sampling interval *k*. It is a vector of dimension $n \cdot p$ by 1.

m_{*k} is a corresponding vector of p *future* values of the m manipulated variables, *i.e.*, a vector $m \cdot p$ by 1. The objective of the control algorithm will be to calculate the first m values in this vector, which will then determine the control signals sent to the plant during the sampling period k.

b_{*k} is a vector of $n \cdot p$ initial conditions for predictions to be made at time k.

G_* is the pulse response matrix, dimension $n \cdot p$ by $m \cdot p$, which is constant and can be determined directly from the continuous-time transfer functions.

It must be emphasized that the subscript "*" on the above variables denotes the use of a diagonal factorization to account for process dead-time, if any (Ricker, 1985). For example, suppose the path between sampled input/output pair $m_j(t_k)$ and $y_i(t_k)$ exhibits the *least* pure delay (with respect to all m manipulated variables). Let this delay be τ_i sampling periods. Similarly, find τ_i for all n outputs. We define m_{*k} such that the first m elements correspond to the m manipulated variables at time interval k, the next m to interval $k+1$ and so on. Then y_{*k} is defined such that the first n elements correspond to the n output variables at the time intervals $k+\tau_i$, the next n to intervals $k+1+\tau_i$, and so on. The effect of this is to ensure that the first n outputs in the prediction horizon, y_{*k}, can be influenced by variations in the manipulated variables, m_{*k}. Note that it is assumed that $\tau_i \geq 1$.

The value of b_{*k} at any sampling interval, k, depends on the *sum* of: 1) the estimated effect of process disturbances on the future outputs and 2) the estimated effect of *past* variations in the manipulated variables (*i.e.*, prior to time interval k) on *future* values of the output variables. The disturbance at time interval k is taken to be the difference between the *measured* output variables and their *predicted* values, *i.e.*,

$$d_k = y_k - y_{m,k} \tag{5}$$

where

d_k is the vector of n estimated disturbances at time interval k.

y_k is the vector of n measured values of the process outputs at time k.

$y_{m,k}$ is the vector of n model-predicted values of the process outputs at time k, based on the *measured* values of the manipulated variables prior to time k.

In our implementation of QP IMC, $y_{m,k}$ is calculated using a *discrete transfer function* form of the internal model. A transfer function model is used for this rather than a pulse response model to minimize the number of past values of $m_j(t)$ that need to be kept for making model predictions. The alternative pulse-response model is an infinite series and a large number of terms may be needed for good accuracy, depending on the chosen sampling time and the process dynamics.

Although there are other possibilities the disturbance vector at times $t > k \cdot h$ is assumed to be equal to d_k. In other words, the disturbance is assumed to have been a step change at time k. Then define

$$b_k = [\ (d_k)^T \quad (d_k + y_{m,k+1})^T \ \dots \ (d_k + y_{m,k+p-1})^T\ ]^T \tag{6}$$

where

$y_{m,k+q}$ is the model-predicted value of the process output vector (length n) at a *future* sampling interval, $k+q$, based on measured values of the manipulated variables for times $< k$, *and zero values of the manipulated variables* for intervals $\geq k$. Again, the transfer-function model is used.

Thus, for example, suppose $k = 1$ and the manipulated variables have been constant and equal to zero for $k < 1$. Then the model prediction of the outputs at $k = 1$ would be $y_{m,1} = 0$. If the actual plant output vector were y_1, then the estimated disturbance for $k \geq 1$ would be $d_k = y_1$. Similarly, in this case $y_{m,k+q} = 0$ for $q \geq 1$. Then we would get $b_1 = [\ y_1^T\ y_1^T\ \dots\ y_1^T]^T$.

Note that b_{*k} is determined in the same manner as b_k but with each predicted output shifted forward in time by τ_i time units to account for the diagonal factorization of dead time. For example, assume that $\tau_i = 1$ for all outputs (which would be the case when only sampling delay were present). Then b_{*k} would be

$$b_{*k}=[(d_k + y_{m,k+1})^T\ (d_k + y_{m,k+2})^T \dots (d_k + y_{m,k+p})^T]^T \tag{7}$$

A general expression for b_{*k} appears in Ricker (1985).

2. Constraints on process variables

We assumed that all constraints on the manipulated and output variables can be expressed as linear inequalities:

$$m_{*k,L} \leq m_{*k} \leq m_{*k,U} \tag{8}$$

$$y_{*k,L} \leq y_{*k} \leq y_{*k,U} \tag{9}$$

Note that in general the upper and lower bound vectors can vary with time. They are assumed to be independent of m_{*k} and y_{*k}, however, and must be defined at each time interval k. This definition can be a mathematical function or values in a data base that are modified occasionally by the process operator.

For convenience in the QP algorithm we define

$$u = m_{*k} - m_{*k,L} \tag{10}$$

Then using the model, eq. 4, to eliminate y_{*k} from eq. 9, we can combine eqs. 8 and 9 in the general form

$$C\,u \leq r \tag{11}$$

where C is a constant matrix of appropriate dimension and r is, in general, time varying. Note that even if all the upper and lower bound vectors in eqs. 8 and 9 are constant, r will be time-varying if there are constraints on the outputs (because it then depends on b_{*k}).

3. Definition of optimality

We have used a typical linear-quadratic objective function:

$$J(u) = \frac{1}{2}[\ (s_{*k} - y_{*k})^T\Gamma(s_{*k} - y_{*k}) + \Delta u^T\beta\Delta u\ ] \tag{12}$$

where

s_{*k} is a vector of setpoints, *i.e.*, the desired values of the outputs y_{*k}.

Γ is a weighting matrix which penalizes deviations of y_{*k} from s_{*k}. In our implementation it is assumed to be diagonal with non-negative elements.

β is a diagonal, non-negative weighting matrix that can be used to penalize large changes in the manipulated variables from one sampling interval to the next (see Ricker, 1985).

4. Solution procedure

Suppose we have just sampled the variables at time $t = k \cdot h$ and we want to calculate the settings of the manipulated variables to be used until $t = (k+1)h$. The procedure is to solve for the vector u that minimizes $J(u)$ while satisfying the inequality constraints, eq. 11. The special QP algorithm used for this is described in Ricker (1985). Note that as is the usual practice for such algorithms, only the first m of the $m \cdot p$ values in u are used. The whole procedure is then repeated at the next sampling interval (including a re-evaluation of the $\mathbf{b}_{*k}$ vector in the internal model). This provides an integral action and improves the estimate of future disturbances (Ricker, 1985; Garcia and Morari, 1984).

5. Blocking of manipulated variables

It can be advantageous to impose additional constraints during the calculation of **u**. One used here is to *assume* that, beginning with the current sampling interval, k, the m manipulated variables will be set and then held constant for κ_1 sampling intervals, then changed and held constant for κ_2 sampling intervals, and so on, where the "blocking factors", $\kappa_j \geq 1$, are a set of n_b specified integer constants where $1 \leq n_b \leq p$ and $\kappa_1 + \kappa_2 + ... + \kappa_{n_b} = p$.

The two main advantages of this are: 1) the number of independent variables in the QP is reduced from $m \cdot p$ to $m \cdot n_b$, and 2) the blocking factors can be used as tuning parameters to improve the robustness of the control system (Ricker, 1985). In general, the use of $n_b << p$ improves robustness but results in a more sluggish control action. Note the blocking factors are simply an assumption used in the QP. Since **u** is recalculated at each sampling interval the signals going to the plant are *not* "blocked".

6. Modified prediction horizon

In some problems it is advantageous to use an unusually long prediction horizon. However, since all the matrix dimensions are proportional to p this can result in a formulation that exceeds the capacity of computer memory. Our software allows one to specify that only n_p selected sampling intervals along the prediction horizon will be given non-zero weights in the objective function, J(**u**), where $1 \leq n_p \leq p$. Thus $n(p - n_p)$ rows can be eliminated from $\mathbf{G}_*$. Of course, the selection of which sampling intervals to give a zero weighting is equivalent to a specification of certain elements of Γ, and hence affects performance and robustness.

7. Robustness filter

Our formulation also includes the possibility of specifying a low-order digital filter that is located in the feedback path. The parameters of this filter can be adjusted to affect performance and robustness as discussed by Garcia and Morari (1984).

8. Summary of "tuning" parameters

The parameters that must be specified and that have an important effect on performance and robustness are:

- The internal model of the process
- The length of the prediction horizon, p, and the number of intervals receiving a non-zero weighting, n_p
- The order and parameters in the robustness filter (if used)
- The number of blocking factors, n_b, and their values, κ_j, $j = 1,n_b$
- The weighting matrices Γ and β
- The upper and lower limit vectors in the inequality constraints, eqs. 8 and 9

Fortunately, even though there are a large number of specifications to be made, guidelines for their use have been developed (Ricker, 1985; Garcia and Morari, 1984; Erickson and others, 1986) and the "tuning" procedure is simplified considerably in practice.

Considerations for systems with constraints on output variables

1. Problems caused by the presence of output constraints

The use of eq. 11 to define constraints on the *output* variables introduces complications that do not exist when one considers only upper and lower bounds on the manipulated variables (as in eq. 8). Two obvious but important considerations are:

- With constraints in the form of eq. 8 it is always possible to find a *feasible* value for each manipulated variable (unless one or more constraints has been defined incorrectly). This is not the case once one adds constraints in the form of eq. 9. For example, in a simple level control problem with an unregulated flow entering a tank there is always a feasible setting for the outlet valve. If the inlet flowrate is high enough, however, the tank will overflow -- regardless of the setting of the outlet valve.

  Thus the possibility of an infeasible QP (or other constrained optimization) problem is very real, especially for complex systems. An algorithm that is to be used in a plant must be designed with this in mind. One way of overcoming this is to redefine the mathematical problem until a solution exists. This is the approach taken by us in this work (see section on softening of output constraints).

- Even when there is a feasible solution to the QP problem, the constraints are on the *model* output variables, not the process outputs. Since there is always model error there is no guarantee that the process outputs will satisfy the constraints. Thus, if a particular constraint should never be violated one must introduce a "safety factor" by, for example, specifying an artificial bound that is more restrictive than the "true" value. Only trial and error on the plant will show whether such safety factors are large enough (or too large).

  On the other hand, *no* controller can prevent constraint violations under all circumstances. Thus it is more appropriate to consider whether this kind of formulation is less likely to cause problems in installation and use than the alternatives.

The evaporator study gave us the chance to evaluate the seriousness of these issues for a specific problem. We doubt that one can generalize from such experience, but it may be useful to others in any case.

2. Use of output constraints as tuning parameters in a non-linear controller

We also wanted to test the use of *tunable*, time-varying upper and lower bounds on output variables, as has been suggested by Cutler (1982) and compared with alternative methods by McDonald, McAvoy and Tits (1986) for level-control problems. The idea is illustrated in Fig. 3 for a single-output system. Suppose that one would like to have the output within a certain region of the setpoint at all times, but that there is no reason for the output to track the setpoint exactly, a common situation. Then instead of specifying a setpoint to be used in the QP objective function one could set Γ equal to zero, β non-zero, and define constraint equations to keep the output in desired region.

A potential advantage of this approach, especially for level-control problems of the type considered by McDonald and his co-workers is that the controller takes *no* action unless a constraint violation is predicted. In that event the severity of the control action depends on current state of the system and the definition of the process model, *i.e.*, it is a completely non-linear control. Variations on this basic theme, involving, for example, constraints that converge, as in Fig. 3, to force the system toward a desired state (Cutler, 1982), or a combination of such constraints with a setpoint (and tuning of the Γ matrix) are also possible and may be advantageous.

To test these ideas our version of QP IMC provides for the definition of output constraints that are a linear function of the setpoint and the time interval within the prediction horizon, as shown in Fig. 3. Thus the elements in the upper and lower bound vectors in eq. 9 may be defined by

$$y_{*i,q,L} = (a_{i,q,L})(s_{*i,q}) \qquad (13)$$

$$y_{*i,q,U} = (a_{i,q,U})(s_{*i,q}) \quad (14)$$

where $y_{*i,q,L}$ and $y_{*i,q,U}$ refer to lower and upper bounds for the ith output at the qth time interval of the prediction horizon, $s_{*i,q}$ is the corresponding setpoint value, and $a_{i,q,L}$ and $a_{i,q,U}$ are specified constants. Constraints of this type will be called *adjustable output constraints* in the remainder of the paper. The constants $a_{i,q,L}$ and $a_{i,q,U}$ are considered to be tuning parameters.

In addition, it is possible to define *fixed output constraints* for any of the output variables. These are independent of the setpoint and the sampling interval and are intended to represent the "true" upper and/or lower bounds for the outputs (*e.g.*, a tank capacity) -- with a margin of safety, if desired. These will be ignored, however, unless they become more restrictive than the adjustable output constraints at a particular sampling interval.

3. *"Softening" of output constraints*

The problems caused by hard constraints on the output variables may be avoided by "softening" them by minimizing any allowed constraint violations. A new independent variable,"e", is defined as the *maximum weighted constraint violation*. The weights referred to here, are those placed on the *output constraints* and are different from Γ and β defined earlier. The weights allow us to give more importance to some constraints over others. The objective function is then modified as shown below (note that e is a scalar):

$$J(u,e) = \frac{1}{2}[\ (s_{*k} - y_{*k})^T\Gamma(s_{*k} - y_{*k}) + \Delta u^T\beta\Delta u + e^2] \quad (15)$$

The constraints may be rewritten in the same form as eq. 11 though **C** and **r** take on different forms to account for softening and any constraints that may placed on the changes in the manipulated variables. If there are no constraint violations e is zero, otherwise it takes some positive value. The weights should be chosen carefully: if they are too small then the optimal solution to eq. 15 may violate some constraint (i.e. e is non-zero) even though there may exist a solution to eq. 12 that does not violate any of the constraints. On the other hand if the weights are too large it may result in some of the stability problems mentioned by Zafiriou (1988). The details of our formulation of the softening problem will be presented in a subsequent paper.

CONTROL OF THE FULL-SCALE EVAPORATOR

Installation and tuning of PI and QP IMC controllers

The scope of the plant experiments was limited to the control of the evaporation process as shown in Fig. 1, *i.e.*, control of the feed and product tank levels *was not* included. For PI control the weak liquor flowrate, F.112, was set by the operators (*i.e.*, held constant unless otherwise noted). The steam flowrate, F.101, was manipulated by feedback to regulate the refractometer signal, D.149, for which a constant setpoint was specified.

The QP IMC algorithm was designed to manipulate both the weak liquor and steam flowrates. In this case the output variables were the refractometer signal (as for PI control) and the evaporator throughput (defined as equal to the weak liquor flowrate -- see *internal model* section).

The plant supervisor requested that we tune the controllers to minimize rapid changes in manipulated variables while keeping the product concentration within ±1% solids of its setpoint. The PI and QP IMC algorithms were thus first tuned off-line using Sewell's non-linear model of the evaporator to obtain this type of response. When the PI controller was subsequently tested at the plant its performance was very oscillatory and it had to be re-tuned on-line. Final settings for the PI controller were a sampling time of 0.25 min., a gain of -200000 lb. steam/hr - % solids, and a reset of 0.036 min^{-1}.

For QP IMC the sampling time was 1 min. The 11/34 needed about 2 seconds of CPU time to complete one cycle of QP IMC calculations. The elapsed time was longer (10 to 30 s) since other tasks such as data acquisition (including archiving of all signals for off-line analysis) and process displays were executing concurrently.

The prediction horizon was 62 minutes. Output weights (*i.e.*, the elements of Γ in eq. 12) were 1.0 for the refractive index and 0.1 for the throughput *at the following 15 intervals along the prediction horizon:* 1, 2, 11-13, 22-25, 41-43, 61, and 62. For all other prediction intervals the weights were zero for both outputs. This use of alternating zero and non-zero weights was necessary because of memory constraints in the 11/34 and would not be necessary in general. Weights on the *rates of change* of the manipulated variables (see eq. 22 of Ricker, 1985) were 0.05 for the steam flowrate and 0.01 for the weak liquor flowrate. Blocking factors were 0, 10, 20 and 30 (see eq. 36 of Ricker, 1985). No robustness filter was used. It is worth noting that these initial tuning parameters *did not* need to be modified on-line, in contrast to the PI controller.

Results for QP IMC

1. *Regulation of product concentration*

Figures 4 and 5 show 3.5 hours of regulation of the strong liquor concentration with a constant target value of the weak liquor flowrate 5.72 m^3/min (1525 gal./min). This is the most common mode of plant operation. The maximum steam flowrate was *specified* to be 1000 kg/min (132000 lb/h). The actual maximum flowrate was from 5 to 10% higher than this. Thus the effect of this specification was to eliminate the possibility of steam valve saturation.

The controller was put into closed-loop operation at 930 minutes. Since the throughput was always below its target value (see Fig. 5) QP IMC held the steam flowrate at the specified maximum value (graph not shown) at all times and manipulated the WL flowrate in order to regulate the product concentration. It quickly brought the product concentration back to the setpoint of 51.5 % solids and held it within ±0.2 % for the next hour (see Fig. 4).

Figure 5 shows the corresponding variations in the weak liquor flowrate. Note that given the constraint on the steam flowrate, it was impossible to achieve the targets on both the product concentration and the throughput. Our weighting of the output errors (given in the previous section) gave higher priority to the product concentration; QP IMC automatically allowed the throughput to deviate from its target as shown in Fig. 5.

At about 1020 minutes the solids concentration setpoint was decreased step-wise from 51.5 to 51 %. In contrast to the setpoint variations to be discussed next, QP IMC was given no advanced warning of this change. Its response was to attempt to increase the WL flowrate rather rapidly back toward the target value of 1525 gal/min, while simultaneously decreasing the steam flowrate. During this period, which lasted about 10 minutes, the estimated disturbance in solids concentration (defined as the measured value minus the estimate from the process model) began to decrease rapidly. Consequently, the controller was forced to decrease the WL flowrate to a value *below* that prior to the change in the solids setpoint.

Since a decrease in the solids setpoint would normally allow an increase in the WL flowrate, the observed behavior suggests either the presence of a large unmeasured disturbance in the evaporator operating characteristics (no sign of which could be found in the operating records), an error in the signal coming from the refractometer, or an error in the internal model. In any case, QP IMC corrected for the apparent disturbance so that the solids concentration was at its setpoint within 1.5 hours of the setpoint change. The maximum deviation from the setpoint was less than 0.5 % solids.

2. Programmed changes in throughput

We also wanted to see whether QP IMC could implement needed changes in plant throughput while keeping the product concentration close to its setpoint. This might be necessary, for example, if the operator were to observe that the feed tank level was too high and rising. As shown in Fig. 6, we specified a ramp increase in throughput of 5% over 45 minutes, an hour at this new target, followed by a ramp decrease over 45 minutes to the original target and a final hour at this value. This was programmed one hour before the initial ramp was to start. Note that the prediction horizon was 62 min. Thus QP IMC had the entire prediction horizon to prepare for the impending setpoint changes -- ideal conditions for setpoint tracking.

Figure 7 shows that the steam flowrate increases *prior* to the start of the WL ramp -- the result of pre-programming the ramp. At the end of the initial ramp, however, the steam flowrate hits the specified upper limit on the steam flowrate of 1000 kg/min (132 Klbs/hr). Meanwhile, at 810 minutes the *actual* steam flowrate (which is under flow control as noted previously) begins to lag the value requested by QP IMC. Between 825 and 875 minutes, the steam flow control valve is saturated, yet the flowrate delivered is well below the expected maximum.

This period illustrates two powerful features of the QP IMC algorithm. First, the *actual* steam flowrate is used as the input to the internal model (rather than the requested steam flowrate). Consequently, the model predicts that the product solids concentration will decrease due to the lower-than-requested steam flowrate. (The product concentration does in fact drop during this period as shown in Fig. 8.) In effect, there is a built-in feed-forward compensation for the difference between the requested and delivered steam flowrate.

Second, since the ***requested*** steam flowrate is at its upper limit during most of this time, QP IMC automatically uses the WL flowrate to maintain accurate regulation of the product solids concentration, even though this means dropping below the throughput target (see Fig. 6). Fig. 8 shows that the maximum deviation from the setpoint is less than 0.4 % solids during the entire 3-hour period -- well within the operator's target of 1 %. Requested changes in WL and steam flowrate are acceptably smooth.

Results for PI control

Figures 9 and 10 are the results of PI control of the product solids concentration during a ramp change in throughput identical to that described above. The ramp increase in throughput begins at the initial time in Figs. 7 and 8. As mentioned previously, in this scheme the WL flowrate is held at its target value at all times, *i.e.*, it tracks the prescribed ramp changes perfectly (not shown).

Figure 7 shows that the solids concentration deviates by more than 1% in both directions from the desired setpoint, which is clearly inferior to the performance of QP IMC shown in Fig. 8. The performance of the PI algorithm could have been improved, however, by adding feed-forward compensation for the variation in the WL flowrate, which is common practice in evaporator control.

SIMULATED CONTROL OF EVAPORATOR WITH STORAGE TANKS

An example of the simulated control of the evaporator with the feed and product tanks is shown in Fig. 11. The internal model used is the one shown in Table 2 discretized for a sampling interval of 5 minutes. The prediction horizon is 35 minutes, and there is a first-order robustness filter on the product concentration with a time constant of 7 minutes. Penalties on the outputs (elements of Γ) are 0, 0.01 and 0 for the feed-tank level, product concentration and product-tank level respectively (these are constant over the prediction horizon). The weights on the output constraints are 10, 10 and 1 for the three outputs (in the same order). Penalties on the rates of change of manipulated variables are 0 and 0.001 for the feed-tank level setpoint and steam flowrate respectively. The plant conditions are listed below:

- the tank levels are within 5% of their maximum values at time zero
- a maximum deviation of 10% is allowed in the product concentration though it should be maintained at its setpoint (at zero).
- constraints on all manipulated variables are at -0.5 and +0.5.
- there is a 20% increase in the flow to the feed tank starting at time zero. This is fed forward to the controller.
- there is an unmeasured disturbance of 10% in the product concentration. This lasts for 260 minutes before it dies out.

The disturbances are larger than would normally be encountered and were chosen just to illustrate the behaviour of the controller under extreme conditions. As shown in Fig. 11 the unmeasured disturbance tends to push the product concentration beyond its lower limit at -0.1 but the output is brought back to its setpoint at the next sampling instant. The temporary deviation of the product concentration from its setpoint at around 300 minutes is caused by the 10% step increase in the additive output disturbance at 260 minutes. The setpoints for the levels are at the half full position (corresponding to -0.35). However, since there is no weight on setpoint tracking for the levels, the controller only attempts to maintain them within their constraints. Figure 12 shows the inputs to the plant. The flow to the evaporator is shown for clarity but it should be noted that the manipulated variable is actually the level setpoint (see Fig. 2). Since the product concentration has a larger penalty on it, the controller initially decreases the flow to the evaporator and increases the steam flowrate, both of which have the effect of increasing the product concentration. Subsequently, the flow to the evaporator is increased (actually the level setpoint is decreased) in an attempt to satisfy the level constraint for the feed tank.

Figure 13 shows the same case run with modeling error. The plant has a 15 minute deadtime in the response of the product concentration to a change in steam flowrate. There is a +10% error in the gain relating the feed-tank setpoint to the level and the product concentration, and a -10% gain error relating product concentration to the steam flowrate. All tuning parameters have the same values as before. The response is more oscillatory at the beginning but smooths out and there is little difference between this case and the one with a perfect model after about 200 minutes.

We also tried using time-varying output constraints with the hope that it may have better robustness properties. After much effort we have seen no significant advantage, here or in other process control problems, for using them. For example, in averaging level control problems of the type considered by McDonald and others (1986) we have been able to achieve good flow filtering with a QP IMC controller having fixed output constraints, a setpoint, and a heavy penalty on movements in the outlet flowrate relative to those on level deviations. The use of time-varying output constraints gave only slightly better flow filtering and was more difficult to tune.

SUMMARY

We have demonstrated that a QP IMC strategy can be used to achieve good control in a full-scale, multi-variable application in which constraints play an important role. Off-line tuning using simulations was sufficient, whereas the PID controller tuned using simulations had to be modified on-line. No robustness problems were evident. Our simulation studies of the larger problem of simultaneous control of the evaporator and storage tank levels shows that there is also considerable promise in this approach. The output constraint softening technique allowed us to maintain the outputs within bounds without the risk of encountering an optimization problem having no feasible solution. However, in general the use of output constraints can lead to stability problems discussed by Zafiriou (1988) and more work needs to be done before they can be used routinely.

ACKNOWLEGEMENTS

This material is based upon work partially supported by the National Science Foundation under grant CPE-8113056 and the Weyerhaeuser Company. Greg Golike, Pierre Leroueil, Barbara Crowell, and Ray Harrison of Weyerhaeuser were instrumental in carrying out the plant tests.

LITERATURE CITED

Brosilow, C., G. Q. Zhao, and K. C. Rao (1984). Proc. ACC 667.

Chang, T. S., and D. E. Seborg (1983). Int. J. Control **37**(3), 583.

Cutler, C. R. (1982). ISA Trans. **21**(1), 1.

Cutler, C. R., J. J. Haydel, and A. M. Morshedi (1983). paper presented at the AIChE Annual Meeting, Washington, D.C., paper 44c, Nov. 2.

Erickson, V. B., and others (1986). AIChE Symp. Ser.

Garcia, C. E., and M. Morari, (1984). Ind. Eng. Chem. Process Des. Dev. **24**, 484.

McDonald, K. A., T. J. McAvoy, and A. Tits (1986). AIChE J. **32**(1), 75.

Mehra, R. K., R. Rouhani, J. Eterno, J. Richalet,and A. Rault (1982). Chemical Process Control 2, Edgar, T. F.; Seborg, D. E., eds., Engineering Foundation, New York.

Prett, D. M., and R. D. Gillette (1980). Proc. JACC San Francsico, paper WP5-C.

Richalet, J. A., A. Rault, J. D. Testud, J. D., and J. Papon, (1978). Automatika **14**, 413.

Ricker, N. L., and T. Sewell (1984). AIChE Symp. Ser. **80**(232), 87.

Ricker, N. L. (1985). Ind. Eng. Chem. Process Des. Dev. **24**, 925.

Sewell, T. (1984). MS Thesis, U. of Washington.

Skogestad, S., and M. Morari (1987). AIChE J. **33**(10), 1620(1987).

Zafiriou, E. (1988). IFAC Workshop on Model Based Process Control, .

TABLE 1 Internal Model Used in Plant Tests

	$\frac{y_1}{m_1}$	$\frac{y_1}{m_2}$	$\frac{y_2}{m_1}$	$\frac{y_2}{m_2}$
a_1	-2.621276	-3.328457	0	0
a_2	2.516800	3.957536	0	0
a_3	-1.151470	-1.763568	0	0
a_4	0.321239	-0.023084	0	0
a_5	-0.064222	0.157688	0	0
b_1	-6.052×10^{-3}	-1.4×10^{-5}	0	1
b_2	2.23×10^{-4}	-2.0×10^{-5}	0	0
b_3	1.0639×10^{-2}	-1.58×10^{-4}	0	0
b_4	-1.272×10^{-3}	-3.7×10^{-5}	0	0
b_5	-1.539×10^{-3}	-7.6×10^{-5}	0	0

where the transfer functions are defined as:

$$\frac{y_i}{m_j} = \frac{b_1q^{-1} + b_2q^{-2} + \ldots + b_5q^{-5}}{1 + a_1q^{-1} + \ldots + a_5q^{-5}}$$

TABLE 2 Internal Model Used in Simulations With Level Control

Outputs	Manipulated Variables: Feed Level Setpoint	Manipulated Variables: Steam Flow
Feed Level	$\frac{1}{30s+1}$	0
Product Conc.	$\frac{648\ s}{(30s+1)(20s+1)}$	$\frac{2.7(-6s+1)}{(20s+1)(5s+1)}$
Product Level	$\frac{-\ 90s}{(30s+1)(30s+1)}$	$\frac{-\ 0.1375\ (-4s+1)}{(30s+1)(2.6s+1)}$

Note: Assumes continuous level control of both tanks

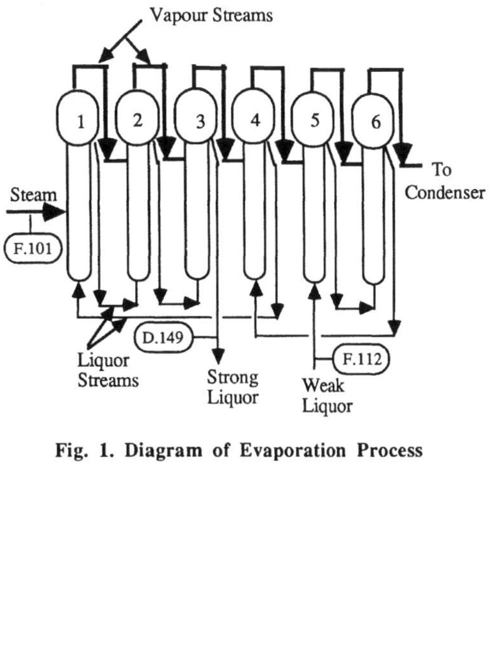

Fig. 1. Diagram of Evaporation Process

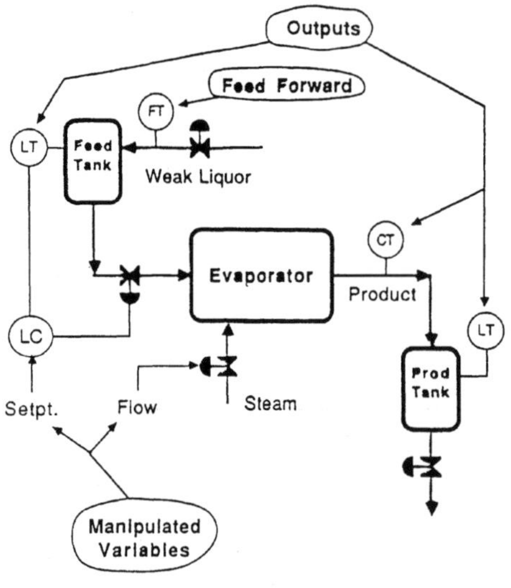

Note: LT is a level transmitter
LC is a level controller
FT is a flow transmitter
CT is a composition transmitter

Fig. 2. Control Structure for Process Including Storage Tanks

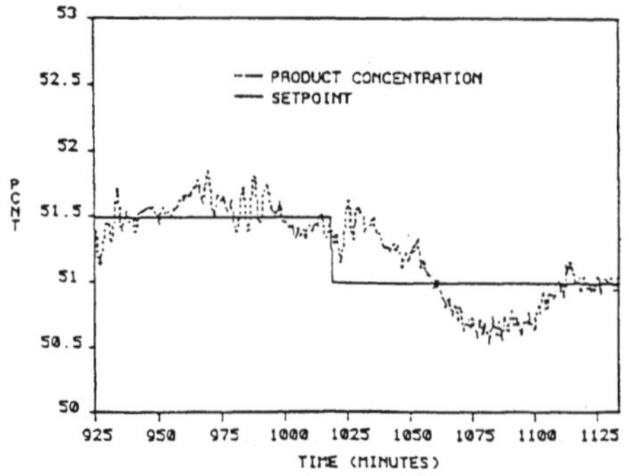

Fig. 4. Reguation of Product Concentration by QP IMC

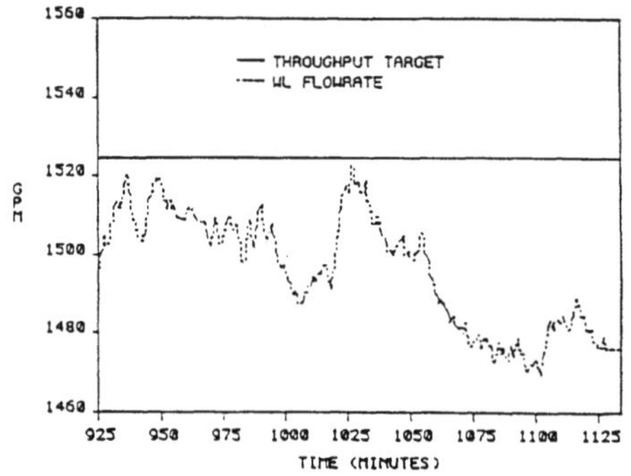

Fig. 5. Adjustment of Weak Liquor Flowrate by QP IMC

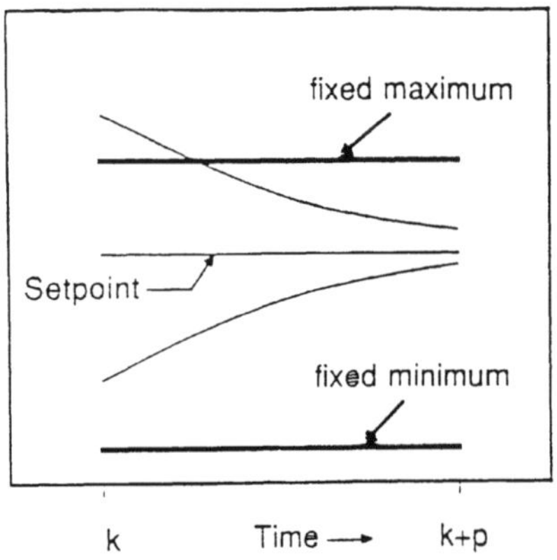

Fig. 3. Illustration of Time-Varying Output Constraints

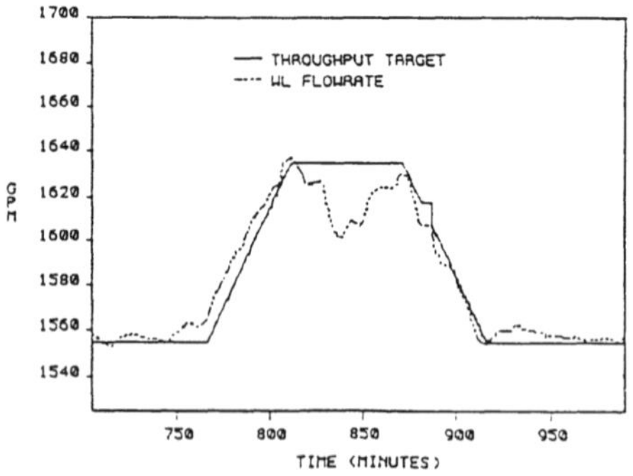

Fig. 6. Programmed Throughput Adjustment by QP IMC

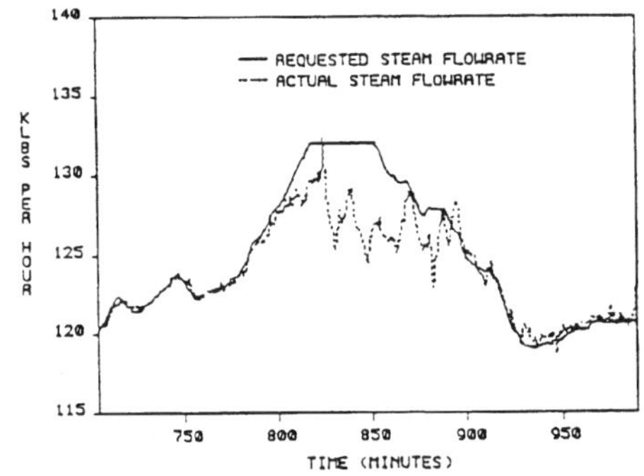

Fig. 7. Steam Flowrate Variations During Programmed Throughput Change by QP IMC

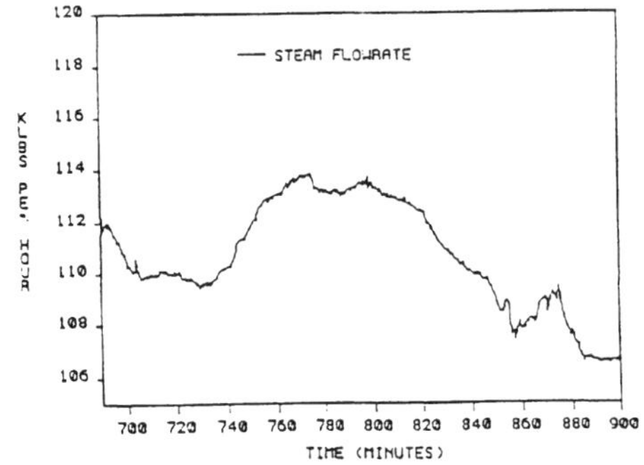

Fig. 10. Adjustment of Steam Flowrate During PI Control

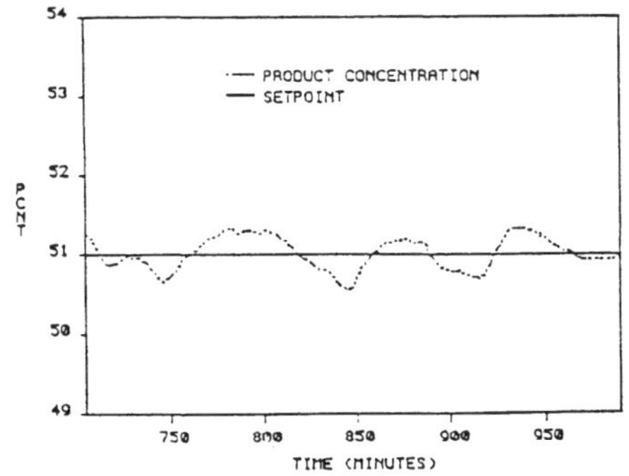

Fig. 8. Regulation of Product Concentration During Programmed Throughput Changes by QP IMC

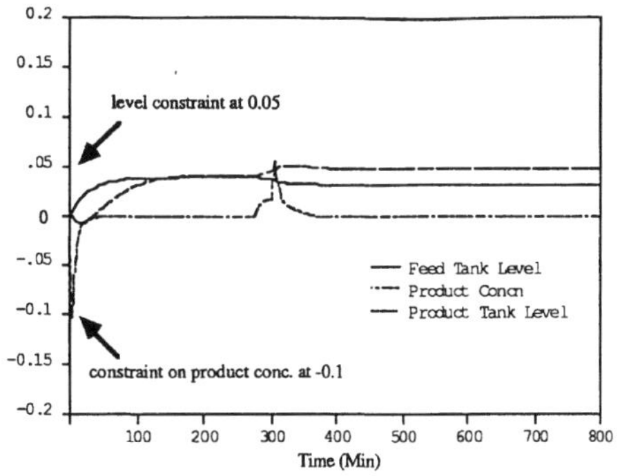

Fig. 11. Outputs for Control of Evaporator and Storage Tanks With a Perfect Model

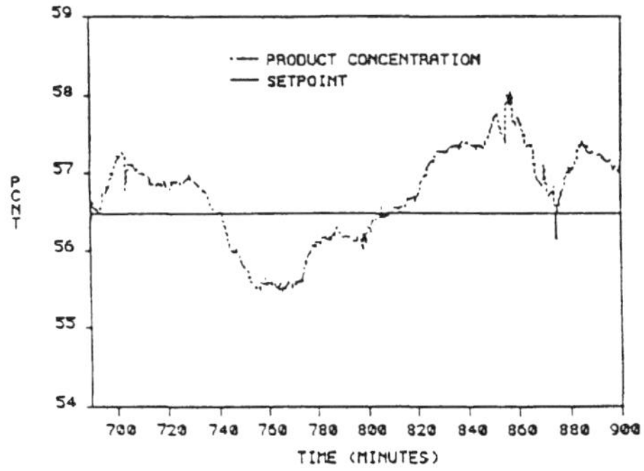

Fig. 9. PI Control of Product Concentration

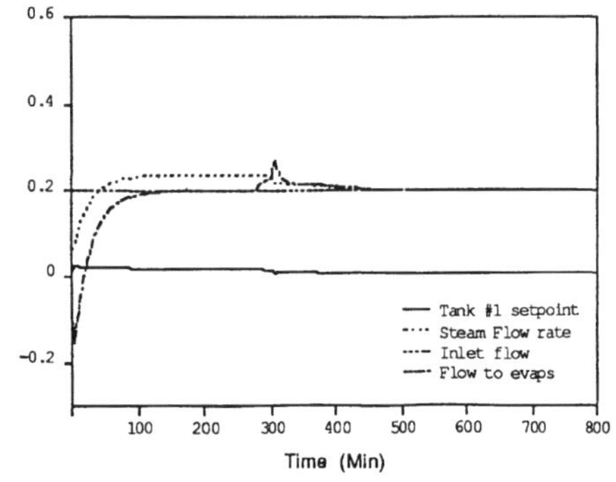

Fig. 12. Inputs for Control of Evaporator and Storage Tanks With a Perfect Model

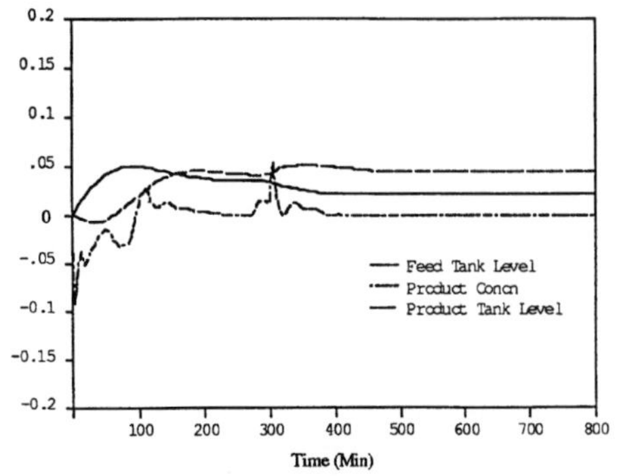

Fig. 13. Outputs for Control of Evaporator and Storage Tanks With Modeling Error

DESIGN CONSIDERATIONS FOR A HYDROCRACKER PREFLASH COLUMN MULTIVARIABLE CONSTRAINT CONTROLLER

C. R. Cutler* and S. G. Finlayson**

**DMC Corporation, P.O. Box 800873, Houston, Texas, USA*
***SUNOCO Inc., P.O. Box 307, Sarnia, Ontario, Canada*

INTRODUCTION

In 1984, Sunoco's Sarnia Refinery commissioned a new Hydrocracker Complex. The start up was extremely successful with the plant running stably and at high throughput in relatively short order. The Hydrocracker Preflash Column is the first distillation column after the Hydrocracker Reactor and is therefore most susceptible to swings in the reactor severity. Since the startup of this Hydrocracker Complex operations has been unable to consistently operate this column successfully. Hydrocracker Reactor Severity swings affect the proportion of light material in the feed to the extent that the normal response of the tower is dominated almost completely by these severity disturbances. This paper discusses the key design aspects considered during the identification of dynamic responses of this system and the design of a multivariable Dynamic Matrix Controller (DMC) for this tower.

The DMC controller built for this Preflash Column uses four manipulated variables and two disturbance variables to control four dependent variables. The controller has significantly reduced the disturbances to the downstream Gas Plant and Main Fractionator. Further, it has made feasible the constrained multivariable control of the second stage of the Hydrocracker Reactor by handling the increased number of disturbances that occur when the reactor system is operated against its constraints. The Hydrocracker Reactor controller is designed to maximize feed by consuming all the hydrogen available. Riding the hydrogen availability constraint passes the disturbances in the hydrogen feed system to the Hydrocracker and then into the Preflash Column.

Traditional PID control schemes have proven to be inadequate due to the highly interactive nature of the Preflash Column manipulated variables on the controlled variables. The design of PID based control schemes was further impeded by the dominant influence of the disturbance variables on the Preflash dependent control variables. The disturbances had obscured the effects of the manipulated variables when traditional identification techniques were employed. An identification technique which is the inverse of the DMC controller permitted the dynamics of the column to be obtained by sorting out the effects of the manipulated variables from the disturbance variables.

PROCESS DESCRIPTION

Figure # 1A shows a simplified process flow of the Hydrocracker Preflash Column. This column is the first column in the fractionation train downstream of the Hydrocracker Reactor. A feed containing methane through to recycle oil enters the tower in the upper section after being preheated in a reactor effluent exchange train. Butanes and lighter are taken overhead as a gas stream with pentanes as the heavy key in this stream, no liquid products are taken from the top of the column. This overhead gas stream is then sent to a light ends system for distillation. The Preflash bottoms stream contains the C5 and heavier material. This stream is sent to a fractionator where recycle oil is separated from the products and returned to the reactor system. Unstable operation in the Preflash Column upsets both the downstream gas plant, as well as, the reaction system by way of the fractionator.

FIGURE # 1A: PREFLASH TOWER PROCESS FLOW

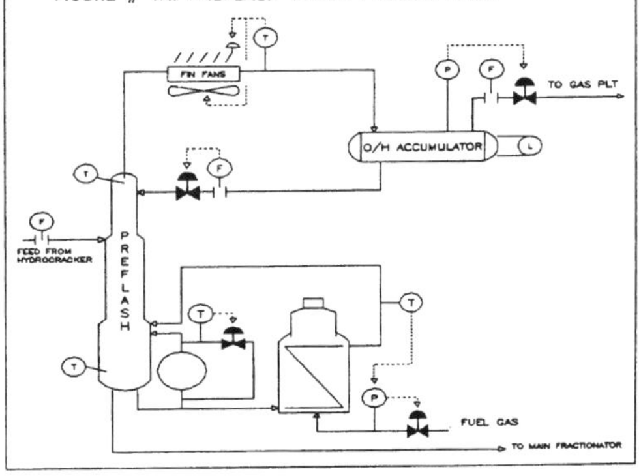

The column has several non standard features which make it difficult to control. Overhead cooling is accomplished by air coolers that have both variable pitch fans and variable position louvers to regulate the air flow and control the outlet temperature of the hydrocarbon. As with most systems of this type, the range of the control by pitch and louver adjustments is limited and the fans must be turned on and off manually to permit operation over a wide range. Control of the overhead accumulator level is particularly difficult due to the absence of a liquid draw stream and the small percentage of the total material entering the column represented by the overhead gas stream. Reboil energy is supplied to the tower bottom via two separate reboilers, a combination gas and oil fired heater and a reboil exchanger using reactor effluent as a source of heat. The column is subjected to several large disturbances. Variations in the reactor temperature have a significant effect on the light gas production. In addition the feed preheater for the column is the last exchanger in the reactor effluent exchange train making it susceptible to variations in the heat input to the feed and recycle streams in the upstream exchangers. The effect of these disturbances on the level are such that the standard level control on the accumulator which sets the reflux rate back to the column is not capable of keeping up with the disturbances. Several attempts had been made to build PID based control schemes to handle the level control problem with only limited success.

The DMC controller designed to handle the Preflash level control problem takes advantage of all four of the manipulated variables which directly affect the level. These four manipulated variables are the temperature out of the overhead air coolers, the reflux flow rate, the heat input to the bottom of the column from the reactor effluent, and the fuel to the Preflash reboil furnace. The actual manipulated variables are the setpoints for the temperature out of the overhead condensers, the reflux flow, and the reboil furnace fuel gas header pressure. The position of the bypass on the flow of the heat medium from the bottom of the column to reactor effluent exchanger is used as the fourth manipulated variable.

In addition to the control of the accumulator level, three other dependent variables are included in the controller. The output signal from the overhead temperature controller is used as a controlled variable to permit the controller to recognize the limited range of the pitch and louver positions. A temperature at the top of the column is used which correlates with the pentanes in the overhead stream to the gas plant. Another temperature at the bottom of the column is used to reflect the butanes left in the bottom stream. The DMC controller was designed to hold the top temperature to a specific setpoint while the bottom temperature is controlled between an upper and lower limit rather than at a specific setpoint.

The feed forward variables for the DMC controller are the feed rate to the column and the weighted average bed temperature for the second stage reactor. The feed temperature was not included directly as a disturbance variable since changes in the feed rate also caused the feed temperature to change. In effect, the feed temperature is not independent of changes in the feed rate.

DESIGN CONSIDERATIONS

The first step in the design of a multivariable controller is to understand the independent and dependent variables in the system to be controlled. From a mathematical point of view, any variable can be an independent or dependent variable; however, from a control point of view the choice is not arbitrary. The origin of an independent manipulated variable is a physical device in the system that can be changed, i.e. a valve position, the speed of a machine, a fan on or off. The basic independence of a variable can be transferred to a dependent variable by placing a controller inside the system. The independence of a valve position may be transferred to the set point of a flow controller. In this case the flow controllers dynamics are introduced into the system and the valve position becomes a dependent variable. If a temperature in the system is used to change the setpoint of the flow controller, then the temperature controller's dynamics are also embedded in the system. The temperature setpoint becomes the independent variable and the flow controller's setpoint becomes a dependent variable along with the valve position of the flow controller.

It is usually desirable to maintain the basic regulatory system intact and superimpose the multivariable DMC controller on top of the regulatory control system. Two circumstances may exist where such a choice is not desirable. If a basic regulator is cyclic and cannot be tuned to be stable without destroying its responsiveness, then the controller should be removed and its output set directly by the DMC controller. In the other case, the controller may be off control much of the time with the manipulated element at one of its limits. The process identification under these conditions is made difficult since part of the time the controller dynamics are in the system and part of the time they are absent. The use of the manipulated element as the independent variable circumvents the problem of identification and provides the DMC controller with the ability to operate the system with the manipulated element at its limit.

Variables that are fed into the system to be controlled are usually independent disturbance variables. Inspection of the variables further upstream may be necessary to determine if the disturbance variables are truly independent. For example, with a simple column where the bottom product exchanges heat with the feed, the temperature of the feed directly into the column is a dependent variable. In this case the temperature of the feed going to the bottoms to feed exchanger is the independent variable. Care must be taken to carefully define the boundaries of the system to prevent the selection a dependent variable as an independent disturbance variable.

For the Preflash Column the Independent variables were selected as follows:

1) The flow of material to the Preflash Column is a function of the feed rate to the hydrocracker, the recycle oil rate, and the hydrogen makeup, which makes the flow an independent disturbance variable.

2) The weighted average bed temperature (WABT) reflects the severity of the reactor which makes its effect on the Preflash feed composition an independent disturbance.

3) The pitch of the fans and the louver positions are the physical devices which are the basic independent variables on the overhead condensing system. A position controller was built with a setpoint which moves the pitch of the fans and the louvers in unison. The setpoint of the position controller was made dependent by including in the system a controller on the overhead condenser outlet temperature. The setpoint for the temperature controller becomes the independent manipulated variable in the DMC controller.

4) The control valve position on the reflux out of the overhead accumulator is the physical device for the independent effect of reflux. The control valve was made dependent by choosing the setpoint of the reflux flow controller as the independent manipulated variable.

5) A circulating stream drawn from the bottom of the Preflash column is heated by the reactor effluent and then returned to the column. The basic regulatory system uses a controller on the return temperature to manipulate a bypass valve around the reactor effluent to reboiler exchanger. Many times the controller has the bypass valve shut; therefore, the valve position itself was chosen as the manipulated variable for the DMC controller.

6) The last independent physical device in the Preflash system is the control valve on the fuel to the reboil furnace. The pressure on the fuel gas header is used to set the fuel valve making the valve position dependent on the setpoint of the pressure controller.

It should be pointed out that the flow of fuel to the Preflash reboiler furnace was chosen as the independent manipulated variable, which made both the valve position and the setpoint for the pressure controller dependent variables. In this case a flow controller did not exist, only a flow measurement. The dynamic analysis of the system used the flow measurement as the independent variable which further required that a relationship be developed between a change in the flow and a change in the fuel gas header pressure. After the calculation of a flow change by the DMC controller, a change in the fuel gas header pressure setpoint is determined. The advantage of the fuel gas flow as the independent variable is the elimination of the nonlinearity associated with different numbers of burners being in service.

The choice of independent manipulated variables is made easy by considering the physical devices which can be changed within the envelope drawn around the system to be controlled. From this basic knowledge of the independencies, the number of controllers imposed on the physical devices can be evaluated for appropriateness.

The dependent variables are chosen basis the objectives set by operations management and by the rangeability of select dependent variables which constrain the movement of the manipulated variables.

The dependent variables for the Preflash tower are as follows:

1) Overhead fan pitch and louver position; controlled within limits.

2) Overhead drum level; controlled within limits.

3) Upper tray temperature; controlled to a setpoint.

4) Bottom temperature; controlled within limits.

The position setpoint of the controller on the fan pitch and air cooler louvers is a dependent controlled variable for the DMC controller. Frequently the temperature setpoint for the overhead condenser effluent is limited by the position controller being overranged. Further, the other independent variables have a significant influence on the output of the temperature controller with the temperature setpoint held constant.

All but one of the dependent variables in the DMC controller are held between upper and lower limits. If the dependent variable is between the limits, it does not have an error. To generate a specific setpoint, the upper and lower limits are set to the same value. In the case of the dependent position setpoint, the limits were set wide, i.e. 10% and 90%. With the position setpoint as a dependent variable, the DMC controller recognizes the limited rangeability of the top temperature controller and moves the other manipulated variables to keep the temperature controller functioning.

The overhead accumulator level is set between upper and lower limits to minimize the fluctuations in the feed to the gas plant. The gas plant feed is the vapour leaving the accumulator. The steady state response for the level for a step change in an independent variable is a ramp, i.e. the level continues to change at a constant rate. The nature of a ramp variable is that if it is not changing with time, but is not at its setpoint, then to return it to its setpoint requires the system be thrown out of balance to permit the level to change to its setpoint, then the system must be moved again to put the level back in balance. The effect on the gas production of moving the level is to change the production rate for a while and then put it back to its original value. Since the scenario of changing the level puts a disturbance into the gas plant, it is preferable to design the controller to stop the level from changing, then leave it unless it is too high or too low.

An upper tray temperature is used as a dependent controlled variable to indicate pentanes in the overhead stream to the gas fractionation plant. This is the only dependent variable controlled to a setpoint to prevent hexanes from going overhead and to prevent the loss of butanes out the bottom. Maintaining a constant separation is extremely important in this tower since this is where the primary separation takes place for the Hydrocracker Complex. Therefore, the tray temperature is controlled to a specific setpoint to stabilize the column separation.

Finally a bottom column temperature is used as a dependent variable with its limits specified to prevent butanes from going out the bottom and to prevent hexanes from going out the top with the gas product. Figure # 1B is a simplified process diagram showing the preflash column DMC controller, it's feedforward variables, manipulated variables and controlled variables.

PROCESS IDENTIFICATION

The key to the success of the DMC Preflash Controller was the quality of the data obtained during the testing of the column. Figure # 2 shows the movement of the independent variables during the test. As can be noted, the movement of the independent variables was spread across the entire test period. A more common practice is to pulse each manipulated variable for a specific period and move on to the next one. The objective in spreading out the movement of the manipulated variables was to reduce the risk that a large unmeasured disturbance would impact on a single independent variable while it was being perturbed. This is an important consideration when working on large multivariable problems. Also it can be noted that the change in the manipulated variables did not follow a prescribed up and down movement. The reason for this behaviour was the personnel conducting the test had the operators change the manipulated variables at each point in time to keep the column moving towards the desired operating range. For example, if the overhead accumulator level was falling too fast, the reflux rate would be reduced. Operators were allowed to make any changes in the variables while the process was being tested. The only condition was that multiple moves were not made in synchronization. Usually this is not an issue, since very few operators follow the same pattern.

The process identification technique used to find the step response curves between the independent and dependent pairs was the inverse formulation of the DMC controller. Since the DMC controller rigorously handles all the interactions between variables, its inverse permits the identification of the step response curves when all the variables are changing at once. Further, the Dynamic Matrix Identification (DMI) technique does not require that the system be at steady state to begin the test or come to steady state any time during the test.

The weighted average bed temperature was found to dominate the other independent variables in the model. For example a change in WABT of one degree C had the same effect on the level in the overhead accumulator as a 15 percent change in the top reflux. The magnitude of the WABT on the response of the dependent variables explains the inconsistent results obtained with the traditional identification techniques used earlier to design the PID controllers.

Examination of the response curves shown in Figure 3 demonstrates the unusual dynamic behaviour of many of the independent dependent pairs. Another advantage of the inverse DMC identification (DMI) is that the form of the differential equations describing the model do not have to be known. The only assumption made is that the process response either comes to steady state or to a constant imbalance. The model is represented by a vector of numbers which can have any form. The method is valid for any system which can be described or approximated by a system of linear differential equations.

The major difficulty in using the DMI method is finding the time to steady state. For the inverse DMC formulation to be valid, requires that the last coefficients in the numerical vector representing the response approach a constant value. The time for the Preflash Column dependent variables to come to steady state was found to be between one and two hours.

The DMC controller is a constrained multivariable controller with a built in linear program. The linear program uses the controller's prediction of where the system will be at steady state to find the most economic set of changes in the manipulated variables to eliminate the steady state error. The nature of the linear program is that it will be at the same number of constraints as there are independent variables. For the Preflash Controller there are four independent manipulated variables or four degrees of freedom, which means the controller will ride four of the sixteen possible constraints. The controller may be at an upper or lower limit on any of the four manipulated and four dependent controlled variables. One degree of freedom is required to balance the ramp response of the accumulator level. The three remaining degrees of freedom are used to find the minimum energy cost for operating the column inside the limits specified for the independent and dependent variables.

SUMMARY

Experience with the controller has been excellent with complete operator acceptance. Historically the control of the Preflash Column has consumed a large portion of the operator's time. Control of the gas plant has improved significantly with the reduction in disturbances that result from the improved level and cut point control. Figure # 4 shows before and after results for the Preflash accumulator level and gas flow to the gas plant. The economic incentives for the controller are derived from the its ability to attenuate disturbances. The reduced disturbances in the down stream columns permits other controllers to move closer to their constraints where significant gains can be realized. Further, on the upstream side it permits the reactor controller to take full advantage of the opportunity to consume hydrogen and increase feed rate without creating additional problems in the gas plant.

FIGURE # 1B: PREFLASH TOWER COMPUTER CONTROLS

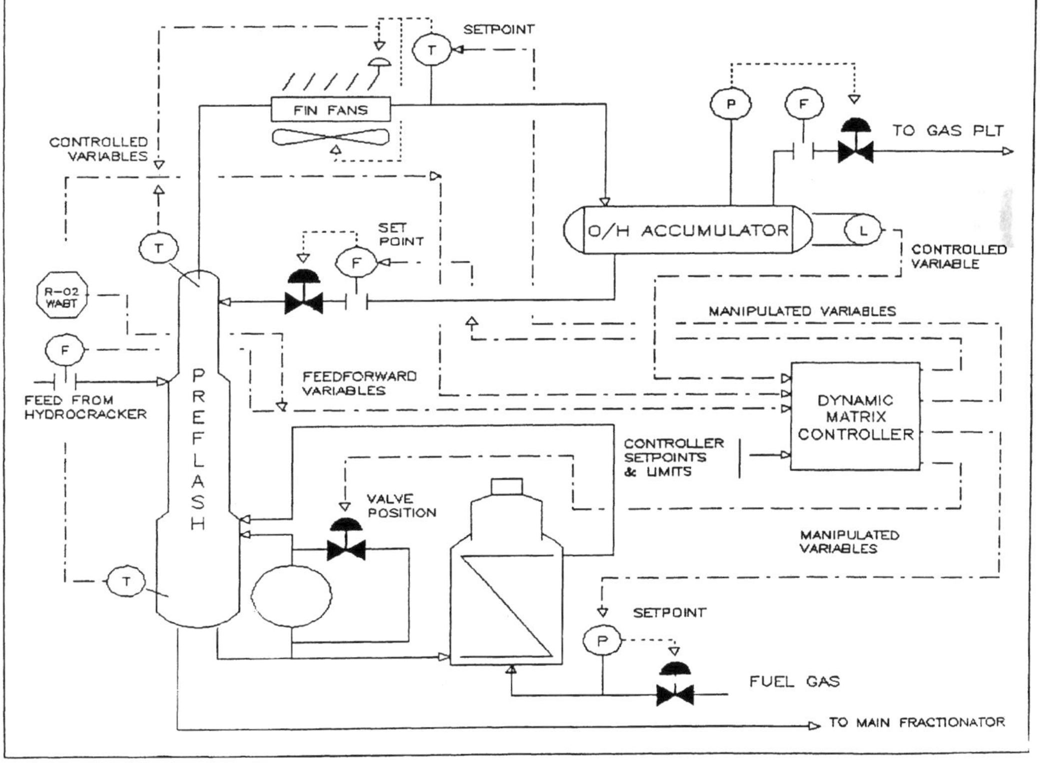

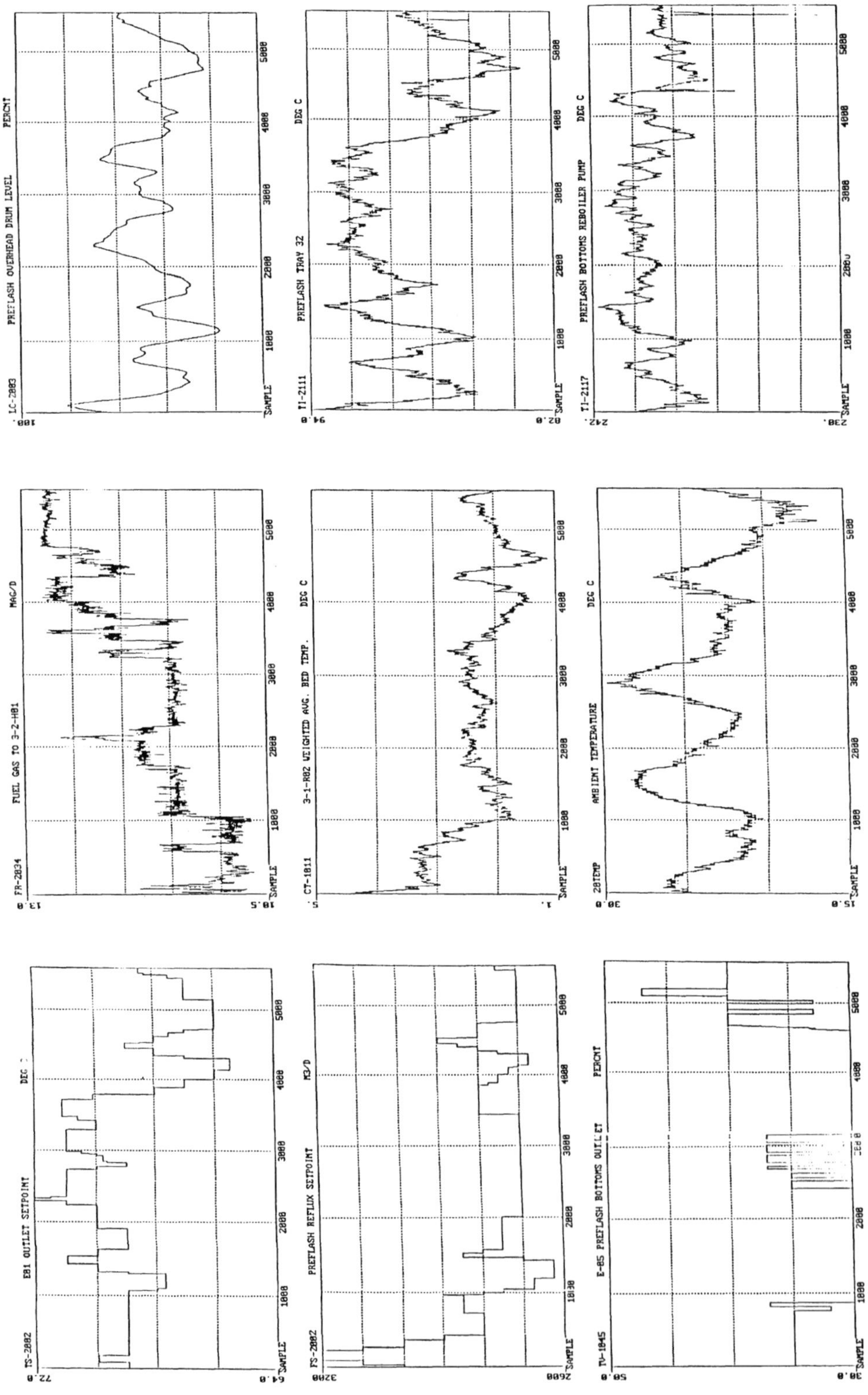

FIGURE 2 DYNAMIC TEST OF PREFLASH COLUMN

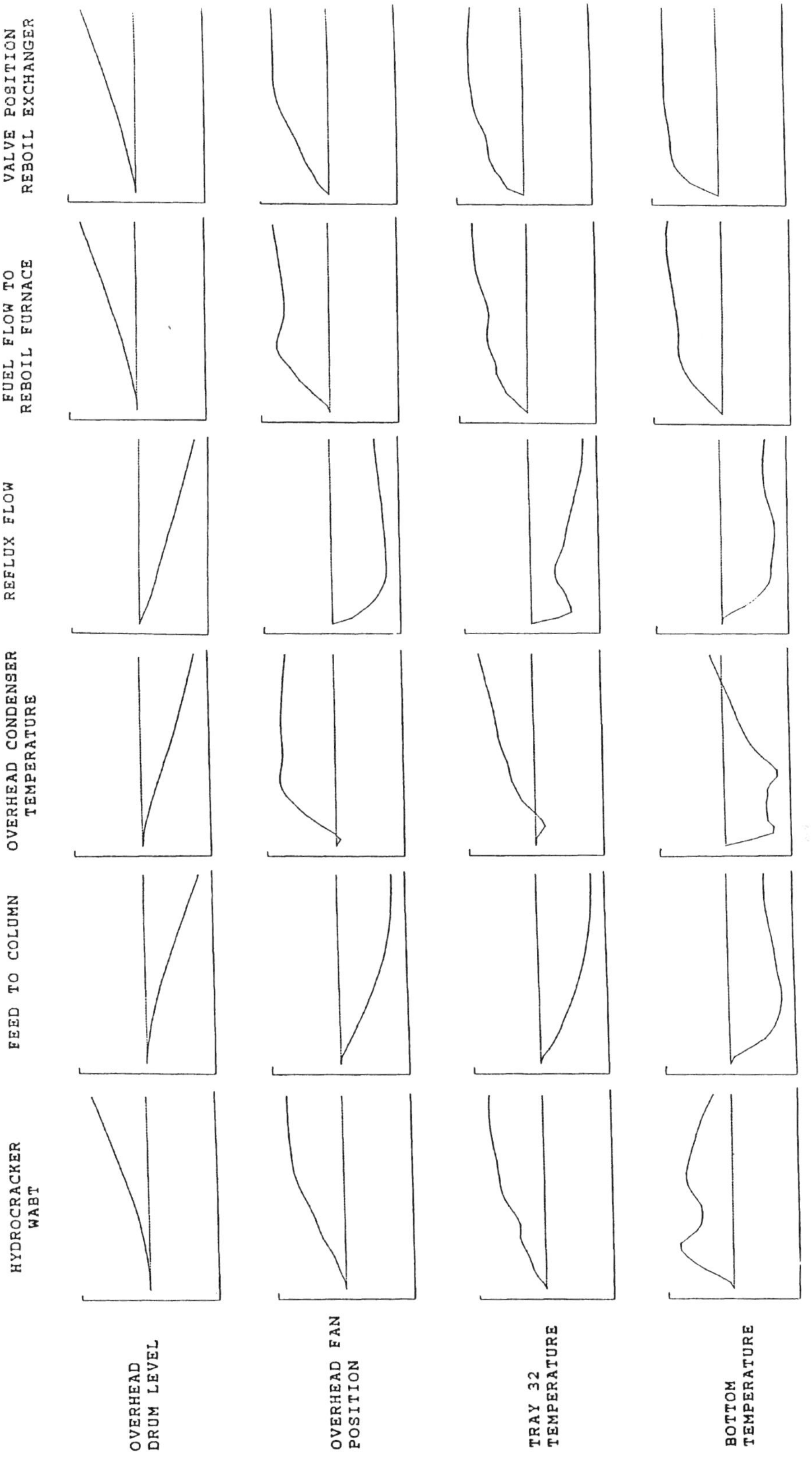

FIGURE 3 DYNAMIC MODEL OF PREFLASH COLUMN

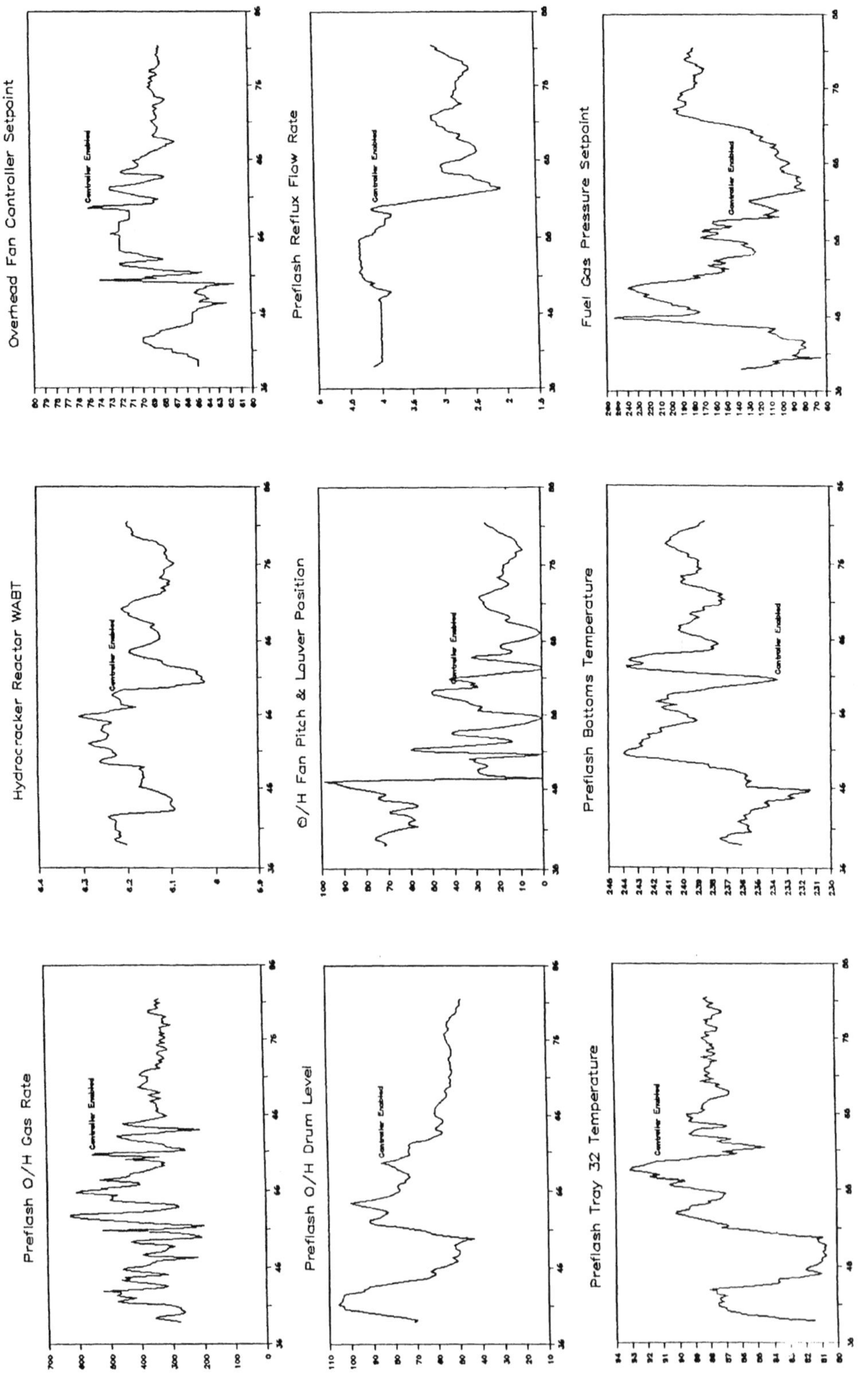

FIGURE 4 CONTROL OF PREFLASH BEFORE AND AFTER DMC

THE IDOCOM–M CONTROLLER

P. Grosdidier, B. Froisy and M. Hammann

Setpoint Inc., 14701 St. Mary's Lane, Houston Texas 77079, USA

Abstract. Multivariable controllers will be applied on a wide-scale basis in the refining and petrochemical industries if they satisfy key requirements that will ensure both their technical and financial success. This contribution attempts to list these requirements and shows how they have been achieved in SETPOINT's own multivariable controller (IDCOM-M).

Keywords. Computer control; controllability; industrial control; least-squares control; multivariable control systems.

INTRODUCTION

Multivariable controllers are being applied more and more widely in the refining, chemical and petrochemical industries. In most, but not all applications, these controllers rest at the highest level of a hierarchy of simpler controllers and regulate the overall operation of a single process unit. This hierarchy of controllers is commonly known as the "advanced controls" for the unit and is distinct from all the "decentralized" controllers which regulate process valves and other basic pieces of equipment. This high level of automation has only been reached in the last few years and has been made possible by the introduction of digital instrumentation and the availability of inexpensive computing power.

The technical and financial success of a multivariable controller will depend mostly on the ease with which it can be tuned and commissioned and on its robustness with respect to changing unit operating conditions and operating objectives. This communication shows how these conditions translate into specific requirements and shows how these have been achieved in SETPOINT's multivariable controller, IDCOM-M.

REQUIREMENTS FOR INDUSTRIAL MULTIVARIABLE CONTROLLERS

A multivariable controller physically takes the form of a high-level language program which resides in the process control computer. Requirements for this program and their motivations are as follows.

Generic Code

The same software must be used for all applications. The amount of time and the level of technical competence necessary to write and test multivariable control software mandate this requirement. This implies that the controller must offer many features, not all of which need be used in any specific application. The resulting code complexity and redundancy will be compensated by standard user training and simpler software maintenance.

Constraints

All process manipulated variables are bound by position and rate-of-change (velocity) constraints. These constraints generally reflect equipment capacity, but can change considerably along with operating conditions. An air blower's maximum throughput, for example, decreases between dawn and the early afternoon with rising ambient air temperature. A multivariable controller must take these constraints into account in an explicit manner.

Special Dynamics

Zero-gain and integrating transfer functions are regularly encountered in chemical processes. A multivariable controller must include provisions to handle these special dynamics.

Communication

Like any other controller, a multivariable controller must be able to communicate with the process as well as with other controllers located above or below in the control hierarchy. In practice, this means the controller must have access to the process database.

Fault tolerance

Provisions must be made for the controller to remain on-line whenever some of the manipulated and/or controlled variables are disabled. This can occur either as a result of instrument failure or simply by operator intervention. Fault tolerance is essential, for without it high controller uptime cannot be expected.

Non-square Systems

This term refers to processes where there are fewer or more manipulated variables (MVs) than controlled variables (CVs). The transfer function matrix for these processes has fewer or more columns than rows, respectively, hence the terminology. A multivariable controller must have the ability to control non-square systems since these may occur spontaneously as a result of sensor or actuator failure. Multivariable controllers generally need this ability since non-square control problems occur in practice far more frequently than commonly thought.

Control problems with fewer MVs than CVs arise, for example, when one of the MVs in a multivariable control scheme is deactivated or reaches saturation. In this case, it may be preferable to maintain control on all the CVs, at the expense of offset between setpoints and measurements, rather than remove one of the CVs from control. This choice can be made simply because it best reflects operating objectives.

Control problems with more MVs than CVs arise whenever extra manipulated variables are available, providing opportunity for optimization and trade-offs. This is the generic situation for some processes such as Fluidized Catalytic Cracking (FCC) units. These extra handles on the process can be put to use either for providing a link between the controller and a process optimizer or simply for improving closed-loop performance. In the latter case, the additional manipulative action compensates for the performance limitations imposed by the velocity clamps on the manipulated variables.

Controllability Supervisor

It is by now well understood that not all multivariable processes are controllable from a practical point of view.[1] A multivariable controller faced with the task of regulating closely coupled variables will invariably be susceptible to stability problems stemming from plant/model mismatch and may attempt manipulative moves of unrealizable magnitude - the so-called "large inverse" effect[3]. In this communication, we will say the process is ill-conditioned whenever these two symptoms appear under feedback. If the symptoms do not appear, the process will be considered well-behaved.

Fault tolerance and design requirements which call for the ability to deactivate any of the MVs or CVs at the operator's will, impose a special burden on a multivariable controller. Although a process with given MVs and CVs may form a well-behaved system, the subsystems created by the loss of some of the MVs and/or CVs may be ill-conditioned. A multivariable controller must therefore be able to identify such ill-conditioned systems on-line and take corrective action to avoid poor closed-loop performance, if not instability.

Tuning and Commissioning

A multivariable controller will be installed on a widespread basis in the process industries only if tuning and commissioning can be done with trained but unspecialized control engineers and in a reasonable amount of time.

Compatibility

Ultimately, a multivariable controller should be viewed in the same terms and should be made to "look" and "act" as much as possible like all the other control blocks in the advanced control hierarchy. Compatibility will not only enhance software uniformity and ease its maintenance, but it will also increase the multivariable controller's chances of acceptance by the operators. The latter is an essential condition for the long-term survival of any control loop in an operating environment, whether multivariable or not.

THE IDCOM-M CONTROLLER

The IDCOM-M controller is a multivariable controller developed by SETPOINT in cooperation with Adersa of France. It is based on Adersa's model-predictive control technology[2] and relies on process models which are identified from dynamic tests performed on the process prior to commissioning. SETPOINT includes this proprietary technology in many of its process unit advanced control package licenses.

The IDCOM-M controller is a FORTRAN program which runs on process control computers and can tie into various databases such as SETPOINT's SETCON*, Texaco's TACS/TDACS*, Honeywell's TDC 3000*, Foxboro's IMPAC* and IBM's ACS*. The program can be scheduled much like any other calculation block on the process control system and differs from other blocks only by its size and complexity. The program is the same for all applications, and the specifications for each of these are made by configuring a fill-in-the-blanks text file which the program reads as part of its initialization routine. This file, called a definition file, identifies the process inputs and outputs and points to the impulse response model files that link each input/output pair. Model files representing self-regulating, integrating and zero-gain transfer functions can each be specified. The definition file also specifies control objectives (setpoints, constraints and tuning parameters) for each of the variables and identifies their tag names in the database. The construction of this file is essentially the design step that sets the overall strategy for each application. Beyond this step is the specification of numerical values for the tuning parameters and the testing and commissioning of the controller. In the following developments, we will present the controller's basic inputs and outputs and its overall flowchart. For the sake of simplicity, many of the controller's detailed features are omitted.

* Registered trademarks.

Controller Inputs and Outputs

The controller recognizes two types of process inputs (MVs and DVs) and one type of process output (CVs). The primary control system objectives placed on manipulated variables are positional and velocity constraints as well as Ideal Resting Values (IRVs). An IRV acts like a MV setpoint. It specifies the desired MV operating point. However, the attempt to satisfy IRVs occurs only after the primary CV control objectives are (predicted to be) satisfied. The IRV concept proves useful when the control strategy has extra degrees of freedom (more MVs than CVs).

The settings for IRVs are motivated by operational or economic considerations. Typically, these settings are made by a process optimizer which is located above IDCOM-M in the advanced control hierarchy. IRVs are therefore part of the link between IDCOM-M and the process optimizer.

Disturbance variables are process inputs that can be measured but not adjusted by the controller. Changes in DVs are used to predict future values of the process outputs.

Feedforward control action counteracts the predicted upsets as DVs vary.

Two different and mutually exclusive control objectives can be placed on CVs. First is the setpoint specification wherein the controller always attempts to maintain the CV at the specified value. Second is the zone specification, meaning that the objective is to keep a process output within some region defined by maximum and minimum zone limits. The process output remains uncontrolled so long as it remains inside its zone limits. The controller takes corrective action whenever the output ventures outside a zone limit and treats the latter as the CV setpoint. The zone concept proves useful when the control strategy calls for keeping some process outputs "under control," which means keeping the variable within a range where drifting is acceptable.

Controller Objectives

The controller's overall objective is first to bring the CVs to their setpoints subject to constraints on manipulated variables. The controller then tries to bring MVs to their IRV subject to "control equalities." These equalities ensure that whatever objective was obtained in the CV optimization is still satisfied in the MV optimization.

A more detailed flowchart of the controller's objectives is shown in Fig. 1 and is discussed below. The controller addresses these objectives sequentially at each execution.

Objective 1: The controller reads all required numerical values from the database. This includes process measurements as well as control objectives such as constraints and setpoints.

Objective 2: The controller determines which MVs, DVs and CVs are available for manipulation, feedforward and control, respectively. This decision is reached by checking the database status of the variables as well as an overriding operator switch.

Objective 3: The controller assesses the controllability of the process and selects which CVs can and should be controlled. The selection is based on process operating conditions, models and a list of "controllability" priorities. The decisions made here reflect fundamental limits of feedback control, regardless of the actual control algorithm used.

Objective 4: The controller finds the future MVs which minimize a sum of squares between the predicted process outputs and their setpoints. Mathematically, the controller solves the following CV setpoint optimization problem:

$$\min_{u_j} \sum (y_i - s_i)^2 W_i^2 \tag{1}$$

subject to

$$u_{jmin} \le u_j \le u_{jmax} \tag{2}$$

and

$$\Delta u_{jmin} \le \Delta u_j \le \Delta u_{jmax} \tag{3}$$

In these expressions u_j denotes the future MVs, y_i the predicted CVs and s_i the setpoints. Δu_j denotes the MV's rate-of-change, and W_i denotes weighting factors which are used to emphasize some of the error terms.

Objective 5: The outcome of Objective 4 may or may not yield a unique solution. In the latter case the controller finds the MVs which minimize a sum of squares between MVs and their IRVs subject to control equalities. Mathematically, the controller solves the following MV IRV optimization problem:

$$\min_{u_j} \sum (u_j - r_j)2\, W_j2 \tag{4}$$

subject to (2), (3) and the control equalities

$$y_i = t_i. \tag{5}$$

In (4) r_j denotes the IRVs or the previous value of a MV if the latter has no IRV. Here too, the W_js are weighting factors. In (5), t_i denotes the predicted values of the CVs which would be obtained with the solution of Objective 4.

Objective 6: The calculated MVs are written back to the database, along with other general maintenance information such as MV and CV statuses and alarm messages.

Discussion

The implications of Objectives 4 and 5 are worth exploring in terms of the relative number of MVs and CVs available for a given process. Three cases can be identified depending on whether the process under control has a smaller, equal or greater number of MVs (N_m) than CVs (N_c), respectively.

Case 1 ($N_m < N_c$): The controller has an insufficient number of MVs to control all the CVs at their setpoints at steady state. The controller is said to lack degrees of freedom. In this case, the IDCOM-M controller controls the CVs in a "least-squares" sense, i.e., IDCOM-M finds the MVs that, at steady-state, minimize the weighted sum of squares of the error terms between the CVs and their setpoints (Equation (1)).

This situation is used to advantage, for example, when two process outputs are strongly coupled with respect to the available MVs but not with respect to process disturbances. In this case, a 2 x 2 controller is not practically feasible and, because of disturbances, one cannot assume that by controlling one of the outputs, the other will be controlled as well by virtue of the coupling effect. The remedy is to control both process outputs in the least-squares sense with only one MV.

In general, a square, ill-conditioned process can become well-behaved and controllable in the least-square sense with the deletion of one or more MV. The price paid in terms of steady-state offset is usually quite acceptable from a unit operation's point of view. Note that the choice of which MV should be ignored is an important one since not all non-square ($N_m < N_c$) processes are well-behaved.

Case 2 ($N_m = N_c$): The controller has exactly enough MVs to control the CVs at their setpoints at steady state but still has no degrees of freedom. This is the standard situation for multivariable control and needs no elaboration here. Note that in this case, as in Case 1, the steady-state solution of equation (1) is unique and therefore Objective 5 does not apply.

Case 3 ($N_m > N_c$): The controller has more MVs than CVs and therefore the steady-state solution of equation (1) is not unique. The controller is said to have $N_f = N_m - N_c$ degrees of freedom and it can address Objective 5. The outcome of this objective depends on whether the number of degrees of freedom is smaller or equal to the number of IRVs. In the first case, the MVs will be manipulated to their IRVs in the least-squares sense at steady state. In the latter case, the MVs will be exactly equal to their IRVs at steady state. A third solution which occurs when the number of degrees of freedom is greater than the number of IRVs is economically wasteful but may be of practical interest in some applications.

The IRV concept can also be used beneficially when attempts are made to control ill-conditioned processes. Square, ill-conditioned processes can become well-behaved and controllable in an offset-free manner with the addition of one or more MVs. Objective 4 ensures that the CVs reach their setpoints with no offset and Objective 5 ensures that the burden of manipulative action is evenly distributed among the MVs. The choice of which MV to add is again critical since not all non-square ($N_m > N_c$) processes are well-behaved.

Current Status

In practice, the hierarchy of control objectives outlined in Fig. 1 has proved extremely flexible in providing a framework for addressing diverse multivariable control problems.

In the recent past, the IDCOM-M controller has been applied in a variety of ways and with considerable success in FCC and hydrocracking units. Other applications are in various stages of development on fractionation and distillation columns. Experience shows that the controller can be tuned and commissioned with adequately trained but otherwise unspecialized control engineers. Operator acceptance has been outstanding and will be discussed in a future publication.

This next section will use the example of a FCC unit to illustrate the capabilities of the IDCOM-M controller. Although the process is fictional and the closed-loop responses are the result of simulations, both closely reproduce actual process behavior observed and achieved with the IDCOM-M controller.

EXAMPLES

The examples discussed in this section are based on a FCC unit with the regenerator running in a full-burn mode. A simplified flow diagram for this unit is shown in Fig. 2. Table 1 contains a list of the manipulated, disturbance and controlled variables as well as their nominal process values. The open-loop transfer function matrix for this process is shown in Table 2, with time delays and time constants expressed in minutes.

Example 1: In this first example we consider the case of an IDCOM-M controller manipulating combustion air and recycle oil flows and controlling regenerator bed temperature. One of the MVs has an IRV (recycle oil flow) and bed temperature is controlled to a setpoint. This control structure is summarized in Table 3.

Figure 3 shows the system's closed-loop response when a setpoint change of + 10 DEGF is imposed on regenerator bed temperature. This response shows the benefits of using more MVs than are needed to bring the CV to its setpoint: the temporary decrease in recycle oil flow helps push bed temperature in the right direction. This effect compensates for the relatively slow response of bed temperature to combustion air flow and to the velocity clamps on the latter variable. Note that in this example a heavy weighting factor brings the oil flow quickly back to its IRV.

Example 2: We now consider the case of an IDCOM-M controller with the same MVs as in Example 1 but controlling flue gas O_2 to a setpoint and maintaining flue gas temperature below a maximum zone limit. This control structure is summarized in Table 4.

Figure 4 shows the system's closed-loop response after a measured feed flow disturbance of + 50 BPH. At the time the disturbance occurs, flue gas O_2 is controlled to its setpoint but since flue gas temperature is below its zone limit it is ignored by the controller. The disturbance raises the temperature above its zone and at this time the controller acts to return it to its maximum zone limit. This explains the inverse response displayed by recycle oil which is first decreased to help combustion air keep flue gas O_2 at its setpoint. Note that at steady state the recycle oil flow does not return to its IRV since flue gas O_2 and temperature must each be controlled to a setpoint and therefore the controller has no degree of freedom.

Example 3: This last example will illustrate the functionality of the controllability supervisor. The control structure is identical to that of Example 2 with bed temperature additionally controlled to a setpoint. This control structure is summarized in Table 5.

Figure 5 shows the system's closed-loop response after a measured feed flow disturbance of + 50 BPH while the controllability supervisor is turned on. At the time the disturbance occurs, both flue gas O_2 and bed temperature are controlled to their setpoints. The controllability supervisor subsequently elects to control flue gas O_2 and flue gas temperature once the latter ventures outside its zone. Bed temperature is left uncontrolled.

This situation lasts until 50 minutes after the onset of the disturbance when the oxygen analyzer fails and the failure is recognized by the controller. At this point the controllability supervisor elects to control only flue temperature and both bed temperature and flue gas O_2 are left uncontrolled. Observe that subsequent to the oxygen analyzer failure, the controller returns the recycle oil flow to its IRV.

An altogether different situation prevails if the same closed-loop response is simulated with the controllability supervisor turned off, as shown in Fig. 6. In this case IDCOM-M controls all three CVs during the initial part of the transient and attempts to control both temperatures after the oxygen analyzer failure. Figure 6 shows that the controller is successful in controlling the three CVs simultaneously - in the least-squares sense, of course, since only two MVs are available. However, the controller is unable to control the two regenerator temperatures as witness the diverging movement of the MVs on Fig. 6. The reason for this behavior is obviously the near singularity of the 2 x 2 system formed by these four variables.

This example illustrates a generic situation which occurs with multivariable controllers: as process operations change, MVs and CVs are turned on or off, either by operator intervention or by failure. The net effect is that at each new execution, a multivariable controller can be faced with a different set of MVs and CVs which may or may not form an ill-conditioned process. In the IDCOM-M controller, the controllability supervisor ultimately decides which CV to control in such a manner that ill-conditioned systems are always avoided.

CONCLUSION

The IDCOM-M controller is a flexible tool for addressing and solving most multivariable control problems encountered in the process industries. In designing this controller, provisions have been made to account for all the requirements expected of a multivariable controller in an operating environment: IDCOM-M is fully integrated into a variety of process databases, fault tolerance and non-square systems are treated in a generic manner and protection is provided against controlling ill-conditioned processes.

Initial applications of IDCOM-M in fluid catalytic cracking and hydrocracking units have been met with considerable success and other applications are in progress on fractionation and distillation columns.

TABLE 1 Process Variables and Their Nominal Values

Variable	Label	Type	Value
Combustion air flow	M_1	MV	300 MLB/HR
Recycle oil flow	M_2	MV	80 BPH
Feed flow	L_1	DV	2000 BPH
Flue gas O_2	Y_1	CV	2.5 PCT
Flue gas temp.	Y_2	CV	1355 DEGF
Regen. bed temp.	Y_3	CV	1320 DEGF

TABLE 2 Process Transfer Function Matrix (Time Units = min)

	M_1	M_2	L_1
Y_1	$\frac{.07e^{-2s}}{5\ s+1}$	$\frac{-.01e^{-2s}}{7\ s+1}$	$\frac{-.01e^{-2s}}{8\ s+1}$
Y_2	$\frac{-.82e^{-36s}}{11\ s+1}$	$\frac{-1.0}{50\ s+1}$	$\frac{.32e^{-2s}}{12\ s+1}$
Y_3	$\frac{-.78e^{-38s}}{8\ s+1}$	$\frac{-.90e^{-8s}}{44\ s+1}$	$\frac{.02}{10\ s+1}$

TABLE 3 Control Structure for Example 1

Variable	Type	Specification
Combustion air flow	MV	No IRV
Recycle oil flow	MV	IRV = 80 BPH
Regen. bed temp.	CV	SETPOINT = 1320 DEGF

TABLE 4 Control Structure for Example 2

Variable	Type	Specification
Combustion air flow	MV	No IRV
Recycle oil flow	MV	IRV = 80 BPH
Flue gas O_2	CV	SETPOINT = 2.5 PCT
Flue gas temp.	CV	MAX ZONE = 1360 DEGF

TABLE 5 Control Structure for Example 3

Variable	Type	Specification
Combustion air flow	MV	No IRV
Recycle oil flow	MV	IRV = 80 BPH
Flue gas O_2	CV	SETPOINT = 2.5 PCT
Flue gas temp.	CV	MAX ZONE = 1360 DEGF
Bed temp.	CV	SETPOINT = 1320 DEGF

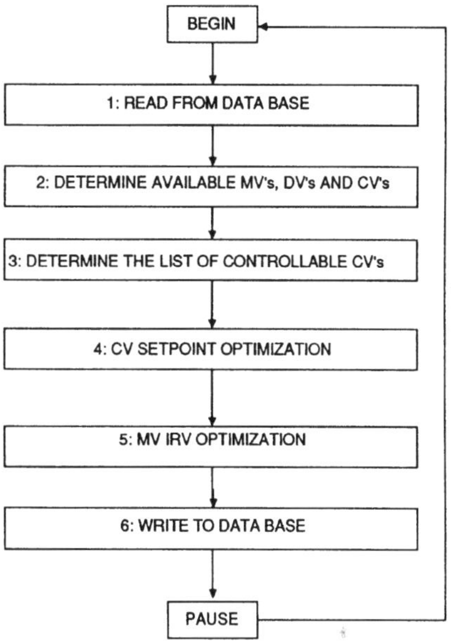

Fig. 1. Controller Objectives Flowchart

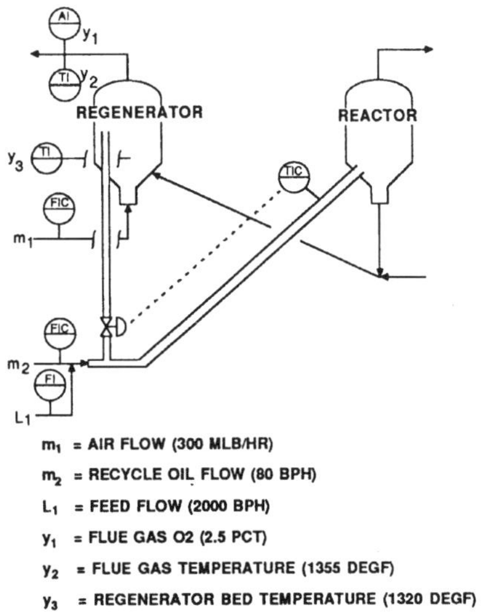

Fig. 2. FCCU Regenerator Control

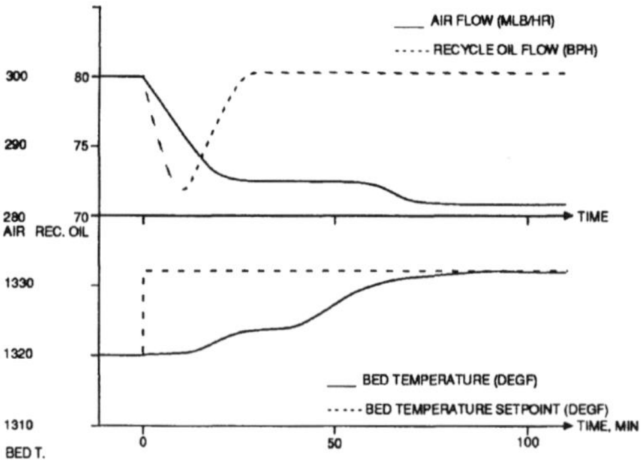

Fig. 3. (Example 1). Closed-Loop response for bed temp. setpoint change (+ 10 DEGF)

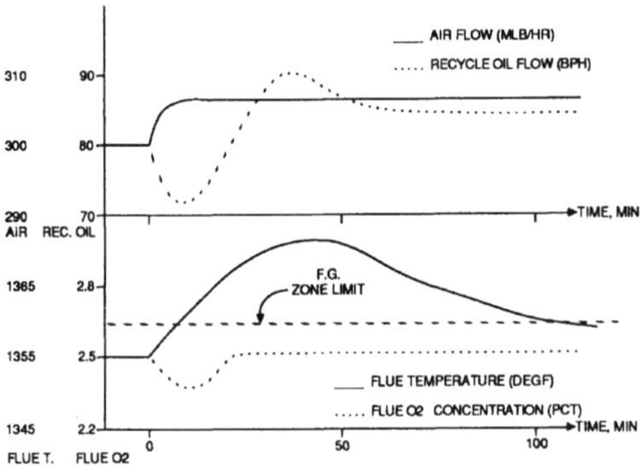

Fig. 4. (Example 2). Closed-loop response for feed flow disturbance (+ 50 BPH)

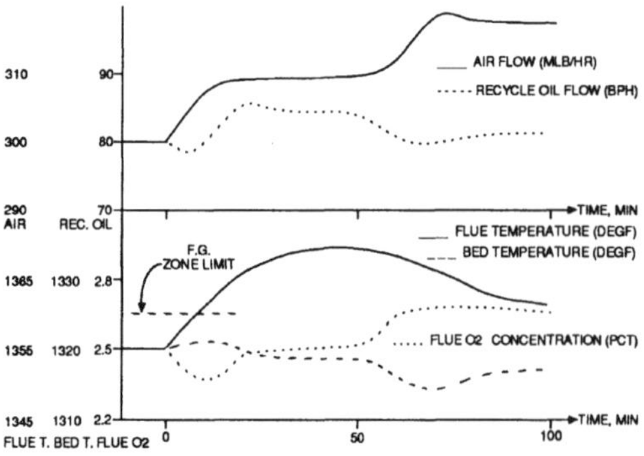

Fig. 5. (Example 3). Closed-loop response for feed flow disturbance (+ 50 BPH) - with controllability supervisor

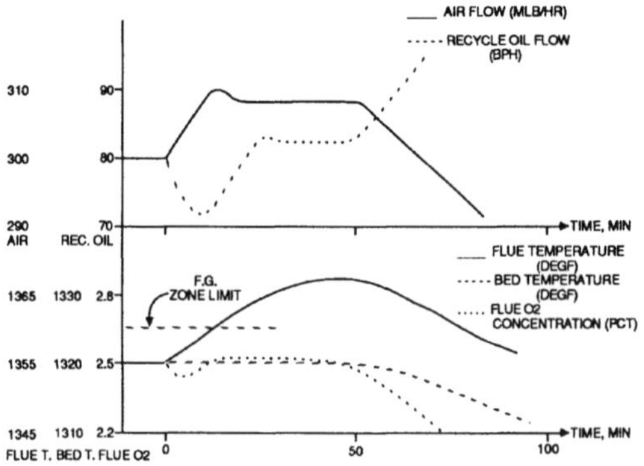

Fig. 6. (Example 3). Closed-loop response for feed flow disturbance (+ 50 BPH) - without controllability supervisor.

REFERENCES

(1) Grosdidier, P., M. Morari and B. R. Holt, "Closed-Loop Properties from Steady-State Gain Information," Ind. Eng. Chem. Fundam. 24, 221 (1985).

(2) Richalet, J., A. Rault, J. L. Testud and J. Papon, "Model Predictive Heuristic Control: Applications to Industrial Processes," Automatica, 14, 413 (1978).

(3) Skogestad, S., and M. Morari, "Effect of Disturbance Directions on Closed-Loop Performance," Ind. Eng. Chem. Research, 26, 2029, 1987.

SMOC, A BRIDGE BETWEEN STATE SPACE AND MODEL PREDICTIVE CONTROLLERS: APPLICATION TO THE AUTOMATION OF A HYDROTREATING UNIT

P. Marquis and J. P. Broustail

Shell Recherche SA, 76530 Grand-Couronne, France

Abstract

A well established advantage of Model Predictive Controllers over State Space Controllers is their ability to on-line take inequality constraints into account.

On the other hand, State Space methods provide a rich theoretical framework for dealing with complex feedback configurations, through the concept of observer ; they offer possibilities which go far beyond those of "model in parallel to the process" structures.

It is therefore attractive to combine these two techniques in order to benefit from their respective strengths. This was our motivation for developing SMOC (Shell Multivariable Optimizing Control).

The example presented here is related to a Hydrotreating Unit and illustrates the practical interest of both constraints handling and closed loop observers.

Keywords : Linear Multivariable control, constraints, disturbance model, closed-loop observer.

1. From Model Predictive Control and Linear Quadratic Gaussian control to SMOC

Multivariable control techniques are now widely applied within the Shell Group. In Europe, in particular, a control method was derived from that developed by Prof. Bornard and Gauthier, from the Laboratoire d'Automatique de Grenoble. This method was then elaborated and has now been operating, on an industrial scale, for over 6 years.

This method, called SMOC (Shell Multivariable Optimizing Control) [1], is based upon a state-variable modelling. It can take constraints into account, just as well on manipulated variables as on process outputs. It is easily tunable on line by means of a set of filter time constants specifying the regulatory and servomechanism behaviour independently.

This method is of the Model Predictive Control (MPC) type, like most methods in operation today [2][3][4][5].

Let us, first, briefly recall what MPC controllers are.

In essence, they are discrete time controllers defined in algorithmic (rather than structural) terms, where the following sequence is computed at each control period :

- step 1 : one iteration is made on the process model by feeding in the latest actions and, possibly, some feedforward measurements. The difference between the current process outputs and the updated model outputs $(Y - \hat{Y})$ is here materialised.

- step 2 : a control planning is computed over a number of forthcoming steps so that the predicted process outputs are as near as possible to a specified scenario. This scenario is, in general, a trajectory going gradually from the current measured value towards the current set value. The prediction of process outputs is based on the same model as in step 1 and assumes that $(Y - \hat{Y})$ will remain constant. Owing to the presence of constraints on U or Y, this step involves a QP resolution.

Presenting MPC this way emphasizes an essential feature of model based control, ie the duality and complementarity between.

- a backward-looking activity, involving the processing of feedback measurements in order to estimate current disturbances (step 1) ;

- a forward-looking function, relying upon the same model, and implying the computation of actions so that the future behaviour of the plant complies with some control objectives. (step 2) .

Should the control problem of step 2 be unconstrained, the MPC would be equivalent to an Internal Model Controller (IMC) [6]. Indeed, an analytical solution to the QP problem would exist and the overall algorithm would amount to a structurally defined IMC scheme.

In other words, MPC techniques subsume IMC ones via an extended control problem formulation. Their constraints handling capability has large practical implications ; indeed, several aspects of automation, which are often considered as separate issues, can now be dealt with in an integrated manner. As the example of the Hydrotreating Unit will show it, MPC can cover, on the basis of a single overall model, the following functions :

- control in the classical sense, ie keeping some process outputs on their setvalue ;

- safety, ie preventing other variables from crossing some threshold ;

- optimisation of the steady state within the safety and control constraints.

However, whereas MPC clearly generalizes IMC with respect to step 2, the philosophy for disturbance estimation and prediction is rigourously the same in both techniques. They share a "model in parallel to the process" structure.

While remaining in this framework, the use of a state space model instead of the usual finite impulse responses only brings minor changes :

- step 1 involves the computation of a state vector prior to that of the model outputs :

$$\underline{\bar{X}}(k) = A.\underline{\bar{X}}(K-1) + B.\underline{U}(k-1) \qquad (1)$$

$$\underline{\bar{Y}}(k) = C.\underline{\bar{X}}(k)$$

- step 2 can make direct use of this state vector, which somewhat reduces the amount of calculations.

This computation-saving effect is also reflected in the block scheme corresponding to the unconstrained case : the only difference with an IMC scheme is the existence of a state feedback link between the model (parallel to the process) and the controller block (see figure 1).

Figure:1 Unconstrained SMOC

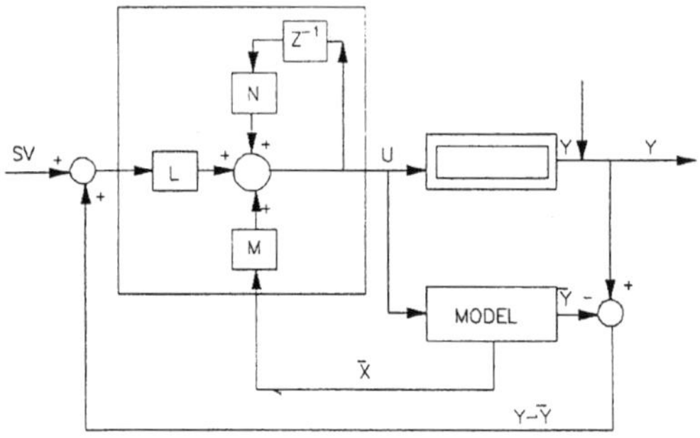

The state space formalism actually starts bearing fruit when step 1 is looked at as an observation procedure. Indeed, when compared with a classical closed loop observer, step 1 can be considered as a state and disturbance estimator relying upon the following stochastic model :

$$\underline{X}(k) = A.\underline{X}(k-1) + B.\underline{U}(k-1)$$

$$\underline{Y}(k) = C.\underline{X}(k-1) + \underline{V}(k) \qquad (2)$$

$$\underline{V}(k) = \underline{V}(k-1) + \underline{\xi}(k-1)$$

where "chsi " is a vector of centered independent white noises.

The corresponding closed loop observer being :

$$\begin{bmatrix} \hat{X} \\ \hat{V} \end{bmatrix}(k/k-1) = \begin{bmatrix} A & 0 \\ 0 & I \end{bmatrix} . \begin{bmatrix} \hat{\underline{X}} \\ \hat{\underline{V}} \end{bmatrix}(k-1/k-1) + \begin{bmatrix} B \\ 0 \end{bmatrix} . \underline{U}(k-1)$$

$$\hat{Y}(k/k-1) = [C \quad I] . \begin{bmatrix} \hat{X} \\ \hat{\underline{V}} \end{bmatrix}(k/k-1) \qquad (3)$$

$$\begin{bmatrix} \hat{X} \\ \hat{V} \end{bmatrix}(k/k) = \begin{bmatrix} \hat{X} \\ \hat{\underline{V}} \end{bmatrix}(k/k-1) + \begin{bmatrix} K_x \\ K_v \end{bmatrix} . (Y(k) - \hat{Y}(k/k-1))$$

where

It can easily be shown that, provided that $x(0/0) = \bar{x}(0)$

$$K_x = 0$$

$$\qquad (4)$$

$$K_v = I$$

$$\hat{\underline{X}}(k/k) = \underline{\bar{X}}(k)$$

$$\qquad (5)$$

$$\hat{\underline{V}}(k/k) = \underline{Y}(k) - \underline{\bar{Y}}(k)$$

The "model in parallel to the process "structure is, thus, a special case of closed loop observer based on the assumption that unmeasured disturbances are independent integrated white noises added to each output (figure 2).

figure 2: MPC uses a special case of OBSERVER

A model in parallel to the process structure

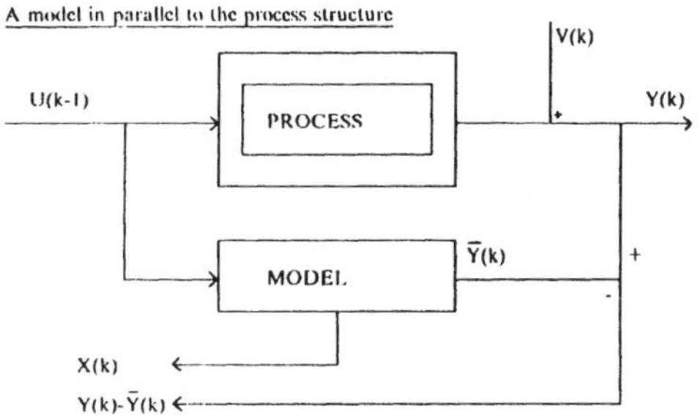

An equivalent layout

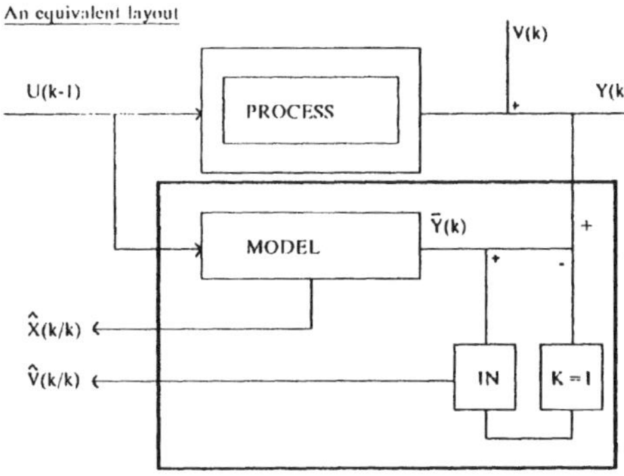

Having reformulated step 1 in this way enables us to generalize MPC simply by adopting other stochastic hypotheses. The case of unstable processes can, for instance, be dealt with by assuming the simultaneous presence of state, measurement and integrated output noises, leading to values of K different from (eq 4) and warranting the observer stability.

Another example of generalization, which proved to be a key factor of success in the Hydrotreater application, will be presented in section 2.3.

2. Application of SMOC to a hydrotreater

The application of SMOC which is now presented has been operational since March 87. It concerns a hydrotreating unit and provides a good illustration of the practical interest of :

- formulating control objectives in terms of combined setvalues, constraints and steady state optimization criterion ;

- processing all the measurements available through the concept of extended closed loop observer.

2.1 Description of the unit

An HTU unit is a four catalyst beds reactor ; after being heated in a furnace, the feed undergoes a hydrogenation reaction which improves the product quality.

The reaction is exothermic, but a minimum temperature must be reached to obtain it ; hence the furnace at the inlet of the reactor.

At each input of beds 2, 3, 4, quenches can be injected through manipulable valves ; these quenches with a high hydrogen content enable the product to be cooled down through the reactor.

It must be emphasized that sensors, distributed all along the process, allow us to measure the inlet and outlet temperatures of each bed.

The unit scheme is presented in Figure 3a.

2.2 Control problem formulation

As has been pointed out in section one, SMOC can deal with three kinds of functionalities.

In the case of the HTU :

- SMOC first has to control the Weighted Average Bed Temperature (WABT), which is a linear combination of elementary temperatures, versus a set-value.

- Meanwhile, and this objective is ever more important than controlling the WABT, safety constraints must be satisfied, namely :

. Outlet temperatures of each bed $T_o(i)$ must be kept below a maximum value ($T_{max}(i)$)

Last, the two previous types of objectives must be achieved with a minimum energy consumption in the furnace (hence with minimum quenches flows when the steady state is reached for a given WABT).

The realisation of this SMOC controller has been based upon a state space model with :

- 4 manipulated inputs (SV inlet temperature, valve position on each quench VQi).

- 12 outputs :

. 8 subject to upper limits : TO4x, TO3x, TO2x, TO1x (2 outlets temperature per bed).

. 1 subject to set-value : WABT (combination of individual temperatures).

. 3 subject to minimization (unreachable low set-values) : the valves position on each quench.

In order to illustrate how SMOC handles the different functions in the course of time, we can face the 3 following scenarios :

Scenario 1 (figure 3c)

Starting from a situation where :

- WABT is on its set-value
- the inlet temperatures of each bed are the same (Christmas tree profile)

SMOC progressively decreases the inlet temperature and the valves aperture to reduce the energy consumption, while keeping the WABT on its set-value.

The first active constraints are the minimum bound (0%) on the three manipulated valves.

In this scenario, no output constraint becomes active.

Scenario 2 (figure 3e)

The starting situation is :

- Stable unit
- Christmas tree profile
- WABT too low

When SMOC takes over, all the means of action are used to get a proper WABT answer ; then, in the long term, a rearrangement takes place which progressively brings the temperature profile back to the optimum (closed quench valves).

Scenario 3 (figure 3g)

- Same starting situation as in scenario 2.
- This time, the outputs are lower on the outlet temperatures of beds 3, 4. The first part of the scenario is comparable to scenario 2 but the quench minimization is stopped because of the constraints on temperatures.

In the final state, the active constraints are, this time, output constraints. The temperature profile is very different from the "linear" ones of scenarios 1, 2 ; this time, the bottom part of the reactor (beds 3, 4) is the coolest.

Figure 3 summarizes the different steady state temperature profiles :

- starting profile for cases 1, 2, 3 (figure 3b)
- end of case 1 (figure 3d)
- end of case 2 (figure 3f)
- end of case 3 (figure 3h)

2.3 Observation policy

2.3.1 Critique of the classical MPC approach

So far, from a control perspective, the hydrotreating unit can be seen as a 4 inputs/12 outputs system. The run-of-the-mill approach would, therefore, consist in :

- step 1 : running the 4 by 12 model in parallel to the process and computing the 12 process/model differences to estimate the disturbances ;

- predicting, during step 2, that these 12 disturbances will remain constant over the horizon.

Our first implementation of SMOC on the actual unit relied upon this philosophy ; this led to unacceptable oscillations.

An explanation to that lies in the specificities of the HTU problem :

- controlled outputs are not as fully independent as a global black box view would suggest ; for instance, outlet temperature of bed 1 has an impact on any other temperatures ;

- one of the controlled variables, namely the WABT, is a linear combination of elementary measurements, each of them being a measured state of the whole system.

Consequently, predicting that the overall effect of disturbances will be constant on each controlled output is inconsistent. Moreover, for the elaborated WABT, an efficient disturbance prediction should rely upon an estimation/prediction made for each component variable.

2.3.2 Towards a new approach

Further to these remarks, the following principles have been adhered to :

- in order to avoid inconsistencies, an explicit stochastic model must be established first.

- to make an extensive use of the available measurements, the outputs of this model must be all the relevant elementary measurements, whether subject or not to a control objective. This introduces an important distinction between the outputs relative to the control problem (the controlled variables) and those relative to the observation problem (the feedback variables).

- the estimation made at step 1 and the prediction of disturbances affecting controlled outputs made during step 2 must be coherent with the stochastic assumptions.

In order to give some substance to these general statements, it is now proposed to apply them to a simplified version of the HTU process.

A simple example

Let us consider the following system :

$$T_i = H_i(u_1)$$
$$(6)$$
$$T_o = H_{io}(T_i) + H_o(u_2)$$

where :

u_1 and u_2 are the manipulated variables

T_i and T_o are the 2 measured variables (feedback variables)

H_i, H_{io}, H_o are global notations for SISO dynamic systems ; corresponding state vectors and matrices being :

$$\underline{X}_i, A_i, \underline{b}_i, \underline{c}_i^T$$
$$\underline{X}_{io}, A_{io}, \underline{b}_{io}, \underline{c}_{io}^T \quad (7)$$
$$\underline{X}_o, A_o, \underline{b}_o, \underline{c}_o^T$$

Let us, also, define an elaborated variable :

$$WABT = \frac{T_i + T_o}{2} \quad (8)$$

The controlled variables and their corresponding objective are :

- T_o subject to a setvalue

- WABT subject to a maximum bound

In addition to that, u_2 must be minimum when the steady state is reached.

At this stage, this system can be seen (with some minor inconsistencies) as a single bed hydrotreater where :

u_1 is the setvalue of a slave PID controlling T_i

T_i is the inlet temperature of the reactor

u_2 is the valve controlling the quench

T_o is the outlet temperature of the reactor

Akin to the HTU case, we are here in a situation where :

- the feedback variables (T_i and T_o) do not fully coincide with the controlled variables (T_o and WABT).

- the feedback variables are not completely independent of each other because of the H_{io} transfer.

A possible set of stochastic assumptions is embodied in Figure 4, where "chsi$_i$" and "chsi$_o$" are centered independent white noises. The contribution of "chsi$_o$" to T_o is the classical integration which generates non stationary effects.

STOCHASTIC MODEL FOR SIMPLIFIED HTU

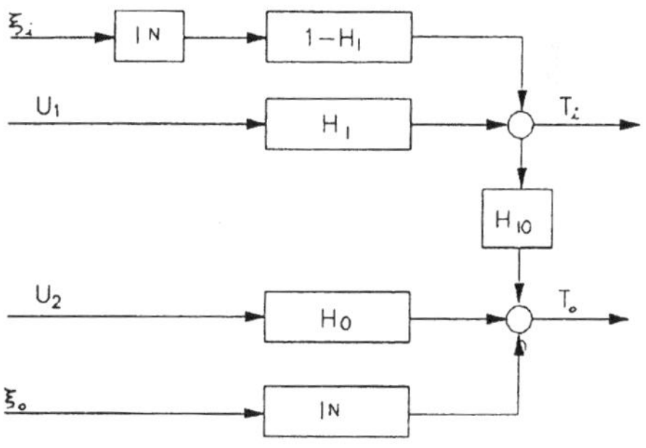

Figure 4 IN: stands for INTEGRATOR

As to "chsi$_i$", it must be remembered that T_i is controlled by a PID loop whose setvalue is u_1. If it is assumed that the disturbance affecting T_i inside this loop is an integrated white noise added to the output, then, when seen from SMOC, disturbances become an integrated white noise filtered by the closed loop regulatory transfer.

Note that, due to the triangularity of the system and to the simplicity of noise (no measurement noise, no direct noise on the intra states of the H) there is no need to know noise variances. Of course, in a more general case this would no longer hold.

2.3.3 The extended closed-loop observer

An alternate non minimal representation of the system is made in figure 5,

Equivalent non minimum representation

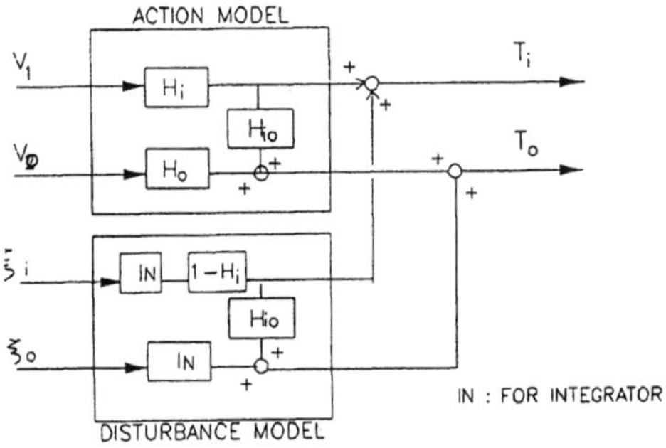

Figure 5

corresponding equations are :

$$\begin{bmatrix} X \\ \tilde{X} \end{bmatrix}(k) = \begin{bmatrix} A & 0 \\ 0 & \tilde{A} \end{bmatrix} \cdot \begin{bmatrix} X \\ \tilde{X} \end{bmatrix}(k-1) + \begin{bmatrix} B \\ 0 \end{bmatrix} . U(k-1) + \begin{bmatrix} 0 \\ \tilde{B} \end{bmatrix} . \xi(k-1) \qquad (9)$$

$$Y(k) = \begin{bmatrix} C & \tilde{C} \end{bmatrix} \cdot \begin{bmatrix} X \\ \tilde{X} \end{bmatrix}(k) + \eta(k)$$

$$Y_c(k) = \begin{bmatrix} C_c & \tilde{C}_c \end{bmatrix} \cdot \begin{bmatrix} X \\ \tilde{X} \end{bmatrix}(k) + D.\eta(k) \qquad (10)$$

where:

Y are the feedback variables

Y_c are the controlled variables

X the state of the deterministic part

$\tilde{X}$ the state of the stochastic (i.e. disturbance) part

η et ξ are white centered independant noises

The corresponding observation policy can be summarized as follows :

Step 1 : Processing of the current measurements (at time k)

$$\begin{bmatrix} X \\ \tilde{X} \end{bmatrix}(k/k-1) = \begin{bmatrix} A & 0 \\ 0 & \tilde{A} \end{bmatrix} \cdot \begin{bmatrix} X \\ \tilde{X} \end{bmatrix}(k-1/k-1) + \begin{bmatrix} B \\ 0 \end{bmatrix} . U(k-1)$$

$$\hat{Y}(k/k-1) = \begin{bmatrix} C & \tilde{C} \end{bmatrix} \cdot \begin{bmatrix} X \\ \tilde{X} \end{bmatrix}(k/k-1) \qquad (11)$$

$$\begin{bmatrix} X \\ \tilde{X} \end{bmatrix}(k/k) = \begin{bmatrix} X \\ \tilde{X} \end{bmatrix}(k/k-1) + \begin{bmatrix} 0 \\ \tilde{K} \end{bmatrix} . (Y(k) - \hat{Y}(k/k-1))$$

where :

y is the measured value of the feedback variables

$\hat{y}$ is the corresponding predicted value

$\tilde{K}$ is the observer feedback matrice

Prediction of the disturbances acting on the controlled outputs

During step 2, the contribution of disturbances to the controlled variables (here, T and WABT) is computed by iterating, over the horizon on the following equations :

$$X(k+i/k) = A.X(k+i-1/k) \qquad (12)$$

$$\tilde{Y}_c(k+i/k) = \tilde{C}_c . \tilde{X}(k+i/k)$$

where:

$\tilde{X}(k/k)$ is obtained from equation (11)

$\tilde{Y}_c(k+i/k)$ is the contribution of disturbances to the controlled outputs at the i-th step of the horizon.

Return to the previous example.

To come to more details about our single bed reactor, adopting the notations of (7) and the stochastic assumptions of figure 4 and 5 :

$$X = \begin{pmatrix} X_1 \\ X_{1\bullet} \\ X_\bullet \end{pmatrix} \quad A = \begin{pmatrix} A_1 & 0 & 0 \\ B_{1\bullet} . C_1^T & A_{1\bullet} & 0 \\ 0 & 0 & A_\bullet \end{pmatrix} \quad B = \begin{pmatrix} b_1 & 0 \\ 0 & 0 \\ 0 & b_\bullet \end{pmatrix}$$

$$U = \begin{pmatrix} u_1 \\ u_2 \end{pmatrix}$$

$$Y = \begin{pmatrix} T_1 \\ T_\bullet \end{pmatrix} \quad C = \begin{pmatrix} c_1^T & 0 & 0 \\ 0 & c_{1\bullet}^T & c_\bullet^T \end{pmatrix}$$

$$\tilde{X} = \begin{pmatrix} v_1 \\ \tilde{x}_1 \\ \tilde{x}_{1\bullet} \\ v_\bullet \end{pmatrix} \quad \tilde{A} = \begin{pmatrix} 1 & 0 & 0 & 0 \\ b_1 & A_1 & 0 & 0 \\ b_{1\bullet} & -b_{1\bullet} . c_1^T & A_{1\bullet} & 0 \\ 0 & 0 & 0 & 1 \end{pmatrix} \quad \tilde{K} = \begin{pmatrix} 1 & 0 \\ 0 & 0 \\ 0 & 0 \\ 0 & 1 \end{pmatrix} \qquad (13)$$

$$Y = \begin{pmatrix} T_1 \\ T_\bullet \end{pmatrix} \quad \tilde{C} = \begin{pmatrix} 1 & -c_1^T & 0 & 0 \\ 0 & 0 & c_{1\bullet}^T & 1 \end{pmatrix}$$

$$Y_c = \begin{pmatrix} T_\bullet \\ WABT \end{pmatrix} \quad \tilde{C}_c = \begin{pmatrix} 0 & 0 & c_{1\bullet}^T & 1 \\ 0.5 & -0.5 * c_1^T & 0.5 * c_{1\bullet}^T & 0.5 \end{pmatrix}$$

$$C_c = \begin{pmatrix} 0 & c_{1\bullet}^T & c_\bullet^T \\ 0.5 * c_1^T & 0.5 * c_{1\bullet}^T & 0.5 * c_\bullet^T \end{pmatrix}$$

2.3.4 Genericity of the approach

Equations 11 and 12 are generic to any disturbance model provided the process is stable.

It can be noted that the case of measured disturbances, in other words when feedforward is possible, fits in this framework of model of action/model of disturbances ; in this situation, the recommended approach is to embed to measured disturbance model into the $\tilde{A}$, $\tilde{B}$, $\tilde{C}$ model of equations 9 and to consider the measured w as the output of a generating process which can be selected freely (eg an integrated white noise, if one wants to predict it to be constant over the horizon).

In the case of unstable systems, the observer must be based upon the minimal state space representation corresponding to figure 8 ; in this case the K matrix of the observer must be computed through a Ricatti equation, which guarantees that the state/disturbance estimator is stable. A consequence of minimality is that it is no longer possible to distinguish between states related to past actions and states related to disturbances.

Of course, a minimal observer can also be used for stable systems ; but, from a practical viewpoint, it is most convenient to keep a separate track of the contribution of disturbances ; this, indeed, enables simple on line tuning of disturbance rejection via a filtering, in step 2, of the y_c values coming from equation 12.

As to the computation of the observer feedback matrix K, it may range from a full Kalman gain calculation to a straigthforward combination of 0 and 1. Such was the case in our example and for the HTU itself.

Conclusion

The objective of this article was to show the complementarity between state space approaches and model predictive control techniques.

The reason for that lies in the essential duality of model based control. In any control where an explicit model is referred to, two types of activities are involved :

- processing of measurements (observation)

- computation of actions (control)

These two functions are not only based upon the same deterministic model, they also share assumptions about the unmeasured disturbances. These are seldom made explicit in classical MPC techniques although it would be a convenient route towards generalization.

Whereas MPC have made a quantum jump when incorporating constraints in the control problem, LQG offer much more flexibility with regard to observation. SMOC, which retains from MPC the receeding horizon approach and from LQG the state space modelling, is an attempt to make the most of both methods (see figures 6 and 7).

REFERENCES

[1] MARQUIS, P. (1987)
Shell Multivariable Optimizing Control
IFAC DS1 (vol. 11) 10-11

[2] CUTLER, CR. and RAMAKER, BL. (1979)
Dynamic Matrix Control - A Computer Control Algorithm
AICHE National Mtg

[3] RICHALET, J. and Al (1978)
Model Algorithm Control
AUTOMATICA, 14, p. 413

[4] MEHRA, RK. - ROUHANI, R. - ETERNO, J. - RICHALET, J. and RAULT, A. (1982)
Model Algorithm Control - Review and Recent Development
Eng. Foundation Conference on Chemical Proc.

[5] GARCIA, CE. and PRETT DM. (1986)
Advances in Industrial Model Predictive Control
Chemical Process Control

Conference - CPC III (M. MORARI and TJ. Mc AVOY, eds.)
CACHE and Elsevier, Amsterdam, 249-294

[6] GARCIA, CE. and MORARI, M. (1982)
Internal Model Control - A Unifying Review and Some New Results.
IEC Proc. Des. Dev. 21-308.

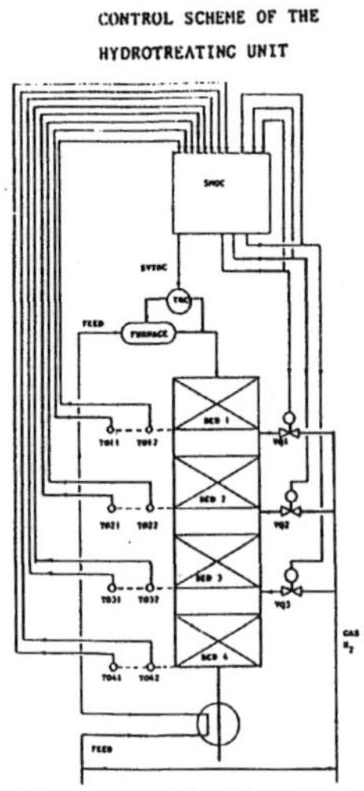

(3a)

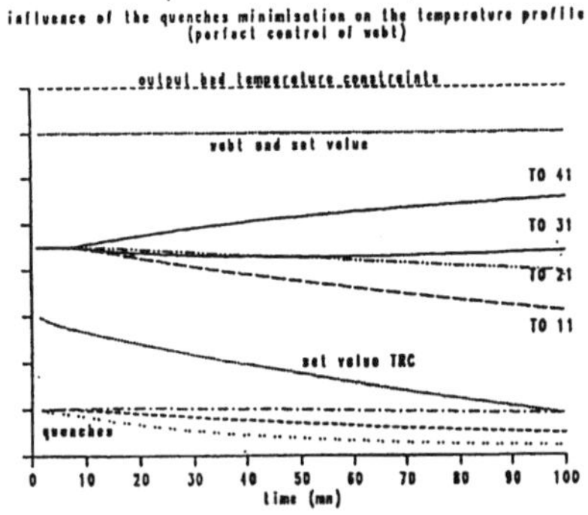

(3c)

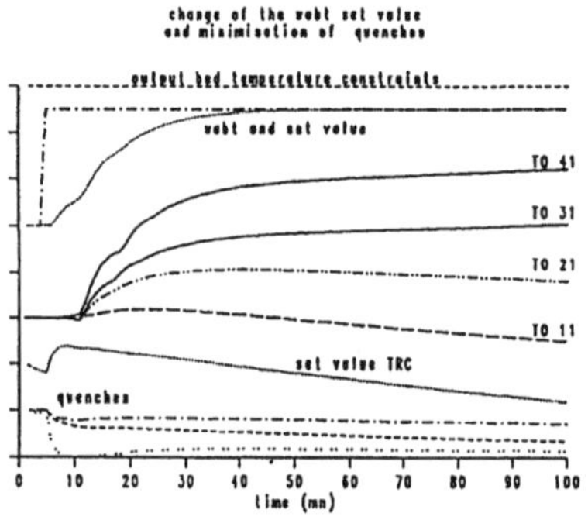

(3e)

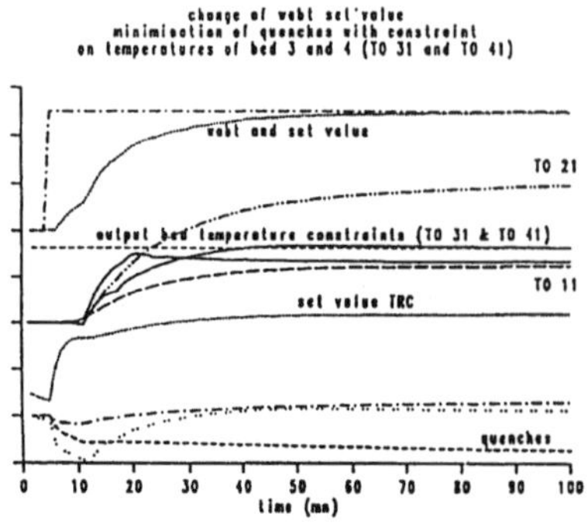

(3g)

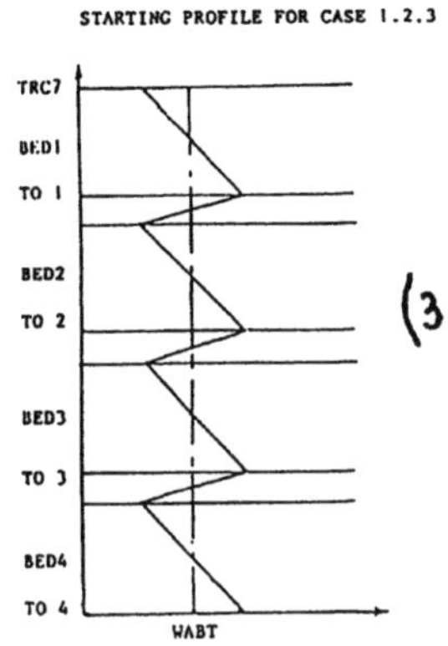

(3b)

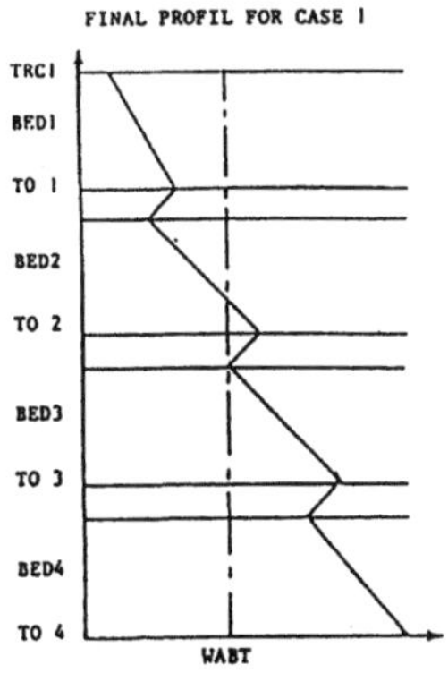

(3d)

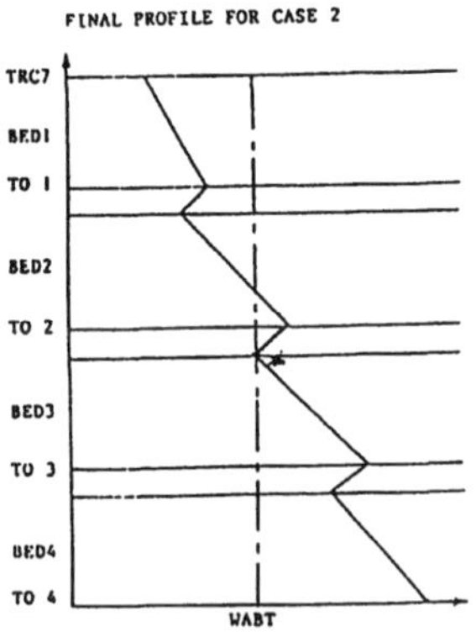

(3f)

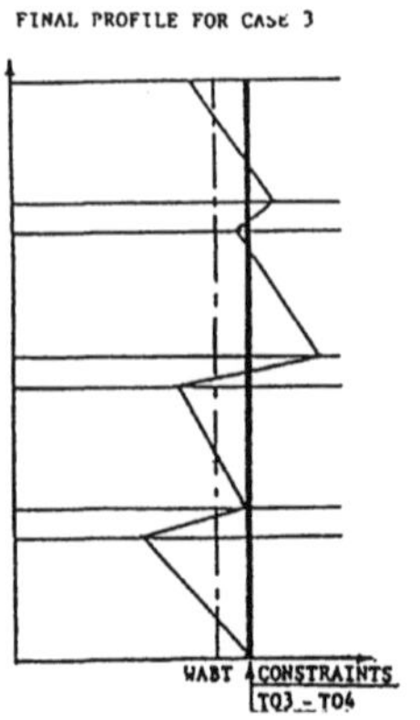

(3h)

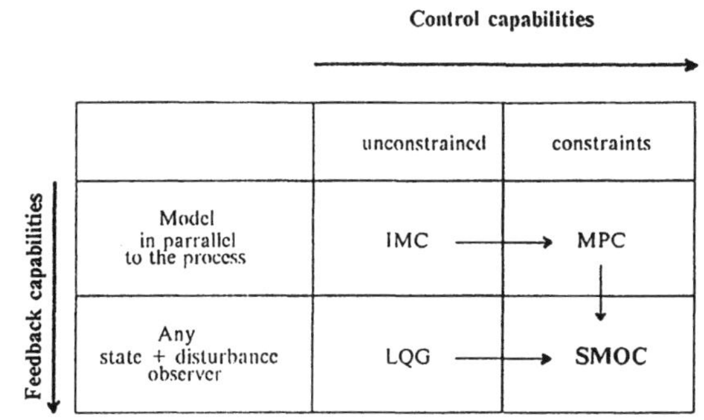

Figure (6)

Figure 7: A functional overview of SMOC

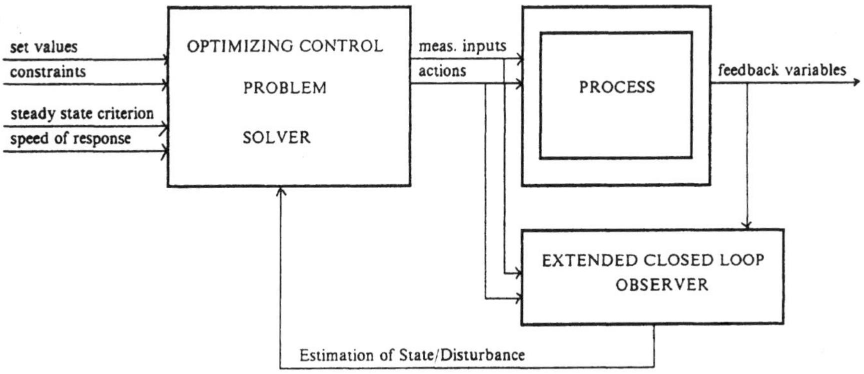

STATE SPACE MODEL PREDICTIVE CONTROL OF A MULTISTAGE ELECTROMETALURGICAL PROCESS

J. G. Balchen, D. Ljungquist and S. Strand

Division of Engineering Cybernetics, The Norwegian Institute of Technology, Trondheim, Norway

Abstract. The paper deals with the proposed application of a novel technique for Model Predictive Control to a multistage electrometalurgical process. The novel control technique is based upon high speed repetitive simulation of a nonlinear state space model of the process including relevant constraints, and searching in a parameterized control space by an efficient optimization routine until an optimal set of control actions has been found. This MPC-technique is not limited to linear processes with quadratic objective functionals and is therefore believed to offer major improvement to the control of many industrial processes where the standard, linear control solutions fail because of nonlinearities and constraints in the process system.

Keywords. Optimal control, predictive control, nonlinear control system, state-space methods, iterative methods, on-line operation.

INTRODUCTION

Since the time of Norbert Wiener, it has been known that optimal feedback control algorithms for dynamic processes will in some way or other, reflect the models which describe the behaviour of the process and its environment and the measure of the performance to be minimized. Optimal control as derived by Pontryagin and co-workers (1957/1962) is an example of this. Later, numerous contributions in the literature present control techniques which are model based . The algorithms for process control which were introduced from the mid 1970's, MAC, DMC and IDCOM are more recent examples of the same fact (Richalet and co-workers, 1978; Cutler and Ramaker, 1979; Reid, 1981; Marchetti, Mellichamp and Seborg, 1983; Mehra and Mahmood, 1985; Garcia and Prett, 1986; Clarke, Mohtadi and Tuffs, 1987). The latter type of control algorithms, often referred to as Model Predictive Control (MPC), computes the process variables over a future time interval. The control variables which will minimize a performance criterion (objective functional) over this time interval, can be determined. If the process is regarded as linear and the objective functional is quadratic, a closed form solution is found which is easy to implement in a computer system. This is not the case if the process is nonlinear, has a nonquadratic objective functional or is constrained. This has been generally known for a long time.

A simple way of deriving optimal control of a general dynamic process (nonlinear and constrained) is to parameterize the control variables in a simple manner and optimize the control variables by searching in this parameter space by simulating the process behaviour over a future time interval using a fast repetitive computer (Strand 1987). This is a promising technique which has been employed in the following. An alternative technique is to apply the Pontryagin Maximum Principle (Ljungquist 1986) which due to its more efficient mathematical technique yields better results in certain cases.

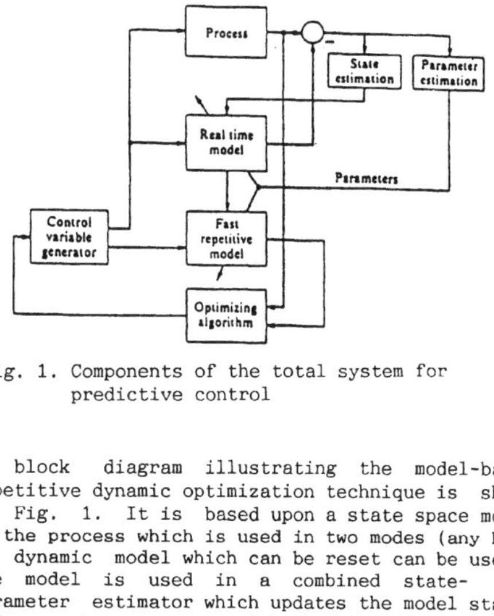

Fig. 1. Components of the total system for predictive control

A block diagram illustrating the model-based repetitive dynamic optimization technique is shown in Fig. 1. It is based upon a state space model of the process which is used in two modes (any kind of dynamic model which can be reset can be used). One model is used in a combined state- and parameter estimator which updates the model states and parameters to agree with the behaviour of the real process. This makes it possible to adjust the other model which is running repetitively at a much higher speed. This second model can be identical to the first or may be reduced in complexity. The major prerequisite for the present method is the availability of a high speed on-line computing capacity. This should allow repetitive simulation and optimization calculations to be done so fast that the search in the control parameter space converges nearly instantaneously compared to the dynamics of the real process. Such computer capacity is available today at a reasonable price, which makes this technique applicable to most processes encountered in the process industry. In case the process dynamics is too fast for the simple searching technique suggested, a modified algorithm is available which employs a LQG-strategy in parallel with the searching strategy (see Balchen, Ljungquist, Strand, 1988).

PROPERTIES OF THE PRESENT METHOD

The properties of the present method compared to conventional process control techniques and MPC strategies based upon a linear model are briefly as follows:

- The process (and its model) may be nonlinear, highly interacting and with constraints on both control variables and state variables provided the optimization algorithm converges. This becomes a matter of the relative bandwidth (speed) of the optimization and the process. A similar statement is out of the question for other available process control strategies.

- The technique is very straightforward and obvious and directly appeals to operators because of its simplicity.

- The objective functional to be employed should be based on economic aspects (profit functions) including the cost of raw materials, energy, labour, maintenance etc. and value of the products.

- The present technique, as well as other MPC techniques, assumes the availability of a reasonably good model of the process which is continuously updated against the real process. In many cases of industrial practice, this requirement is not controversial at all. In other cases it will limit the applicability of the method.

THE APPLICATION OF MODEL PREDICTIVE CONTROL TO A MULTISTAGE ELECTROMETALURGICAL PROCESS

A simplified blockdiagram indicating how the multistage electrometalurgical process operates, is shown in Fig. 2. Each unit is identical and all units in the same row are electrically connected in series. Each unit is being fed with the raw material which appears in two aggregates, A and B. These aggregates are chemically similar, but differ in energy content. The product of this process is a valuable metal and a less valuable gas. Each of the units can be manipulated by the flow of the two different aggregates of feed, by the electric current through the unit and additional heating through a special circuit. The main objective of the control system is to maintain the mass and energy balances in each unit, that means maintaining the concentration of the active material and the temperature at optimal values. Due to operational reasons, the temperature must be kept between rather narrow constraints. The energy balance is influenced by the ratio of the two aggregates of feed, by the electric current and the additional heating. Additional heating is undesirable because it is unproductive. The yield of the process is not constant, but influenced by certain impurities in the feed, and must be regarded as a disturbance. Since all units in one row are connected in series, the electric current may be regarded as both a control variable and a disturbance.

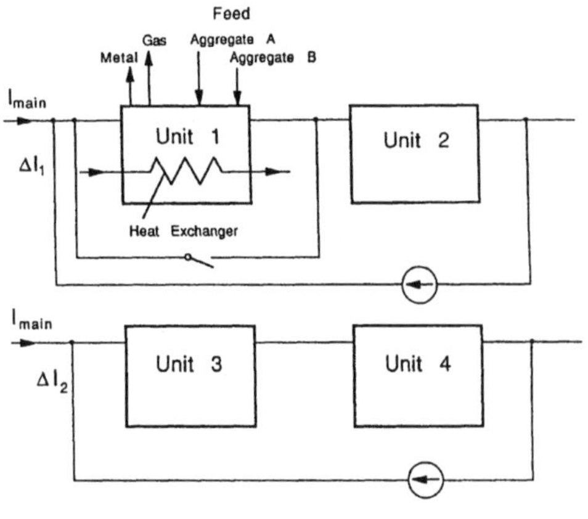

Fig. 2. The multistage electrometalurgical process

Thus this process becomes very interesting for the type of control presented in this paper, since:

- The control actions are discontinuous (quantised).

- There are severe constraints on both control variables and states.

- The units are highly interacting and interdependent.

- All units have identical models, but with different parameters and states.

- The objective functional of the system is easily established in terms of economic factors.

THE MODEL

First a detailed, precise description of the process dynamics was developed. The result was a model containing about ten state variables for each unit and several unknown nonlinear functions and unknown parameters. Model reduction to five or six states per unit could be done by aggregation of related effects. Simulation results showed that even two states represented the dynamics of an undisturbed unit in a satisfactory way. These results seemed to fit perfectly with the control strategy illustrated in Fig. 1.

- A simplified model with three state variables per unit is to be used in the optimization.

- In the identification/estimation algorithm two or three extra state variables are used to make the model fit the real process.

The extra variables may be regarded as time-varying parameters in the simplified model. The almost unpredictable dynamics of the contents of impurities and the other disturbances in the units however, make the assumption of constant parameters during the optimization horizon nearly as good as any other assumption. On-line estimation of these parameters then assures updated estimates before every new optimization. The slow dynamics of the actual process makes this strategy applicable even with a conventional personal computer.

The simplified model may be developed based upon elementary energy and mass balances.

i) Energy balance in unit i

The following factors are taken into account in the energy balance:

- energy due to the electric current that flows through the unit

- additional heating

- energy contents in the feed aggregate A
- energy contents in the feed aggregate B
- energy contents in the produced metal removed at certain points in time
- energy in the escaping gas
- energy loss to the surroundings
- other losses

Aggregate A contains more energy than aggregate B.

ii) Mass balance of the active component in unit i

The rate of change of the active component in the unit equals the difference between the sum of the two aggregates of the feed and the consumption of the component in the reaction of the unit. Moreover, some of the active component is inevitably removed each time the product is removed.

iii) Mass balance, total mass in unit i.

Aggregates A and B of the feed come into the unit, whilst the metal product together with a small amount of feed is tapped off and the gas is continuously escaping.

This leads to the following differential equation (nomenclature listed at the end of the paper):

$$\begin{aligned}
\dot{x}_{1i} &= \frac{1}{b_{1i}}[f_{1i}I_i^2 + u_{3i} - a_1(x_{1i} - a_2) + \\
&\quad a_4(a_5 - x_{1i})u_{1i} - (a_3 + a_4x_{1i})u_{2i} \\
&\quad + (a_6 - a_7f_{2i})I_i] \\
\dot{x}_{2i} &= \frac{(a_{13} - x_{2i})^2}{b_{4i}a_{13}}[a_{14}(u_{1i} + u_{2i}) - a_{15}f_{2i}I_i \\
&\quad - a_{16}x_{2i}f_{3i}] \qquad (1) \\
\dot{x}_{3i} &= a_{14}(u_{1i} + u_{2i}) - a_{15}f_{2i}I_i \\
&\quad - (a_{18} + a_{16}x_{2i})f_{3i}
\end{aligned}$$

where

$$\begin{aligned}
f_{1i} &= b_{2i} + a_8x_{2i} - a_9x_{1i} - a_{10}x_{1i}x_{2i} \\
f_{2i} &= b_{3i} - a_{11}(x_{1i} - a_{12}) \\
f_{3i} &= x_{3i} - \frac{b_{4i}a_{13}}{a_{13} - x_{2i}}
\end{aligned}$$

Each unit may be electrically short-circuited. If this is done to unit number i, the effect is that $I_i \equiv 0$. I_i therefore represents three control variables; the main current being fed through all the units, additional current fed through all units in a series and the possibility of short circuiting each unit individually. The aggregate A of the feed is fed in batches (u_{1i}), whilst the other control variables are continuous. Because of the nature of the process equipment, some of the control actions are implemented manually by the operators.

In the model (1), three different constants appear. The a_j's are constants common to all the units, while the b_{ji}'s may vary from unit to unit. Moreover f_{ji} indicates time-varying functions; f_{1i} equals the internal electrical resistance of the unit, f_{2i} equals the yield factor and f_{3i} is the predicted mass removal. f_{1i} and f_{2i} are very simplified expressions. Therefore b_{2i} and b_{3i} have to be updated in the estimation algorithm, and will typically change from one optimization to the next.

THE OBJECTIVE FUNCTIONAL

The simple model presented in the previous section can only be justified when the concentration of the active component is kept between 7 and 13 percent, and it is desirable to keep the concentration in the middle of this range. For operational reasons the temperature must be kept between narrow limits. Equation (1) is scaled such that the temperature varies around 1.0, and this relative temperature is to stay within ± 3 percent.

These two constraints are taken care of by means of penalty functions in the objective function. The other terms in the objective function reflect economic considerations:

$$J = \int_{t_o}^{t_o+T} [\sum_{i=1}^{n_u} L_i]dt \qquad (2)$$

$$\begin{aligned}
L_i = &- \beta_0f_{2i}I_i + \beta_1((R_{yi} + f_{1i})I_i + e_i)I_i \\
&+ \beta_2u_{1i} + \beta_3u_{3i} + \beta_4u_{2i} + \beta_5\dot{x}_{1i}^2 \\
&+ c_{1i}(x_{1i}) + c_{2i}(x_{2i}) \qquad (3)
\end{aligned}$$

$c_{1i}(.)$ and $c_{2i}(.)$ are the penalty functions which keep the temperature and concentration respectively within the given boundaries. Different functions may be used, but their value should be nearly negligible in the permitted area and increase rapidly when the boundaries are reached (or just before). A penalty function suggested by Balchen (1985) has the desired properties. An additional linear penalty on relative temperatures above 1.0 reflects increased unit maintenance.

The objective functional (2) has two important qualities:

- Each of the objective functions (L_i) represent economic factors (expenses or incomes) or penalties with scaling factors to be chosen.
- The economic scaling factors are obvious whereas the scaling factors of the penalty functions are easily tuned until the constraints are maintained in a satisfactory way.

This results in an optimal control problem which can be solved using an efficient searching algorithm.

SOLUTION STRATEGY

Given an arbitrary, nonquadratic objective functional and a nonlinear state-space model, there are several ways to reach the optimal choice of the control vector. The following outlines the strategy used to obtain the results in this paper.

The original problem may be formulated as:

$$\min_{\underline{u}\in U} J(\underline{u}(t))$$

$$J(\underline{u}(t)) = S[\underline{x}(t_0+T)] + \int_{t_0}^{t_0+T} L(\underline{x}(t), \underline{u}(t), t)dt \quad (4)$$

$$\dot{\underline{x}}(t) = \underline{f}(\underline{x}(t), \underline{u}(t), t)$$

where U is the space of admissible controls, i.e. the control variables may be subject to hard constraints. In the present case, hard constraints are not applied to the state variables, since we are using penalty functions to limit them.

To make this original problem formulation numerically solvable, the control variables have to be parameterized, leading to somewhat different formulations and to suboptimal solutions of the given problem. A piecewise constant parameterization of each control variable is chosen since this is a practical way of implementing the control signal in the process. Thus the problem may be stated as follows:

$$\min_{\underline{\theta}\in\Theta} J(\underline{\theta})$$

$$J(\underline{\theta}) = S[\underline{x}(t_0+T)] + \int_{t_0}^{t_0+T} L(\underline{x}(t), \underline{\theta}, t)dt \quad (5)$$

$$\dot{\underline{x}}(t) = \underline{f}(\underline{x}(t), \underline{\theta}, t)$$

where $\underline{\theta}$ is the set of parameters describing the control variables. Now the original problem (4) has been turned into a finite-dimensional numerically solvable optimization problem.

The dimension of the parameter space has a great influence upon the solution speed, and should be kept to a minimum. However it is known that reduction of the number of parameters describing each control variable will make the corresponding suboptimal solution less satisfactory compared to the optimal solution of the originally stated problem in (4). Hence the choice of parameterization is a trade-off between solution speed and the quality of the suboptimal solution achieved.

In the present example as few as three parameters for each control variable are used. This turns out to be a suitable number in process control cases where trajectory tracking is not required. The discretization intervals do not have to be of equal lengths (Balchen, Ljungquist, Strand, 1988).

Moreover a quite simple optimization technique is used. The gradient of $J(\underline{\theta})$ with respect to $\underline{\theta}$ is computed experimentally, perturbing each parameter in turn and measuring the corresponding change in $J(\underline{\theta})$. The search in the admissible parameter space is performed using a modified Newton algorithm, which aggregates the information retrieved by sequential computation of the gradients at points in the parameter space, in an approximation to the inverse Hessian. Special care is taken for parameters which have reached their constraints, according to the constraints on the control variables. In order to reduce the number of gradient computations, the line search in each direction is done quite accurately.

This optimizing technique is more time-consuming than mathematically more advanced ones like the Pontryagin Maximum Principle. But it is easy to implement in a parallel processing system, since the gradient components may be computed on different processors. Furthermore it is an advantage that process operators are able to understand how optimal control is computed using this technique. Finally, if the solution speed is too slow compared with the process dynamics, which is not the case in the present example, a LQG-strategy with immediate feedback around the computed optimal trajectory is suggested (Balchen, Ljungquist, Strand, 1988).

SIMULATION EXPERIMENTS

As pointed out in the presentation of the model, there are several uncertainties related to the parameters of the simplified model. Simulation experience shows that the inaccuracies related to the flow of feed and the removal of the produced metal only contribute marginally to the process dynamics. Changes in the electrical resistance and the yield factor of each unit, however, represent considerable disturbances as can be seen from Fig. 3a and 3b.

The two curves in Fig. 3a illustrate the temperature responses due to 5 percent reduction in the yield factor (at t=45 hours) and 5 percent increase in the electrical resistance in the unit when the control variables are kept unchanged. Figure 3b shows the corresponding changes in concentration of the active component in the unit. Both temperature and concentration must remain within narrow limits, however, therefore the control variables have to change.

In the simulations, 4 units are studied, units 1 and 2 in row number 1 and units 3 and 4 in row number 2. The electrical current is allowed to be somewhat different in each row (see Fig. 2). Units 1 and 3 are exposed to disturbances in the yield factors, increase and reduction of 5 percent respectively, while units 2 and 4 are disturbed by decrease and increase of 5 percent respectively in the electrical resistance of the unit. The step disturbances occure at the points in time illustrated in Fig. 4a and 4b. It is assumed that the real time estimator described in Fig. 1, is able to give estimates of these disturbances within a period of 2.7 hours. The optimization horizon is 48 hours. Moreover the control parameters are updated every 2.7 hour.

The disturbances indicated above are not supposed to be reasonable from a process engineers point of view, but it is known that the stepsizes are reasonable and that the disturbances will be a challenge to the optimization algorithm. The reader should be aware that the dynamic behaviour of the yield factor and the electrical resistance of the unit, still apply in the simplified model, i.e. just the parameters b_{2i} and b_{3i} in (1) are changed in steps. The exact numerical values of the different prices in the objective functional are not important to a fundamental interpretation of the optimization results.

Many conclusions can be drawn from the simulation experiments. However, we wish to limit the amount of comments by just pinpointing the most important results obtained (See Fig. 5-16).

- Increase in the yield factor in unit 1 calls for an increase of electrical current in this row. The temperature in unit 2, however, increases as the row current increases. Thus the current cannot increase very much if the resistance in unit 2 does not decrease for some reason (Fig. 13).

- The increase of current in units 1 and 2 demands increase in the feed. The responses in the feed of the two different aggregates A and B due to the step changes in yield factor in unit 1 and resistance in unit 2 reflect the importance of the energy balance, and the fact that aggregate A contains more energy than aggregate B (Fig. 9, 11 and 13).

- Similar statements are valid for the control variables in units 3 and 4 (Fig. 10, 12 and 13).

- In the present experiments there is no need for adding auxiliary heat or short circuiting the units, but such conditions may occur under severe disturbances.

- The main peaks in the computational effort (Fig. 14) occure immediately after the disturbances are detected, except the one at t= 150 hours. The latter is caused by the objective functional being flat near optimum and an imperfect searching algorithm. Figure 14 is scaled from 0.0 to 1.0 . With the present algorithm installed on an IBM AT, 1.0 corresponds to 10 min. Updating the control parameters with a 2.7 hours interval, this means that more than 50 units can be controlled by an IBM AT. This updating interval is small compared to the process dynamics, and a parallel LQG-strategy is not required.

- Figure 15 shows the net profit rate of the process. The step disturbances have direct consequences for this rate. The peaks are due to the time-delay in the estimation of the disturbances. As shown in Fig. 16 the sum of penalty functions is nearly negligible.

The following statements are justified by the simulation experiments.

- A reasonable objective functional based upon economic considerations and physical interpretation is easily derived.

- When the state variables are highly coupled through the dynamics and the control variables it is not straightforward to figure out how the control variable should change due to physical disturbances or changes in the operational conditions. It is not difficult, however, to understand the optimality of the new controls derived by the present strategy.

CONCLUSION

From an economic point of view optimal control should be based on nonquadratic objective functionals. The corresponding control variables will drive the process over a wide range of operational conditions. Therefore in many cases nonlinear models and constraints have to be considered. Existing control strategies cannot handle such problems in a satisfactory way. The proposed method, however, permits the flexibility required as illustrated by a simple simulation experiment.

ACKNOWLEDGEMENT

This work has been sponsored in part by The Royal Norwegian Council for Scientific and Industrial Research through the Predictive Control Research Program.

NOMENCLATURE

x_{1i} - temperature in unit number i
x_{2i} - concentration of active component in unit number i
x_{3i} - total mass of unit number i
I_i - the current through unit i
u_{1i} - feedrate of aggregate A to unit i
u_{2i} - feedrate of aggregate B to unit i
u_{3i} - additional heating to unit i
t_o - the initial time for the optimization
J - the objective funtional
L_i - the objective function (cost function) for unit number i
T - the optimization horizon
$S(\underline{x}(t_o+T))$ - a scalar function weighting the final value of the state vector
n_u - number of units
β_i - prices corresponding to the different terms in the objective functional(in Nkr)
a_j, b_{ji} - model parameters, i indicates unit No.
f_{1i} - internal resistance, unit i
f_{2i} - yield factor, unit i
f_{3i} - predicted mass removal, unit i
R_{yi}, e_i - parameters used to compute the voltage across unit i

REFERENCES

Balchen, J.G. (1985). A Quasi-Dynamic Optimal Control Strategy for Non-linear Multi-variable Processes Based upon Non-quadratic Objective Functionals. Modeling, Identification and Control, vol. 5, No. 4.

Balchen, J.G., D. Ljungquist, and S. Strand (1988). Predictive Control Based upon State Space Model, 1988 American Control Conference (in press)

Clarke, D.W., C. Mohtadi, and P.S. Tuffs (1987). Generalized Predictive Control. Part I and II Automatica, Vol. 23, No. 2, pp 137-160,1987

Cutler, C.R., and B.L. Ramaker (1979) Dynamic Matrix Control - a Computer Control Algorithm. AIChE 86th National Mtg, Houston,TX, Apr.1979

Garcia, C.E., and D.E. Prett (1986). Advances in Industrial Model Predictive Control. CPC III, Asilomar, Calif., Jan. 1986.

Ljungquist, D. (1986). Predictive Control Based on Pontryagin's Maximum Principle (in Norwegian), Thesis(sivilingeniør).Division of Engineering Cybernetics, The Norwegian Institute of Technology, Trondheim.Dec. 1986.

Marchetti, L., D.A. Mellichamp, D.E. Seborg (1983). Predictive Control Based on Discrete Convolution Models. Ind. Eng. Chem. Proc. Des. Develop., 1983, 22, 488-495.

Mehra, R.K., and S. Mahmood (1985). Model Algorithmic Control, chapter 15 in Distillation Dynamics and Control, ed. P.B. Desphande, Instruments Society of America, 1985.

Pontryagin, L.S. et al.(1962). The Mathematical Theory of Optimal Processes. J. Wiley & Sons.

Reid, J.G. (1981). Output Predictive Algorithmic Control: Precision Tracking with Application to Terrain Following. AIAA, Journal of Guidance and Control, Vol. 4, No. 5, 1981.

Richalet, J.A.; Rault, A.; Testud, J.D. and Papon, J. (1978). Model Predictive Heuristic Control: Applications to Industrial Processes, Atomatica, 14:413, 1978

Strand, S. (1987). System for Predictive Control Based on Repetitive Simulation of a Dynamic Model with a Parameterized Control Vector (in Norwegian), Thesis(Sivilingeniør),Division of Engineering Cybernetics, The Norwegian Institute of Technology,Trondheim. March, 1987.

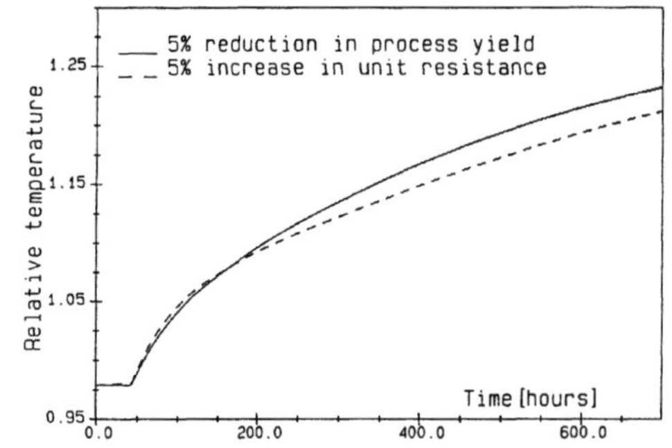

Fig. 3.a. Step responses, control variables unchanged

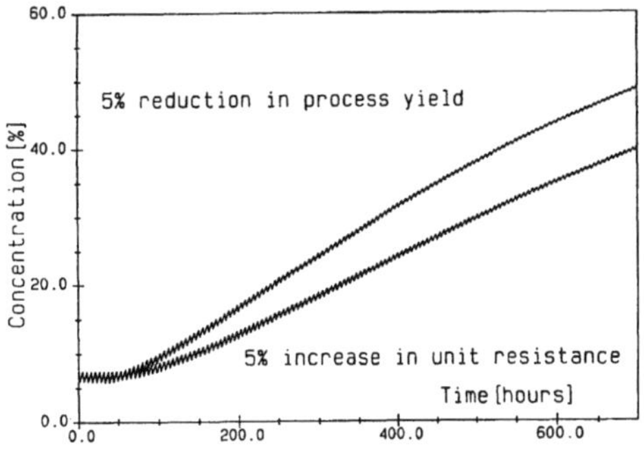

Fig. 3.b. Step responses, control variables unchanged

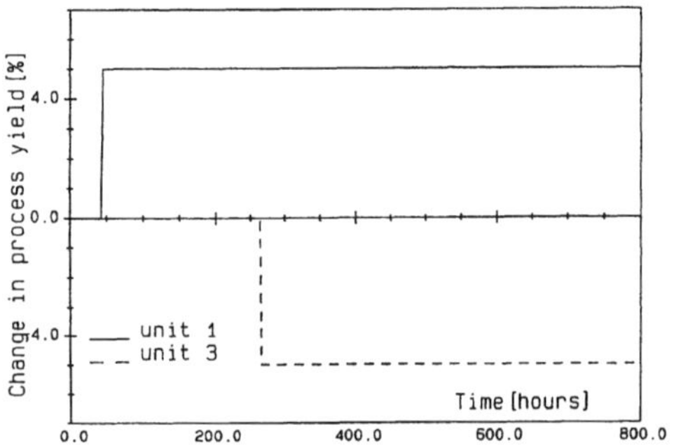

Fig. 4.a. Disturbances in process yields

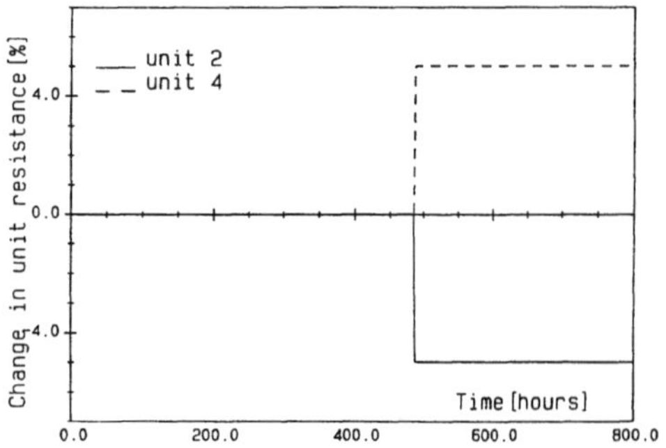

Fig. 4.b. Disturbances in unit resistances

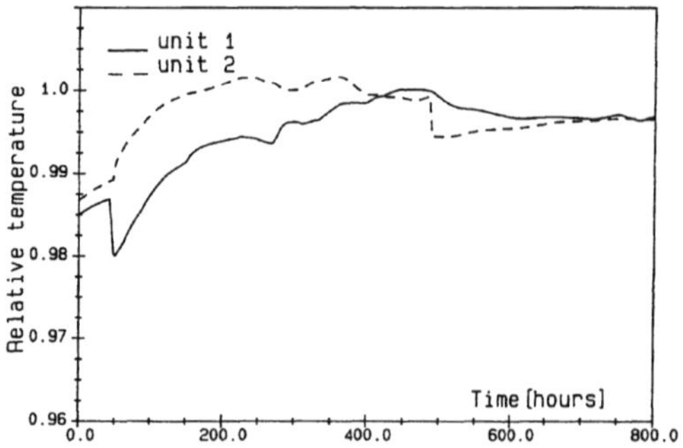

Fig. 5. Unit temperatures

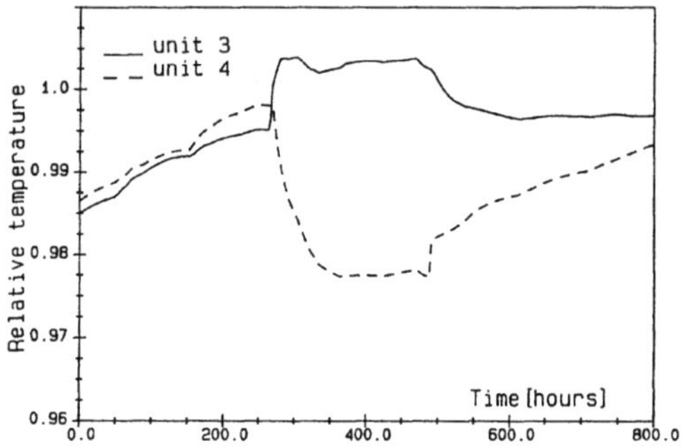

Fig. 6. Unit temperatures

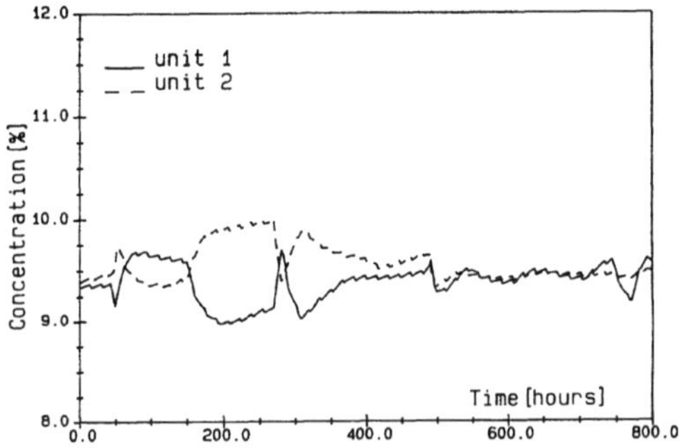

Fig. 7. Unit concentrations

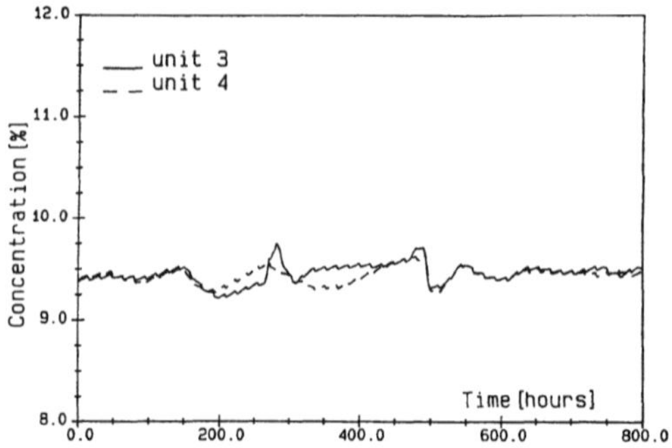

Fig. 8. Unit concentrations

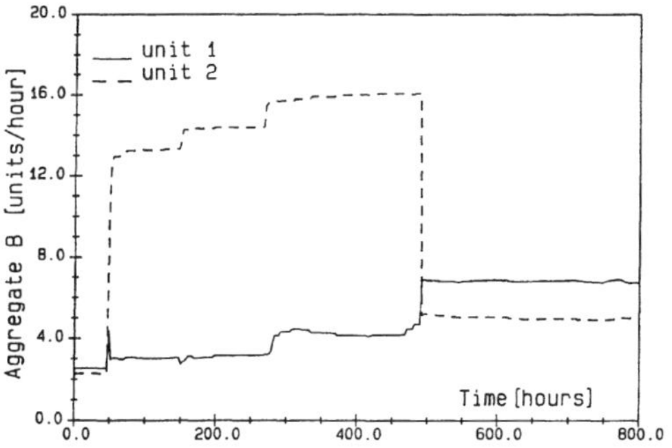

Fig. 9. Use of aggregate B

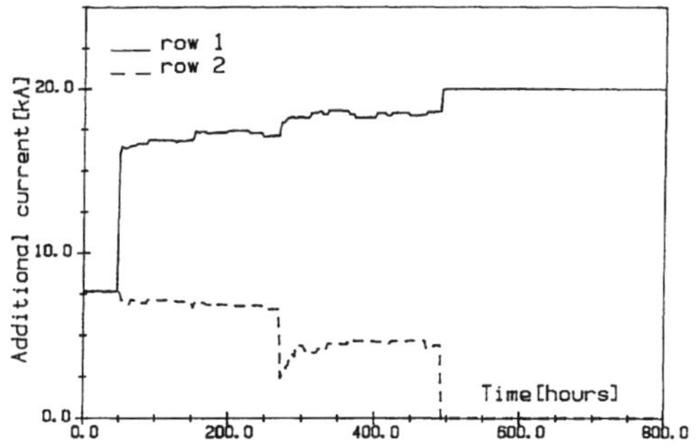

Fig. 13. Use of row current

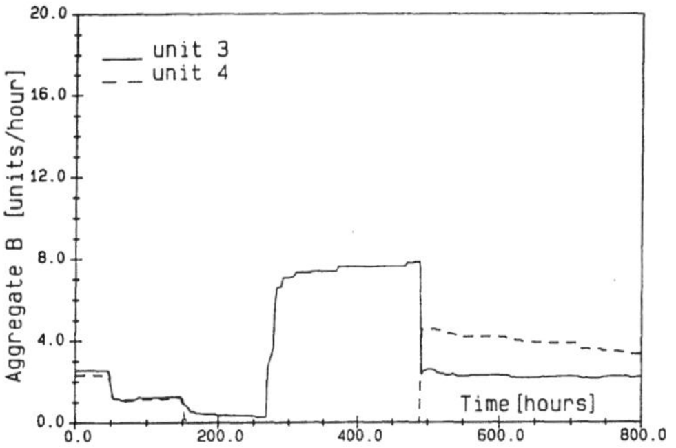

Fig. 10. Use of aggregate B

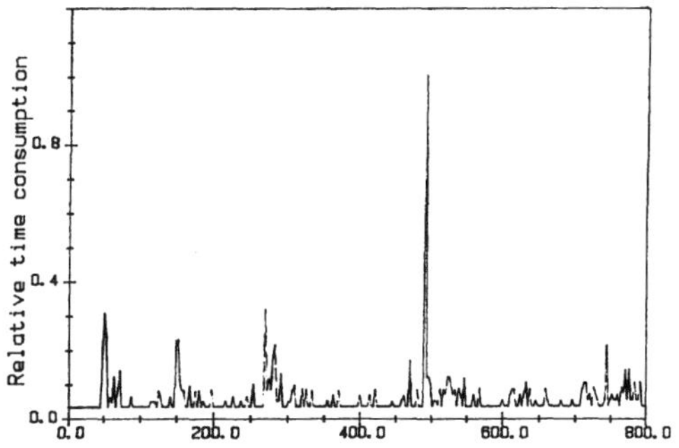

Fig. 14. Computational effort in optimization

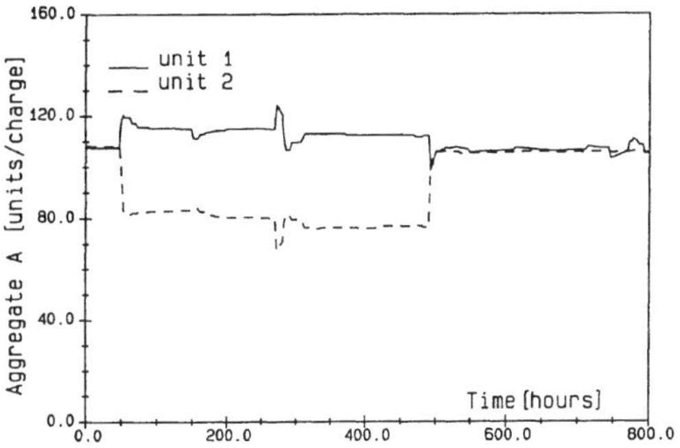

Fig. 11. Use of aggregate A

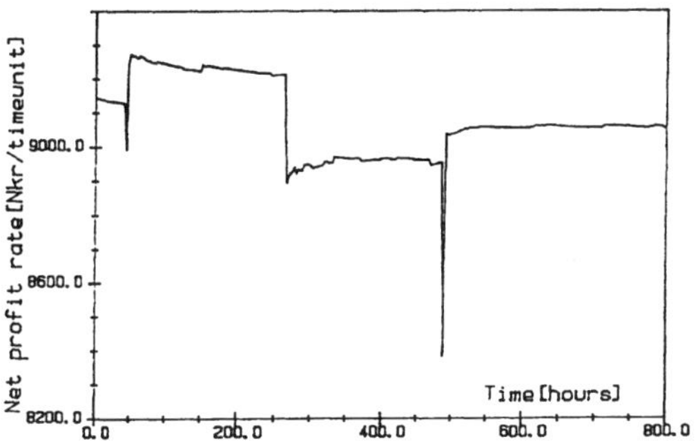

Fig. 15. Net profit rate from 4 units

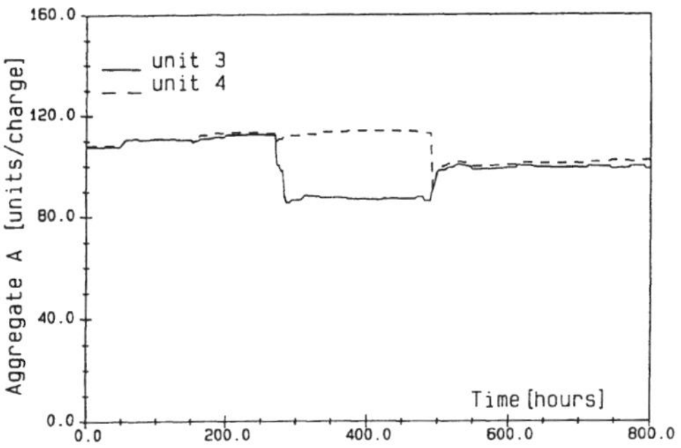

Fig. 12. Use of aggregate A

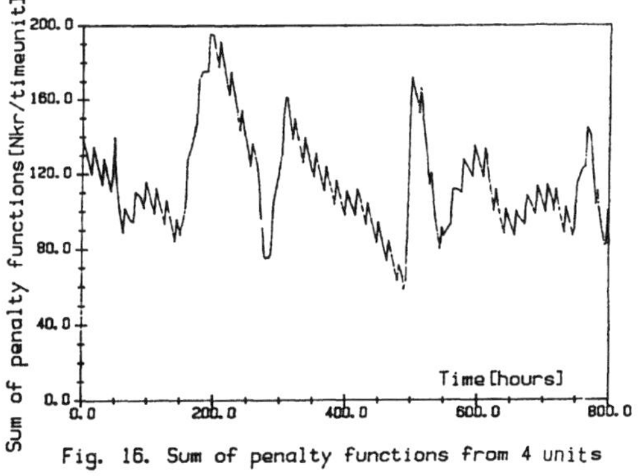

Fig. 16. Sum of penalty functions from 4 units

MODEL-BASED DUAL COMPOSITION CONTROL OF AN INDUSTRIAL PROPYLENE SPLITTER

S. A. Malik

Associate Control Engineer, Process Engineering Department, Polysar, Sarnia N7T 7M1, Ontario, Canada

Abstract. An application of model-based control to the dual composition control of a propylene splitter is described. This is a medium-to-high purity column. The method of control does not rely on linear decoupling to achieve the regulation of tower overheads and bottoms quality. A non-linear model, based on process characteristics, is used to decouple the overhead and bottoms dynamics.

Keywords. multivariable control; model-based; distillation column; high purity distillation; non-linear control

INTRODUCTION

The purpose of this paper is to show the application of model based control to an industrial propylene splitter. This is one component in the combined refinery and petrochemicals complex at Polysar (Corunna) site.

The propylene tower is the final seperation unit, in a reaction/distillation train. It produces medium to high purity propylene for external customers. This unit was recently added to the complex a few years ago, and had been operated under regulatory control for some time. This paper looks at the issues involved both in the design of the regulatory control as well as the decisions behind the selection of model-based control, versus the other multivariable techniques such as DMC(1) and linear decoupling(2).

PROCESS DESCRIPTION

The propylene splitter is designed to split a feed that contains approximately 88% propylene,12% propane, into an overhead product which contains 96% propylene, with less than 5% propylene in the bottoms product.

There is a penalty for making a product that is too pure, as this causes processing problems for a customer. Any propylene in the bottoms of the tower is downgraded from product value, as it is recycled back to the steam cracking heaters.

The control problem therefore is to keep the overhead product purity X_d within the limits, $0.95 < X_d < 0.96$ while maintaining the bottoms propylene loss X_b and keeping it to the criterion $0.015 < X_b < 0.025$.(note that all fractions in the text are mole fractions. Weight fractions will be explicitly indicated)

The condensing capacity, a design value of about 80 MBtu , is provided by cooling water, and the reboiler heat duty of about 78 MBtu, is provided by quench water. This is shown in Figure 1 which also shows the steady state conditions.. There is a curious conflict here between controllability and optimisation, where driving the tower to zero bottoms losses can make it unstable, that is the tower overheads loops reponse becomes non-linear and finally unstable.

The bottoms stream from the tower is typically less than 10% of the feed mass flow rate , and X_b about 0.02. Naturally, subject to constraints, this value should be driven to zero. However, when X_d = 0.96 and X_b tends to zero, the medium purity split, becomes a high purity split. Therefore, the locally linearised values of open-loop process gains changes by a large amount as verified by an RGA about that point (3),(4). The problem also becomes ill-conditioned, in the sense of (5). A further reason is that as the reboiling heat medium is 'free' energy, there is sometimes a overzealous attempt to maximise reboiler heat duty. Further numerical details for this column, and based on the earlier work in (5),(6) are given in (3).

The following graphs show the composition and temperature profiles in the tower. All units are mole fraction. These simulations were necessary to determine the basic control structure of the tower. The temperatures were found to be very insensitive to composition changes, for this column, and analysers were provided on the feed, bottoms and product. The details of the RGA work are not given here but covered in (4).

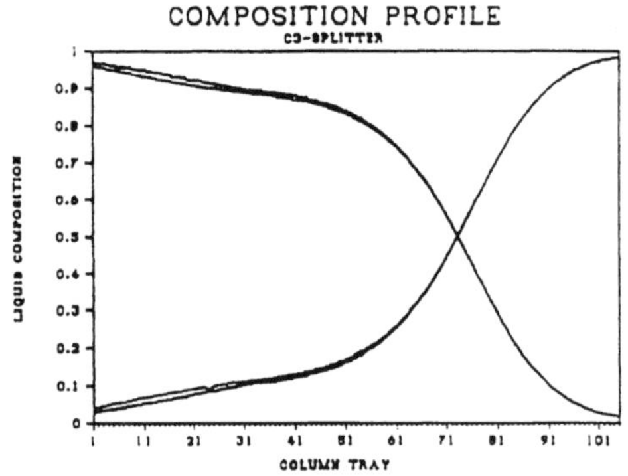

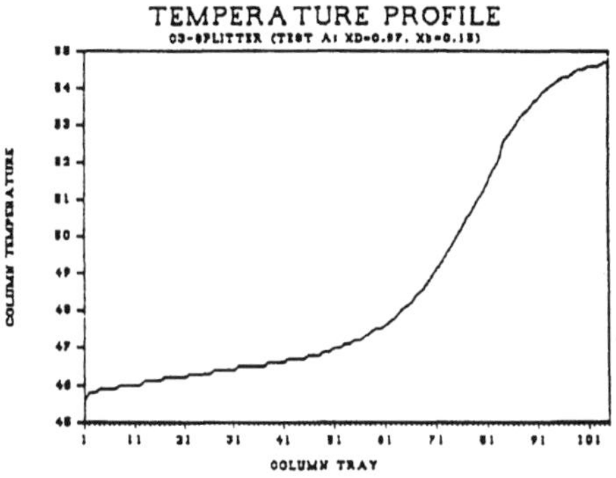

This problem of inventing a high purity problem can be avoided by noting that the propylene concentration in the bottoms (by weight) is controlled in the model-based controller to be less than 0.015 (weight fraction). With a nominal flow rate less than 6 Klbs/hr, this represents a small flow of propylene downgraded to propane cost. (The next step of the control will attempt to keep the propylene downgrading close to zero. The above figures give a daily propylene downgrading of about $100k!) It is important to note that in practice at the plant , due to the changed nature of the problem, the operators tend to keep X_b away from zero, to about .01 .

MODEL BASED CONTROL

The composition control problem in the propylene splitter is a multivariable problem, with constraints. The DMC(1) method was applied to a high purity ethylene splitter, but the method was not robust enough to handle the changing feed composition in this high-purity column, along with the process constraints which limited the control action of the algorithm. The least-squares version of the DMC was applied, and not the more sophisticated LP/QP version which can handle constraints. For this reason, this method was not used on the propylene splitter. The linear decoupling approach of (2) was also tried. In this column, however, there were problems with control initialisation and the magnitude of the constants used in the decoupling. This arose from the very low X_d/Q_r open-loop process gain, which called for a large decoupling controller gain.

Two non-linear algorithms were tested for their suitability for on-line model-based control. These methods were selected based upon the criteria and test results in (6),(7),(8). (4) also examines the extension to very high purity ethylene fractionators. The two methods selected are :

Smith-Brinkley (SB) method (6).
JDM method (7).

There is an obvious difference between these non-linear models based on process correlations, and the linear steady state models used in internal model control IMC. In (3), there is a interesting discussion on robustness and uncertainity which explains the links between these transfer function approaches, and the steady state models. The dominant time constant approach of (3) was also used to directly compute the tower dominant time constant, of about 62.5 minutes. This last work is the first connection between steady state simulations, based on extensive heat and mass balances, and the tower overall dynamics.

These non-linear models are based on the concepts of seperation ,S, and product splits. Note that recovery as defined in (6) is really the product split. The formal definition in (9) is Recovery,W = D / F X_f :

$S = X_d\ (1-X_b)\ /\ X_b\ (1-X_d)$ where S = seperation

$f_1 = B\ X_b\ /\ F\ x_f$

$f_2 = B\ (1-X_b)\ /\ F\ (1-x_f)$

where X_f is the mole fraction propylene in the feed, and B,F are the bottoms and feed mass flow rates respectively.

This makes it convenient to develop the models from detailed mass and energy balance simulations, and tray-to-tray calculations.

The test results which document the work in the comparison between the models is discussed in (4). The details of the application of the Smith Brinkley method are explained very lucidly in (6), where an application on a depropaniser is reported. The outline of the algorithm there is sufficient for coding as an on-line computer control algorithm. The JDM method is somewhat more straightforward and is discussed further in the text.

From the results in (4), there was not an appreciable difference in predictions from the SB and JDM models. Further, an iterative solution of non-linear equations was called for in the SB method at each control interval. Extensive steady state and plant historical data had shown that the tower top and bottom temperatures vary are relatively insensitive to variations both in the feed, and within the tower. Appendix (1) shows that with X_d=0.95 and with X_b varied in the range $0.001 < X_b < 0.100$, the overhead temperature stayed virtually constant at 45.6oC , and the bottoms temperature T_{btm} varied less than 1.5oC.
This simulation data corresponds very closely with actual bottoms temperature variations seen under equivalent conditions.

The SB method goes through an iterative approach to estimate seperation factors for the top and bottom section of the tower.

compute first the top and bottom (avg section) temperatures T_t and T_b.

$X_d\ /\ K_{t1}(\ T_t,\ P\ )\ +\ (1-X_d)\ /\ K_{t2}(\ T_t,\ P)$
$X_b\ /\ K_{t1}(\ T_t,\ P\ )\ +\ (1-X_b)\ /\ K_{t2}(\ T_b,\ P)$

$T_n = (T_t + T_{feed})\ /\ 2$
$T_m = (T_{feed} + T_b)\ /\ 2$

compute the seperation factors for the bottom and top section

$S_{ni} = K\ (T_n,\ P)\ V/L$
$S_{mi} = K\ (T_m,\ P)\ V_p/L_p$ where $i = [1,2]$

where the K_i are the usual K values in vapour/liquid equilibrium, V is the vapour rate = L+D. L (or R) refers to the reflux mass flow rate. V_p and L_p are defined as vapour and liquid flows in the stripping section of the column, assuming a saturated feed, F.

It can be seen from here that in the presence of steady overhead and bottoms temperatures, the attraction of the SB method over the JDM method is slight. This was confirmed in a plant test, over a 1 week period, and through steady state simulations by a commercial package (SimSci). Figure 3 and Figure 4 show the variation of bottoms propylene concentration X_b with bottoms temperature T_{btm}. and T_{btm} with reboiler heat Q_r. The plant data is not represented here but will be included in (4).

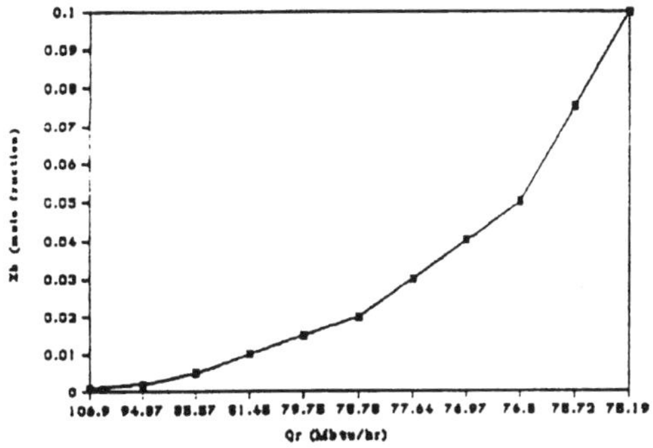

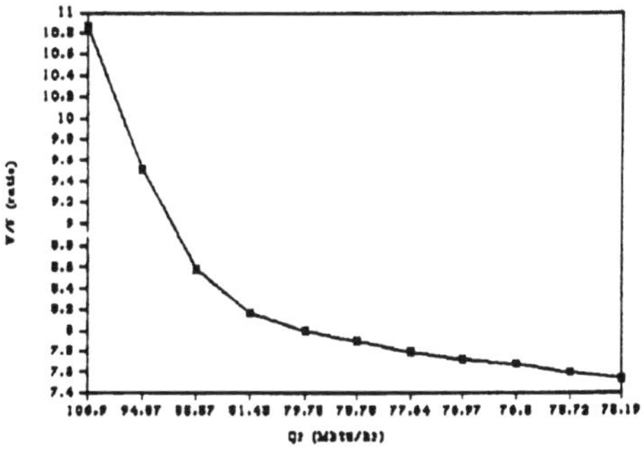

The numerical steady state solution of the tower was obtained while keeping X_d constant. The solution for values of X_b in the range $0.10 < X_b < 0.001$ is shown in Appendix 1. The steady state temperature and composition profiles are shown below. Note that this data was obtained from simulation models where the model had been matched to the operating plant. It was not permissible to take an operating unit through this entire test cycle, to produce the data in the above graphs. The plant data is limited to a narrower range, and there are problems filtering out other operational changes.

However, on another computer control application on the depropaniser, the SB method will be tried in favour of the JDM. This is because there is less analyser information, and greater correlation between mean stage temperatures and mean compositions, in the sense of (3). Also the tower is lower purity, with three feed streams. The rest of the discussion applies only to the JDM method as applied to the propylene splitter tower.

The tower is under material balance control. This method of control was chosen over the energy balance based on the criteria in (9), and through an RGA analysis (2). One of the major factors, was simply that the reflux/distillate ratio is so large, $5 < L/D < 8$, that it would have been difficult to control level in the overhead drum any other way. There are analysers available to measure propylene content in the feed, product and bottoms X_f, X_d and X_b.

The reflux enters the tower on a level to flow cascade, and the bottoms leaves the tower on a bottoms level to bottoms flow cascade also. The product flow, and flow of quench water to the reboiler, is manipulated by the control algorithm.

Therefore to control the pair $[X_{d(sp)}, X_{b(sp)}]$ the JDM algorithm computes setpoints $[D_{(sp)}, V(sp)]$. Note that (sp) refers to the setpoint transmitted to the controllers on the process.

The JDM algorithm is slightly modified in application. The algorithm outputs dimensionless numbers [D/F, V/F]. This makes it easier to fit into a control hierarchy, so that the operator can enable/disable computer control at the appropriate level. It also simplifies the model updating procedure.The operation of the bottoms controller is described first. <u>Bottoms controller for X_b</u>.The variable $V_{(sp)}$ is the desired vapour flow rate in the tower. The controller for this is entirely in software, and the translation between V desired, and the flow rate of heating medium to the reboiler, is done as follows. The actual implementation, as mentioned earlier, is that of a V/F controller. The linearised discrete form is used, and is based on the approach in (5) :

$$Q_{r(sp)} = V_{(sp)} \cdot dH, \text{ where } dH \text{ is latent heat of vaporisation}$$
$$F_{r(sp)} = Q_{r(sp)} / C_p \cdot dT, \text{ where } F_r \text{ is the reboiler flow}$$

C_p and dT are the specific heat and reboiler heating fluid temperature drop, respectively

In the above, data filtering details have been omitted. For the actual control, all variables are dynamically reconciled. It is important to allow for the dynamics of the process, especially when introducing algebraic calculations within a control algorithm.

For the overhead control loop, the main control loop is a one-step ahead predictive controller. Dead time compensation is inherent in the controller design, and the IMC approach is used. The typical analyser deadtime is 10 minutes, but an infrared analyser is used here and the response is virtually instantaneous. So measurement dead time is negligible. Initially the analyser was on tray 12 of the column, but it was later relocated just after the condensors, for operational reasons. The predictive controller drives a D/F controller which resets D, the product flow.

The way in which the JDM controller comes into this loop is through pseudo-linear compensation for disturbances in Xf and V/F. The small perturbation model is used, but <u>the partial derivatives are derived numerically at each control step</u>. Therefore, for a not excessive control interval, the assumptions are valid (this is not unlike the predictor/corrector numerical integration methods for solving differential equations). The problem in the equations below is to derive the incremental change to (D/F) which have to be made to accomodate changes in X_f and V.

$$d(D/F)_k = dU_{1(k)} = d^*/d^*X_f \{(X_f - X_b) / (X_d - X_b)\}(X_{f(k)} - X_{f(k-1)})$$
$$d(D/F)_k = dU_{2(k)} = d^*/d^*V \{(F_1(Ln_e(S)) V/F / (R_m-1)\}(V_{(k)} - V_{(k-1)})$$

$$dU_k = dU_{1(k)} + dU_{2(k)}$$

Note that the term (d^{-}) denotes a partial derivative. The procedure for determining the numerical partial derivative is the usual first order Taylor expansion. Numerical accuracy has not been a problem.

In the above the objective is simply to use the JDM equations to work out what effect the tower internals and feed composition are having on the analyser required D/F target.

For the bottoms control loop, the JDM equations are applied directly :

$$D/F = (X_f - X_b) \,/\, (X_d - X_b)$$
$$V/F = \{(R_m+1)\ D/F\} \,/\, \{ 1 - a_1 (a_2 - a_3)^b \}$$

where

$$a1 = 1.6612,\ b = 1.7643,\ a3 = 0.25$$
$$R_m = (1/(r_v-1))\ ((X_d/X_f) - r_v(1-X_d)/(1-X_f))$$
$$a_2 = \{ ((Ln_e S/Ln_e r_v)+1)/(N+1) + 1 \}$$
$$S = X_d(1-X_b) \,/\, (X_b(1-X_d))$$

where R_m, r_v are the minimum reflux and relative volatility repectively. Note that for this column $1.01 < r_v < 1.14$

The purpose of the JDM controller is more direct in the bottoms loop, than for the top loop.The objective here is to first reconcile the equations with process conditions, calculate a modelling bias term, and then to apply known values of D/F, X_d, X_b and X_f in the right hand side to give a value of the desired (V/F), denoted here as $(V/F)_{sp}$.

The given product specs $X_{d(sp)}$ and optimisation spec $X_{b(sp)}$ determine the seperation S required. The reconciled model predicts the $(V/F)_{sp}$ which will give this ideal. This now is the target for a (V/F) controller, which accounts for the significant dead-time (18 minutes) and nonlinearities to manipulate the heat input to the base of the column. This chain of cascades was described earlier in the text.

This method of control has been used successfully , and good servo performance has been achieved. More importantly, the disturbance rejection for feed composition and rate changes is very acceptable. The model based approach is physically appealing because it allows one to double check modelling changes, with the use of commercially available steady-state simulation packages.

REFERENCES

(1) Chang, T.S — *A new linear programming approach to multivariable control with constraints.* Ph.D Thesis 1983, Univ. of Calif., Santa Barbara

(2) McAvoy, T. — *Interaction Analysis - Principles and Applications* ISA books 1983

(3) Morari, M. — *Dominant time constant for distillation columns* Comp. and Chem. Engng, 1987, Vol.11, 6, 607-617

(4) Malik, S.A — *Non-linear model matching for model-based computer control algorithms* (Report in preparation)

(5) Skogestad, S. — *Control of ill-conditioned plants: High purity distillation* _ Paper 7a A.I.Ch.E Annual Meeting, Florida Nov. 1986

LV control of a high purity distillation column Chem. Engng. Sci. 1988, Vol.43,1, 33-48

(6) Sullivan, G. — *Selection techniques for process model based controllers* Submitted for A.I.Ch.E Meeting 1986

Process model based control and optimisation of binary distillation columns Transcript. Personal Communication. Univ. of Waterloo

(7) Jafarey, A. McAvoy, T. — *Steady state feedforward control algorithms for reducing enegy costs in distillation.* ISA Trans. 1980, Vol.19, 4, 89

Short-cut techniques for distillation column design and control . 1. Column design
Ind. Eng. Chem. Process Des., 1979, Vol.8,2, 197

(8) Tolfo,F. *Group methods for advanced column control compared*
Hydrocarbon Processing May 1985, 93

(9) Shinskey,F.G. *Distillation Control . 2nd ed.*
(Mc Graw Hill) 1984

APPENDIX (1)

The following are results from a steady state simulation package (Simsci). The tower equations were solved for the conditions below :

F = 79.58 klbs/hr
X_f = 0.8758
X_d = 0.9500
N = 105 (theretical)
P = 275 psig

dP = 5 psig
Feed stage = 33
all liquid feed. saturated.
Rel. Voltly $1.09<rv<1.16$

X_b	Q_r	D	B	R/F	B/F	L/D	T_{btm}
mole frac	Mbtu/hr	Klbs/hr	Klbs/hr				°C
0.100	75.19	72.82	6.76	6.63	.0849	7.24	53.7
0.075	75.72	73.00	6.58	6.68	.0827	7.28	54.0
0.050	76.50	73.18	6.40	6.75	.0804	7.35	54.3
0.040	76.97	73.24	6.33	6.80	.0795	7.39	54.4
0.030	77.64	73.31	6.27	6.87	.0788	7.46	54.5
0.020	78.81	73.38	6.21	6.98	.0780	7.57	54.6
0.015	79.75	73.41	6.17	7.08	.0775	7.67	54.7
0.010	81.45	73.44	6.14	7.25	.0772	7.85	54.8
0.005	85.57	73.47	6.11	7.67	.0768	8.30	54.8
0.002	94.87	73.49	6.09	8.59	.0765	9.3	54.9
0.001	106.9	74.5	6.08	9.81	.0764	10.61	54.9

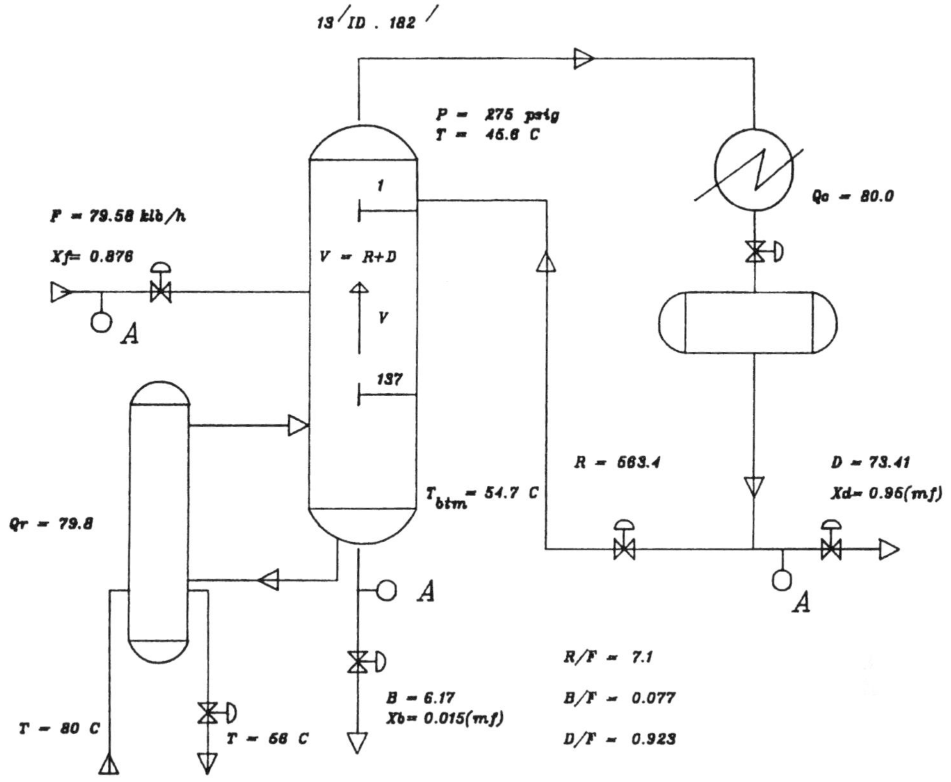

FIGURE (1)

* all flows are klb/hr, duties are mbtu/hr

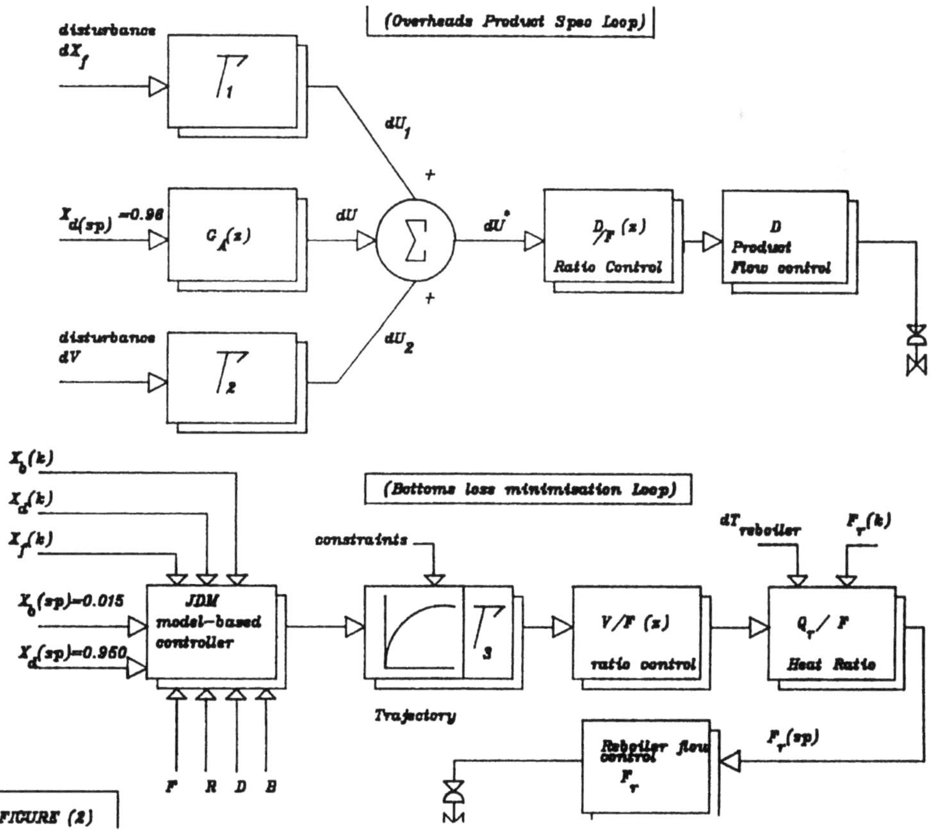

FIGURE (2)

DISTURBANCE FEEDBACK IN MODEL PREDICTIVE CONTROL SYSTEMS

J. P. Navratil, K. Y. Lim and D. G. Fisher*

Department of Chemical Engineering, University of Alberta, Edmonton, Canada

ABSTRACT. Model Predictive Control (MPC) schemes such as MOCCA, DMC, MAC, MPHC and IMC have been shown to provide excellent performance. This is particularly true for servo control applications in which it is assumed that the desired trajectory (setpoint) is known a priori. However, to achieve equivalent performance in regulatory control the future effects of residual values $y(k)-\hat{y}(k)$, which include the effects of disturbances and model process mismatch, must be predicted. In this paper a state space form of the process step response model is used as the basis for implementing state observer forms and a Kalman filter to predict these residual effects. It is shown that the addition of disturbance step response models to the closed loop observer and the Kalman filter predictor improves the prediction of structured residuals. It is also shown that closed loop feedback/prediction can guarantee zero steady state offset in the presence of non-zero mean residual effects.

KEYWORDS. state space; Kalman filter; model predictive control; observers

INTRODUCTION

In this paper the generic term "Model Predictive Control (MPC)" is used to define the class of control techniques which include MPHC (Model Predictive Heuristic Control, Richalet and co-workers 1978); DMC (Dynamic Matrix Control, Cutler and Ramaker, 1980); MAC (Model Algorithmic Control, Rouhani and Mehra, 1982); IMC (Internal Model Control, Garcia and Morari, 1982); MOCCA (Multivariable Optimal, Constrained Control Algorithm, Sripada and Fisher, 1985) plus several others. Each of these control schemes differs in detail but includes the following key features:

1. The future process outputs $\{y_m^{\prime}(k+i|k), i=1,\ldots,P\}$ are predicted using discrete step or impulse response coefficients rather than a typical state space or transfer function model.
2. A "correction term" for each element of the predicted output is usually calculated to account for the difference between the estimated value and the measured plant output.
3. A **predictive** control strategy is used to calculate the control action sequence $\{\Delta u(k+i), i=0,1,\ldots,M; M \le P\}$ which minimizes a user specified performance index.

This paper is primarily concerned with the estimation of the future process output trajectory $\hat{Y}(k)$. As indicated by Fig. 1, this generally involves **prediction** based on the process step response model plus a **correction** term based on the residual value $r(k)$ defined as the difference between the measured and estimated values of the process output. In MPC the correction term requires the prediction of the effect of the residual values on the future output values. The problem of estimating $\hat{Y}(k)$ is solved using state space techniques and theory. The use of state observers or a Kalman filter to generate the correction term based on the residual value is shown to improve system performance in the presence of noise and non-zero mean residuals.

*Author to whom correspondence should be directed.

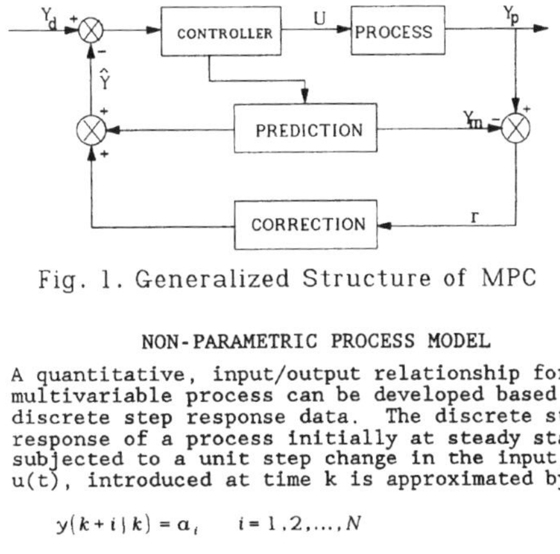

Fig. 1. Generalized Structure of MPC

NON-PARAMETRIC PROCESS MODEL

A quantitative, input/output relationship for a multivariable process can be developed based on discrete step response data. The discrete step response of a process initially at steady state subjected to a unit step change in the input, u(t), introduced at time k is approximated by

$$y(k+i|k) = a_i \qquad i = 1,2,\ldots,N \tag{1}$$
$$= a_{ss} \qquad i > N$$

where $y(k+i|k)$ is defined as the output value at time k+i, calculated at time k, based on all process inputs available up to and including time k-1. The output at any time is assumed to be a function only of N past changes in input and the input (N+1) intervals ago. For example,

$$y(k|k) = \sum_{j=1}^{N} a_j \Delta u(k-j) + a_{ss} u(k-N-1) \tag{2}$$

Predicted Output Trajectory

The non-parametric process model can be used to generate a trajectory of predicted outputs as shown by Sripada and Fisher (1985). Assuming superposition, the trajectory of future predicted output values for $i=1,\ldots,P$ is:

$$y_m(k+i|k+i) = \sum_{j=1}^{N} a_j \Delta u(k+i-j) + a_{ss} u(k+i-N-1) \tag{3}$$

For control purposes Eqn. (3) is divided into two parts involving: 1) present and future (unknown) inputs and 2) Y^*_m past (known) inputs. The effect of future (unknown) inputs, $\{\Delta u(k+i), i=0,1,\ldots,M-1\}$, on the process output trajectory is:

$$\begin{bmatrix} y_m(k+1\mid k+1) \\ y_m(k+2\mid k+2) \\ \cdot \\ \cdot \\ y_m(k+P\mid k+P) \end{bmatrix} = \begin{bmatrix} y^*_m(k+1\mid k) \\ y^*_m(k+2\mid k) \\ \cdot \\ \cdot \\ y^*_m(k+P\mid k) \end{bmatrix} + A_2 \begin{bmatrix} \Delta u(k) \\ \Delta u(k+1) \\ \cdot \\ \cdot \\ \Delta u(k+M-1) \end{bmatrix} \quad (4a)$$

The contribution to the future output trajectory of past input changes is given by:

$$\begin{bmatrix} y^*_m(k+1\mid k) \\ y^*_m(k+2\mid k) \\ \cdot \\ \cdot \\ y^*_m(k+P\mid k) \end{bmatrix} = a_{ss} \begin{bmatrix} u(k-N) \\ u(k-N+1) \\ \cdot \\ \cdot \\ u(k-N+P-1) \end{bmatrix} + A_1 \begin{bmatrix} \Delta u(k-N+1) \\ \Delta u(k-N+2) \\ \cdot \\ \cdot \\ \Delta u(k-1) \end{bmatrix} \quad (4b)$$

In a more compact notation

$$Y_m = A_2 \Delta U_f + Y^*_m \quad (5a)$$

$$Y^*_m = A_1 \Delta U_p + a_{ss} U_p \quad (5b)$$

where the subscripts p and f signify past and future values respectively.

Control Action

The control action $\Delta U_f(k) = \{\Delta u(k+i), i=0,1,\ldots,M\}$ is calculated directly from the predicted $\hat{Y}(k)$ and the desired $Y_d(k)$ trajectories such that a user-specified performance index is satisfied. A simple and commonly used performance index is the weighted least square performance index.

$$J = \frac{1}{2}\Big[\sum_{i=1}^{P} \gamma(i)\big[y_d(k+i) - y_m(k+i\mid k+j)\big]^2 + \sum_{i=1}^{M} \gamma_m(i)\Delta u(k+i-1)^2\Big] \quad (6)$$

where j = min (i-1,M-1), $\gamma(i)$ are weights on the output deviations and $\gamma_m(i)$ are weights on the changes in the input variable. Note that for regulatory control purposes it is necessary to predict the effect of residuals on $\hat{Y}(k)$ as shown in Fig. 1. Without such a prediction there is no predictive control. By defining an error trajectory at time k as:

$$E(k) = \big[(y_d(k+i) - \hat{y}(k+i)), i=1,\ldots,P\big] \quad (7)$$

the performance index, Eqn. (6), becomes

$$J = \frac{1}{2}\big[\,\|E(k) - A_2\Delta U_f\|^2 \Gamma + \|\Delta U_f\|^2 \Gamma_m\big] \quad (6a)$$

Calculation of ΔU_f to minimize J is a standard weighted least squares problem with the solution

$$\Delta U_f = A^* E(k) \quad (8)$$

where A^* (the inverse or pseudo inverse of A_2) is a constant matrix calculated **off-line**. Constraints can be handled by converting the control calculation to an on-line, linear or non-linear, optimization problem e.g. linear or quadratic programming (Prett and Gillette, 1980, Morshedi, Cutler, and Skrovanek, 1985, Ricker, 1985, Morshedi, 1986, Campo and Morari, 1986).

STATE SPACE FORM OF THE PROCESS MODEL

A recursive relationship for estimating current and future values of the process output has been developed by Li, Fisher, and Lim (1988). At any time interval k the predicted value of an output may be calculated from the previous prediction, at k-1, and the most recent change in the input variable as follows

$$y^*_m(k\mid k) = y^*_m(k+1\mid k-1) + a_1\Delta u(k-1) \quad (9)$$

$$y^*_m(k+1\mid k) = y^*_m(k+2\mid k-1) + a_2\Delta u(k-1)$$

Extending Eqn. (9) to the entire trajectory, i=0,1,...,N yields

$$y^*_m(k+i\mid k) = y^*_m(k+i+1\mid k-1) + a_{i+1}\Delta u(k-1) \quad (10)$$

By definition $a_{i+1} = a_{ss}$ for i>N indicating that

$$y_m(k+N+1\mid k-1) = y_m(k+N\mid k-1) \quad (11)$$

Eqns. (10) and (11) expressed in a more compact vector/matrix notation are:

$$X(k) = \Phi X(k-1) + \Theta \Delta u(k-1) \quad (12)$$

where

$$X(k) = \big[y_m(k\mid k) \;\; y_m(k+1\mid k) \;\; \cdot \;\; \cdot \;\; y_m(k+N\mid k)\big]^T_{(N+1)\times 1}$$

$$\Phi = \begin{bmatrix} 0 & 1 & 0 & \ldots & \ldots & 0 \\ 0 & 0 & 1 & 0 & \ldots & 0 \\ \cdot & \cdot & \cdot & \cdot & \cdot & \cdot \\ \cdot & \cdot & \cdot & \cdot & \cdot & \cdot \\ 0 & \ldots & \ldots & 0 & 0 & 1 \\ 0 & \ldots & \ldots & 0 & 0 & 1 \end{bmatrix}_{(N+1)\times(N+1)}$$

$$\Theta = \big[a_1 \;\; a_2 \;\; \cdot \;\; \cdot \;\; a_N \;\; a_{ss}\big]^T_{(N+1)\times 1}$$

The output equation (i.e. the current estimate) is:

$$Y(k) = HX(k) \quad (13)$$

where

$$H = [1 \;\; 0 \;\; 0 \;\; \ldots \;\; 0]_{1\times(N+1)}$$

Reduced Order State Space Form

The dimension of the state vector in Eqn. (12) can be reduced from N+1 to P+1 as shown by Lim (1988). Eqn. (12) is still valid if the following symbols are redefined as:

$$\Delta U(k-1) = [\Delta u(k-N+P-1) \;\; \cdot \;\; \cdot \;\; \Delta u(k-1)]^T_{(N-P+1)\times 1} \quad (14)$$

$$\Theta = \begin{bmatrix} 0 & \ldots & \ldots & 0 & a_1 \\ 0 & \ldots & \ldots & 0 & a_2 \\ \cdot & \cdot & \cdot & \cdot & \cdot \\ \cdot & \cdot & \cdot & \cdot & \cdot \\ h_{ss} & h_N & \ldots & h_{P+2} & a_{P+1} \end{bmatrix}_{(P+1)\times(N-P+1)} \quad (15)$$

Note that the input term becomes a matrix/vector product rather than a vector/scalar product and the model order reduces to P+1. The definitions and forms of the remaining vectors and matrices are unchanged.

Properties of the State Space Model

The recursive form of the process model is a non-minimal state space model. The observability matrix for the system is

$$\big[H, H\Phi, \ldots, H\Phi^{N+1}\big]$$

It can be shown that the rank of the observability matrix is P+1. Thus the system is observable and the state (cf. predicted output trajectory) can be reconstructed from the measured output $y_p(k)$ and the known input.

Delayed States. When a time delay of length d sample intervals is encountered then d delayed states are added to the beginning of the state vector. This is equivalent to increasing N and allowing the first d step response coefficients to be zero. Non-integer values of d do not occur due to the nature of the step response model and discretization. The delayed states can not be altered by control action and thus are not included as part of the trajectory used in the control calculation. It is important to note that the controller design is dependent only upon the contents of the performance index and that any desired subset of the predicted trajectory may be specified in the performance index.

Multi-Input Multi-Output (MIMO) Model

The multivariable state space model with r output and s input variables can be constructed by simply redefining the vectors and matrices as block vector and matrices. The contents of the block matrices are defined by the type of model used, i.e. full or reduced order. The MIMO input vectors for the full and reduced order model respectively are:

$$\Delta U(k-1)=\left[\Delta u^1(k-1),\ldots,\Delta u^s(k-1)\right]^T \quad (16)$$

and

$$\Delta U(k-1)=\left[\Delta U^1(k-1),\ldots,\Delta U^s(k-1)\right]^T \quad (17)$$

PREDICTION AND FEEDBACK OF RESIDUAL CORRECTION TERM

Open Loop State Observer

The most commonly used method of handling residuals in DMC (Garcia and Prett, 1986) can be interpreted as an open loop state observer with unity gain values. This method involves adding the current value of the residual to all future output predictions used in the control calculation. The open loop state observer, shown in Fig. 2, is defined by

$$\hat{X}(k)=\Phi\hat{X}(k-1)+\Theta\Delta U(k-1) \quad (18)$$

$$\hat{Y}(k)=\left[H_C\hat{X}(k)+K\left[y_p(k)-y_m^*(k\mid k)\right]\right] \quad (19)$$

where

$y_m^*(k\mid k)=H\hat{X}(k)$

$H_C=(P+1)x(P+1)$ matrix used to choose the subset of predicted outputs used in the control calculation

$K=[1,\ldots,1,1]^T_{(P+1)x1}$

$\hat{Y}(k)$ is the corrected, model-based, trajectory of future outputs used by the controller.

The observer corresponding to Eqns. (18) and (19) contains the assumption that all future outputs will be effected by the current residual in exactly the same way as the current output. Gains other than unity can not be used in the open loop observer as they will cause steady state error in the prediction of $\hat{Y}(k)$. The predicted output value $y_m^*(k\mid k)$ will not approach the actual output value in the presence of non-zero mean disturbances and thus the residual value will not approach zero.

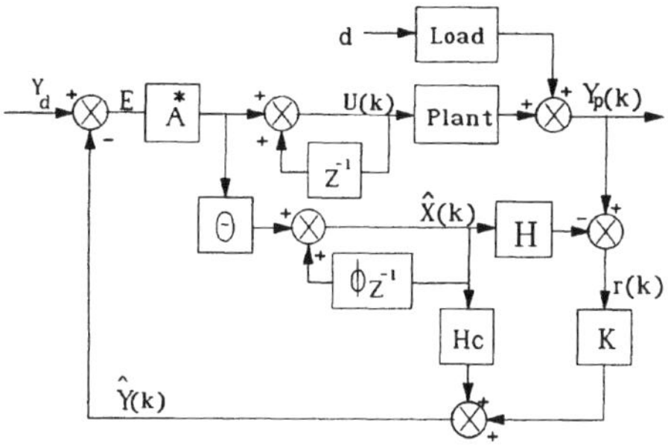

Fig. 2. Open Loop Observer Form

Closed Loop State Observer

In the open loop state observer the states are calculated directly from the process model and are not altered by feedback of the output. Therefore the open loop observer will exhibit steady state prediction offset in the presence of model process mismatch and/or non-zero mean disturbances. This offset is eliminated in the closed loop state observer by adding feedback corrections directly to the state variables. The closed loop observer form is shown, in Appendix A, to guarantee zero steady state offset whenever $K_{N+1}/K_{N+1}=1$ (i.e. whenever the last gain value is non-zero). The closed loop state observer, shown in Fig. 3, is defined by

$$\bar{X}(k)=\Phi\hat{X}(k-1)+\Theta\Delta U(k-1) \quad (20)$$

$$\hat{X}(k)=\bar{X}(k)+K\left[y_p(k)-y_m^*(k\mid k)\right] \quad (21)$$

$$\hat{Y}(k)=H_C\hat{X}(k) \quad (22)$$

where

$y_m^*(k\mid k)=H\bar{X}(k)$

$K=\left[K_1,\ldots,K_N,K_{N+1}\right]^T_{(N+1)x1}$

The closed loop observer gain vector K may be generated by several methods including the use of known disturbance models and closed loop transfer function analysis. Note that the controller input trajectory, $\hat{Y}$, is equivalent for both the open and closed loop observers when the observer gains are set to unity.

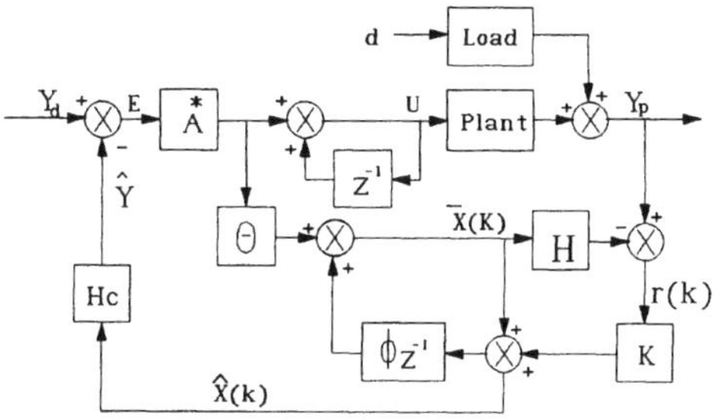

Fig. 3. Closed Loop Observer Form

Closed Loop State Observer Gains from Step Response Data

Disturbance step response data may be used to generate the gains of the closed loop state observer directly. A change in disturbance input at time k-1 will cause a prediction error at time k of magnitude $a_1*\Delta d(k-1)$ and subsequent errors at future time intervals k_i of magnitude $a_i*\Delta d(k-1)$ By defining the closed loop observer gains as a_i/a_1 for $i=2,\ldots,N+1$ prediction errors at all except the first instant can be eliminated in the absence of model mismatch and measurement error. The closed loop observer obtained by this method is sensitive to measurement noise, discretization error, and disturbance model errors. One assumption made in using this method is that the disturbance input signal can be approximated as a series of discrete steps occurring at integer multiples of the sampling interval. When the disturbance step responses are not known an adaptive disturbance model can be used to obtain the closed loop observer gains. This is the subject of current work.

Closed Loop State Observer Gains from Transfer Function Analysis

The closed loop observer gains may be calculated to give a desired closed loop response to residuals by examining the closed loop system equations. The closed loop system transfer functions for the process response to setpoint changes and disturbance inputs when using the closed loop observer form are given by

$$\frac{y(k)}{dG_L}=\frac{1+DG_M\left[1-G_F\right]}{1+DG_M+DG_F\left[G_P-G_M\right]} \quad (23)$$

$$\frac{y(k)}{y_d(k)}=\frac{DG_P}{1+DG_M+DG_F\left[G_P-G_M\right]} \quad (24)$$

where

$$G_F(q^{-1})=\frac{(1-q^{-1})\sum_{i=1}^{N-1}k_iq^{-i}+k_Nq^{-N}}{(1-q^{-1})\left[1+\sum_{i=1}^{N-1}k_iq^{-i}\right]+k_Nq^{-N}} \quad (25)$$

D,G_M,G_P,G_L are discrete transfer function forms in terms of the unit delay operator q^{-1} of the controller, model prediction, process (including zero order hold), and load respectively. G_F is the transfer function form of the closed loop observer feedback method. These equations indicate a two-degree-of-freedom control configuration allowing the design of a control system with independently specified reactions to setpoint and disturbance signals. The controller and the observer each contribute one degree-of-freedom in the design of system reactions to servo changes and residual values. As suggested by Zafiriou and Morari (1987), the setpoint and disturbance responses can be designed independently in the absence of model error. In the presence of model process mismatch the controller and the observer must be designed simultaneously to achieve the desired trade-off between performance and robustness. Note that the open loop observer corresponding to the most common form of DMC must have unity gain values. Therefore only one degree-of-freedom exists and an independent response to disturbances can not be designed. This method of choosing the closed loop state observer gains is not examined further in this paper due to the length of the calculations involved and the number of possible choices of closed loop performance criteria.

Kalman Filter

The block diagram and corresponding state update and output equations for the Kalman filter are identical to those of the closed loop observer. The sole difference between the two feedback schemes is the calculation of the gain values. The property of zero steady state offset in prediction, developed in Appendix A, therefore also holds for feedback via the Kalman filter. The major advantage of using the Kalman filter is that measurement and process noise are incorporated directly into the process model and gain calculation. The Kalman filter is able to take advantage of knowledge of noise covariances to limit the sensitivity of the prediction to noise and to create a more robust system. The state space form of the process model with the noise terms included is

$$X(k) = \Phi X(k-1) + \Theta \Delta u(k-1) + Dw(k-1) \quad (26)$$

$$Y(k) = HX(k) + v(k) \quad (27)$$

If w(k) and v(k) are independent white noise sequences then the standard Kalman Filter design procedure (Åström, 1970, Åström and Wittenmark, 1984, Goodwin and Sin, 1984) leads to

Time Update of the State

$$\hat{X}(k \mid k-1) = \Phi \hat{X}(k-1 \mid k-1) + \Theta \Delta U(k-1) \quad (28)$$

Time Update of Covariance

$$M(k) = \Phi P(k-1)\Phi^T + DR_w D^T \quad (29)$$

Gain Calculation

$$K(k) = M(k)H^T\left[HM(k)H^T + R_v\right]^{-1} \quad (30)$$

Measurement Update of the State and Prediction

$$\hat{X}(k \mid k) = \hat{X}(k \mid k-1) + K(k)\left[y_p(k) - y_m(k \mid k)\right] \quad (31)$$

$$\hat{Y}(k) = H_C \hat{X}(k \mid k) \quad (32)$$

Measurement Update of Covariance

$$P(k) = M(k) - K(k)HM(k) \quad (33)$$

where

R_w = covariances of w(k)
R_v = covariances of v(k)
M(k) = a priori error covariances
P(k) = a posteriori error covariances
K(k) = Kalman gain

The matrix (vector) D contains information on how the state variables are effected by the noise sequence w(k). Since the state variables have been defined as the future output values the elements of D may be obtained from disturbance step response coefficients. In most applications the Kalman filter converges rapidly to steady state gain values. Therefore, the steady state Kalman filter will provide satisfactory performance and the gains may be calculated off line by integration of Eqns. (29), (30), and (33).

It is interesting to note that when assuming all future outputs are effected in the same manner as the current output by w(k) the elements of D become unity. By further assuming zero covariance values for v(k) (no measurement noise) the Kalman gains approach unity and the input signal to the controller approaches that of the open loop observer form equivalent to the most common DMC feedback method.

Although the Kalman filter can not be assumed to be an optimal filter when the noise dynamics do not meet the assumption of zero mean Gaussian noise the noise covariance estimates can still be used as tuning parameters. This can be compared to using the closed loop state observer and filtering the residual values by using ad-hoc filtering techniques.

EXPERIMENTAL RESULTS

The different residual prediction and feedback correction schemes were evaluated using a series of simulated runs based on non-linear experimental step response data from the pilot plant distillation column described in Wood and Berry (1973). The distillation column is controlled via reflux rate and steam. The controlled variables are the top and bottom compositions and disturbances are simulated via changes in feed flow. Measurement noise was zero mean Gaussian. The control action was generated using the MIMO formulation of the MOCCA controller and prediction. The weighted least squares objective function in Eqn. (6) was used with output weighting of $\gamma=1.0$ and input weighting of $\gamma_m=0.001$. The controller design was based on N=50, P=50, M=5 for each output with a sampling time of three minutes.

Closed Loop Observer Comparison

A comparison of two closed loop observers is shown in Fig. 4 for a step input in feed flow with no measurement noise. The first observer has unity gain values, and therefore provides the equivalent performance of an open-loop observer or the common DMC method, while the second observer gains were designed based on discrete step response data. The model based observer clearly provides superior performance to the observer with unity gains by providing better prediction of future outputs. The prediction error for the model based observer is zero after the first measurement of the residual while the prediction error for the observer with unity gains only approaches zero when the disturbance effect approaches a steady state value.

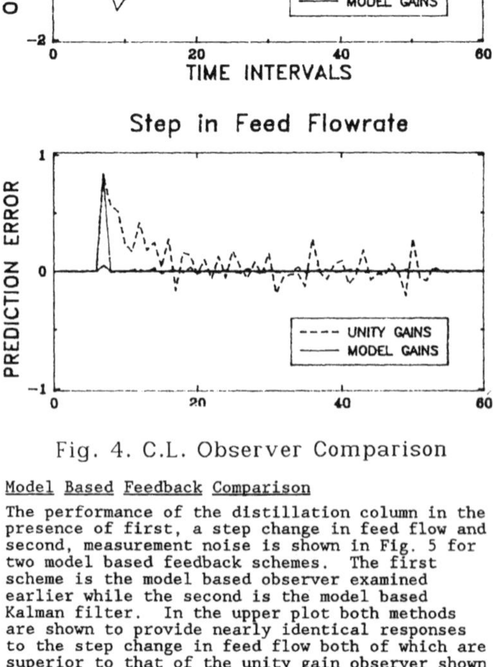

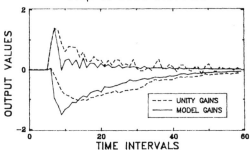

Fig. 4. C.L. Observer Comparison

Model Based Feedback Comparison

The performance of the distillation column in the presence of first, a step change in feed flow and second, measurement noise is shown in Fig. 5 for two model based feedback schemes. The first scheme is the model based observer examined earlier while the second is the model based Kalman filter. In the upper plot both methods are shown to provide nearly identical responses to the step change in feed flow both of which are superior to that of the unity gain observer shown in Fig. 4. However the system with the model based closed loop observer is shown in the second plot to be unstable in the presence of measurement noise. Conversely, the Kalman filter is based on both the model and noise estimates and thus is able to provide good prediction in the presence of both disturbances and measurement noise.

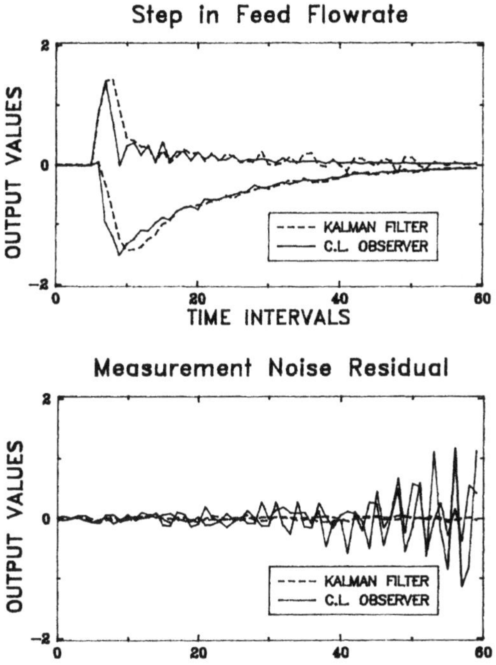

Fig. 5. Model Based Feedback

CONCLUSIONS

1. For regulatory control predictive control schemes require the prediction of the future effects of residual values. Such predictions involve measurement noise, disturbances, and model process mismatch.
2. A state space model provides an excellent basis for techniques such as open loop observers, closed loop observers, Kalman filters etc. for use in the feedback path of MPC.
3. It is often desirable to be able to choose different responses to setpoint changes and residuals (disturbances). This can be accomplished only if there is sufficient design freedom in the feedback path. The feedback gain values can be used for this purpose only if the residual feedback scheme provides zero steady state offset in prediction.
4. Adding the prediction of residual effects directly to the state variables, as in the closed loop state observer and Kalman filter, ensures zero steady state prediction offset in the presence of non-zero mean disturbance effects.
5. Incorporating a disturbance model into the prediction of residual effects can significantly improve regulatory performance.
6. By using a Kalman filter in the feedback path the effects of measurement noise can be reduced and the prediction of the output values appears more robust.

NOTATION

Γ	weighting matrix for output variable
Γ_m	weighting matrix for change in input variable
Δu	change in input variable
ΔU_f	current and future change in input variable
a_i	step response coefficients
A^*	controller gain matrix
$\hat{Y}$	estimate of future output trajectory
Y_m	output prediction based on past and future inputs
$y_m^*(k\|k)$	current output prediction based on past inputs
Y_m^*	future output prediction based on past inputs
$\hat{Y}$	corrected future output trajectory
Y_d	desired output trajectory
r(k)	estimated current residual
E(k)	error trajectory

LITERATURE CITED

Åström, K. J., (1970) Introduction to Stochastic Control Theory, Academic Press, New York, USA.

Åström, K. J. and Wittenmark, B., (1984) Computer Controlled Systems: Theory and Design, Prentice-Hall, Englewood Cliffs, N. J., USA.

Cutler, C. R. and Ramaker, B. L., (1980) "Dynamic Matrix Control - A Computer Control Algorithm", Proceedings of Joint Automatic Control Conference, San Francisco, USA, Paper WP5-B (Also published in AIChE 86th National Meeting, Paper No. 51B, Houston, 1979).

Garcia, C. E. and Morari, M., (1982) "Internal Model Control: 1. A Unifying Review and Some New Results", Industrial and Engineering Chemistry: Process Design and Development, 21, No. 2, 308-323.

Garcia, C. E. and Prett, D. M., (1986) "Advances in Industrial Model Predictive Control", Chemical Process Control, 21, No. 2, 308-323.

Goodwin, G. C. and Sin, K. S., (1984) Adaptive Filtering Prediction and Control, Prentice-Hall, N. J., USA.

Li, S., Fisher, D. G. and Lim, K. Y., (1986) "A State Space Formulation for Model Predictive Control", Pre-prints of the 36th National CSChE Conference, Sarnia, Ontario, Canada.

Lim, K. Y., (1988) "State Space Formulation of MOCCA", MSc. Thesis in progress, Department of Chemical Engineering, University of Alberta, Edmonton, Alberta, Canada.

Morshedi, A. M., Cutler, C. R. and Skrovanek, T. A., (1985) "Optimal Solution of Dynamic Matrix Control with Linear Programming (LDMC)", Proceeding American Control Conference, Boston, USA, 199-208.

Morshedi, A. M., (1986) "Universal Dynamic Matrix Control", Proceedings of the Third International Conference on Chemical Process Control (CPC III), (Morari, M. and McAvoy, T., Eds.), CACHE and Elsevier, Amsterdam.

Prett, D. M. and Gillette, R. D., (1980) "Optimization and Constrained Multivariable Control of a Catalytic Cracking Unit", Proceedings of Joint Automatic Control Conference, San Francisco, USA, Paper WP5-C.

Richalet, J., Rault, A., Testud, J. L. and Papon, J., (1978) "Model Predictive Heuristic Control: Application to Industrial Processes", Automatica, 14, 413-428. (Original version published in 4th IFAC Symposium on Identification and System Parameter Estimation, Tbilisi, Georgian Republic, USSR, 1976)

Ricker, N. L., (1985) "Use of Quadratic Programming for Constrained Internal Model Control", Industrial and Engineering Chemistry: Process Design and Development, 24, No. 4, 925-936.

Rouhani, R. and Mehra, R. K., (1982) "Model Algorithmic Control (MAC): Basic Theoretical Properties", Automatica, 18, 401-414.

Sripada, N. R. and Fisher, D. Grant, (1985) "Multivariable Optimal Constrained Control Algorithm (MOCCA): Part 1. Formulation and Application", Proceedings, International Conference on Industrial Process Modeling and Control, Vol. 1, Hangzhou, China, June 1985.

Tuffs, P. S. and Clarke, D. W., (1985) "Self-tuning Control of Offset: a Unified Approach", IEE Proceedings, 132, Part D, No. 3, 100-110.

Wood, R. K. and Berry, R. W., (1973) "Terminal Composition Control of a Binary Distillaion Column", Chemical Engineering Science, 28, 1707-1717.

Zafiriou, E. and Morari, M., (1987a) "Setpoint Tracking vs. Disturbance Rejection for Stable and Unstable Processes", Proceeding American Control Conference, Minneapolis, USA, 649-651.

APPENDIX A

In order to examine the relationship between feedback gain values and the effect of a disturbance (or other residual) upon the output prediction of the closed loop prediction form of MOCCA a transfer function form of the feedback path is developed. The closed loop relationship between the residual values and the output prediction at any time is governed by the following equations

$$y_m(k) = H\bar{X}(k) \tag{A1}$$

$$r(k) = y(k) - y_m(k) \tag{A2}$$

$$\bar{X}(k) = \Phi\hat{X}(k-1) + \Theta\Delta u(k-1) \tag{A3}$$

$$\hat{X}(k) = \bar{X}(k) + Kr(k) \tag{A4}$$

The prediction of the current output obtained from the expansion of the above equations can then be expressed as

$$y_m(k) = \bar{x}_1(k) \tag{A5}$$

$$\bar{x}_1(k) = \hat{x}_2(k-1) + a_1\Delta u(k-1) \tag{A6}$$

$$\hat{x}_2(k-1) = \bar{x}_2(k-1) + k_1 r(k-1) \tag{A7}$$

where

$$a_1\Delta u(k-1) = a_1[u(k-1) - u(k-2)] \tag{A8}$$

so that

$$y_m(k) = \bar{x}_2(k-1) + k_1 r(k-1) + a_1[u(k-1) - u(k-2)] \tag{A9}$$

continuing in this manner

$$y_m(k) = \bar{x}_{N+1}(k-N) + \sum_{i=1}^{N} k_i r(k-i) + \sum_{i=1}^{N} a_i[u(k-i) - u(k-i-1)] \tag{A10}$$

From the state space model

$$\bar{x}_{N+1}(k-N) = \hat{x}_{N+1}(k-N-1) + a_{ss}\Delta u(k-N-1) \tag{A11}$$

and

$$\hat{x}_{N+1}(k-N-1) = \bar{x}_{N+1}(k-N-1) + k_{N+1} r(k-N-1) \tag{A12}$$

also by definition

$$\bar{x}_{N+1}(k-N) = q^{-1}\bar{x}_{N+1}(k-N-1) \tag{A13}$$

so that

$$\bar{x}_{N+1}(k-N) = \frac{a_{ss}\Delta u(k-N-1) + k_{N+1} r(k-N-1)}{\Delta} \tag{A14}$$

where

$$\Delta = 1 - q^{-1} \tag{A15}$$

Therefore

$$y_m(k) = \frac{k_{N+1} r(k-N-1)}{\Delta} + \sum_{i=1}^{N} k_i r(k-i) + \sum_{i=1}^{N} a_i[u(k-i) - u(k-i-1)] + a_{ss} u(k-N-1) \tag{A16}$$

By expanding the equations using the unit delay operator and using impulse response coefficients

$$y_m(k) = \left[a_1 + \sum_{i=1}^{N} h_i q^{-i}\right] u(k-1) + \left[\Delta\sum_{i=1}^{N} k_i q^{-i} + k_{N+1} q^{-N-1}\right]\frac{r(k-1)}{\Delta} \tag{A17}$$

where

$$h_i = a_{i+1} - a_i \tag{A18}$$

In a more compact form

$$y_m(k) = B(q^{-1})u(k-1) + C(q^{-1})r(k-1) \tag{A19}$$

where

$$C(q^{-1}) = \sum_{i=1}^{N} k_i q^{-i} + \frac{k_{N+1} q^{-N-1}}{\Delta} \tag{A20}$$

From Equation (A2)

$$y(k) = y_m(k) + r(k) \tag{A21}$$

Therefore

$$y(k) = B(q^{-1})u(k-1) + [1 + C(q^{-1})]r(k-1) \tag{A22}$$

and

$$r(k) = \frac{y(k) - B(q^{-1})u(k-1)}{1 + C(q^{-1})} \tag{A23}$$

Equation 16 becomes

$$y_m(k) = B(q^{-1})u(k-1) + \frac{C(q^{-1})}{1 + C(q^{-1})}[y(k) - B(q^{-1})u(k-1)] \tag{A24}$$

Where the term in brackets corresponds to the difference between the model prediction and measured value of the output. This is a discrete signal which can be represented as a series of step changes. Applying the Final Value Theorem in this case results in

$$\lim(\Delta y_m(k)) = \frac{k_{N+1}}{k_{N+1}}\Delta[y(k) - B(q^{-1})u(k-1)] \tag{A25}$$

Which guarantees zero steady state error in prediction for non-zero mean residual effects providing the N+1 gain value is not zero.

DEVELOPMENT OF A MULTIVARIABLE FORWARD MODELING CONTROLLER

K. T. Erickson* and R. E. Otto**

**Department of Electrical Engineering, University of Missouri-Rolla, Rolla, Missouri 65401, USA*
***Monsanto Company, St. Louis, Missouri 63198, USA*

Abstract. The Forward Modeling Controller, a recently developed model predictive digital controller for single input, single output processes, is extended to multi-input, multi-output processes. The multivariable FMC is a promising approach to the control of complex industrial processes with many inputs and outputs. The theory presented in this paper includes stability analysis plus other features necessary for robustness in industrial control. The controller has only two types of adjustments: a robustness/performance setting for each controlled variable and the controller sample interval.

Keywords. Predictive control; multivariable control systems; digital control; industrial control; process control; petro-chemical control; stability.

I. INTRODUCTION

Complex multi-input, multi-output (MIMO) chemical industrial processes such as distillation columns are often difficult to control. Many of these processes have large deadtimes and unusual dynamics and are often affected by persistent disturbances. Automatic control of these processes is usually troublesome due to the interaction inherent in the process, requiring highly skilled operators to maintain acceptable product quality.

Modern control theory, such as optimal control, while used frequently in aerospace and aircraft applications, has been used in only a small number of chemical process applications. The lack of applications appears to be due to the inability of modern control theory to deal with the typically imprecise knowledge of the process and disturbance characteristics. In addition, most process control engineers are unfamiliar with modern control theory and therefore tend to adapt traditional control techniques to solve their control problems.

The limited successful application of modern control theory to chemical process control motivated the development of digital model-based predictive control algorithms such as Dynamic Matrix Control (DMC) (Cutler and Ramaker, 1980), Model Algorithmic Control (MAC) (Richalet and co-workers, 1978; Rouhani and Mehra, 1982) and Internal Model Control (IMC) (Garcia and Morari, 1982, 1985). These multivariable digital control algorithms use an impulse response or step response model of the process to predict the trend of the process outputs and to compute the required change in the process inputs to bring the outputs to their desired values. These model-based predictive control schemes were formulated to deal with the deadtime and unusual behavior of complex MIMO chemical industrial processes. Consequently, these control algorithms have been applied to many chemical processes with favorable results.

Recently, the Forward Modeling Controller (FMC), which circumvents the difficulties of the other model-based predictive controllers was developed for use on single input, single output (SISO) processes by Otto (1986). In this paper, the basic FMC algorithm is extended for use on multivariable processes. The multivariable FMC algorithm is conceptually similar to other model-based predictive digital controllers but is a computationally simpler algorithm. Unlike the DMC or the MAC, the multivariable FMC does not require the on-line solution of an optimization problem. In contrast to the MAC, the FMC does not require the solution of an off-line Ricatti difference equation for processes with nonminimum phase. The multivariable FMC also does not require the derivation of an approximate process inverse as in the IMC. The response models are used directly in the calculation of the controller outputs.

The remainder of the paper is organized as follows. The transfer function form of the controller is developed in the next section and used to examine the system for stability. Robustness issues, feedforward control, and controller tuning are also discussed in this section. In section III, the performance of the multivariable FMC is evaluated on a simulation of the Wood and Berry methanol/water distillation column. Regulatory and servo behavior are evaluated for various values of controlled variable closed loop settling times. The last section of the paper summarizes the advantages of the multivariable FMC and gives suggestions for future work.

II. ALGORITHM DEVELOPMENT

In the multivariable FMC, the process has q outputs and r inputs and is modeled by a discrete impulse response model:

$$\mathbf{y}_i = \mathbf{H}_1\mathbf{m}_{i-1} + \mathbf{H}_2\mathbf{m}_{i-2} + \cdots + \mathbf{H}_N\mathbf{m}_{i-N} + \mathbf{d}_i \quad (1)$$

where

$\mathbf{y}_i = [y_1\ y_2\ \cdots\ y_q]^T$, a q x 1 vector of process output measurements,

$\mathbf{m}_j = [m_1\ m_2\ \cdots\ m_r]^T$, a r x 1 vector of manipulated variables,

$\mathbf{H}_k$ is a q x r matrix of the kth impulse response coefficient; $(h_{ij})_k$ is the kth impulse response coefficient of the response between the jth manipulated variable and the ith process output,

$\mathbf{d}_i = [d_1\ d_2\ \cdots\ d_q]^T$, a q x 1 vector of the current discrepancies between the model and the measurement.

An impulse response model of the process is also employed by others (Cutler and Ramaker, 1980; Marchetti and co-workers, 1983; Richalet and co-workers, 1978; Rouhani and Mehra, 1982), and has the significant advantages:

- No *a priori* assumptions about model order, time delay, etc. are necessary.
- Unusual process dynamics are handled naturally and do not require the specification of model structure.
- The coefficients of the model can be obtained from simple step response data.
- The step response model, obtained by integrating the impulse response model has high intuitive appeal to process operators.
- With prediction error identification methods, multivariable MA models will, in the limit, converge to the true parameter values (Stoica and Söderström,1982).

The chief disadvantages of an impulse response model are

- Non-mimimality of the representation, i.e., the relatively larger number of parameters used in the model, compared with the smaller number of parameters associated with a low-order transfer function model.
- An *a priori* assumption of the time to steady state.

However, we feel the advantages of an impulse response model far outweigh the disadvantages.

It is important to note that the value of $\mathbf{d}_i$ is determined by two separate factors: (1) unmeasured process disturbances and (2) modeling errors. It is *impossible* to separate the two without making further assumptions.

The multivariable FMC computes a prediction of the process output, $\mathbf{T}^j$, into the future with the following assumptions:

1. The manipulated variable is held constant into the future $\mathbf{m}_i = \mathbf{m}_{i+1} = \mathbf{m}_{i+2} = \cdots$,
2. The future discrepancies remain at their current value $\mathbf{d}_i = \mathbf{d}_{i+1} = \mathbf{d}_{i+2} = \cdots$.

The process output predictions are given by the following equations (where $\mathbf{T}^j{}_i$ is the vector of process outputs at the jth sample in the future, given the information of the current, i^{th}, sample) :

$$\mathbf{T}^1{}_i = \mathbf{H}_1\mathbf{m}_i + \mathbf{H}_2\mathbf{m}_{i-1} + \cdots + \mathbf{H}_N\mathbf{m}_{i-N+1} + \mathbf{d}_i$$

$$\mathbf{T}^2{}_i = \mathbf{H}_1\mathbf{m}_i + \mathbf{H}_2\mathbf{m}_i + \mathbf{H}_3\mathbf{m}_{i-1} + \cdots + \mathbf{H}_N\mathbf{m}_{i-N+2} + \mathbf{d}_i$$

$$\vdots$$

$$\begin{aligned}\mathbf{T}^j{}_i &= \left(\mathbf{H}_1 + \mathbf{H}_2 + \cdots \mathbf{H}_j\right)\mathbf{m}_i + \\ &\quad + \mathbf{H}_{j+1}\mathbf{m}_{i-1} + \cdots + \mathbf{H}_N\mathbf{m}_{i-N+j} + \mathbf{d}_i \\ &= \mathbf{A}_j\mathbf{m}_i + \mathbf{H}_{j+1}\mathbf{m}_{i-1} + \cdots + \mathbf{H}_N\mathbf{m}_{i-N+j} + \mathbf{d}_i\end{aligned} \quad (2)$$

$$\vdots$$

$$\mathbf{T}^N{}_i = \mathbf{A}_N\mathbf{m}_i + \mathbf{d}_i$$

where $\mathbf{A}_j = \sum_{k=1}^{j}\mathbf{H}_k$ is the jth "matrix coefficient" of the process step response. Note that

$$\begin{aligned}\mathbf{T}^j{}_i &= \mathbf{T}^{j+1}{}_{i-1} + \mathbf{A}_j(\mathbf{m}_i - \mathbf{m}_{i-1}) + \mathbf{d}_i - \mathbf{d}_{i-1} \\ &= \mathbf{T}^{j+1}{}_{i-1} + \mathbf{A}_j\Delta\mathbf{m}_i + \Delta\mathbf{d}_i\end{aligned} \quad (3)$$

where $\Delta\mathbf{X}_i = \mathbf{X}_i - \mathbf{X}_{i-1}$. Expressed in words, the prediction of the jth sample in the future is updated from the previous prediction of the $j+1^{st}$ sample by the change in the unmeasured disturbances and the changes in the manipulated variables.

We will design a controller which yields at sample i a controller output $\mathbf{m}_i$ which, if it were kept constant from here on, would minimize V, the Euclidean norm of the error between the prediction and the setpoint at P sample intervals in the future,

$$V = \left[\mathbf{s}^P{}_i - \mathbf{T}^P{}_i\right]^T\left[\mathbf{s}^P{}_i - \mathbf{T}^P{}_i\right] = \left[\mathbf{e}^P{}_i\right]^T\mathbf{e}^P{}_i. \quad (4)$$

The details of the actual computation of $\mathbf{m}_i$ are given in the Appendix. Here, we assume that the controller output is calculated to drive the prediction after P sample intervals in the future to the setpoint $\mathbf{s}^P{}_i$ minus the error vector $\mathbf{e}^P{}_i$:

$$\begin{aligned}\mathbf{T}^P{}_i &= \mathbf{s}^P{}_i - \mathbf{e}^P{}_i \\ &= \mathbf{A}_P\mathbf{m}_i + \mathbf{H}_{P+1}\mathbf{m}_{i-1} + \cdots + \mathbf{H}_N\mathbf{m}_{i-N+P} + \mathbf{d}_i\end{aligned} \quad (5)$$

For systems where the number of manipulated variables is greater than or equal to the number of controller variables ($r \geq q$) it is theoretically possible to calculate a controller output that makes the error $\mathbf{e}^P{}_i$ zero. However, for systems where $r < q$, or for constrained systems where $r \geq q$, it may not be possible to achieve the desired setpoint. For convience, we define

$$\mathbf{s}_i = \mathbf{s}^P{}_i + \mathbf{e}^P{}_i. \quad (6)$$

The value of $\mathbf{s}$ thus represents the setpoint that can be achieved, given the limitations of the system. Substituting for $\mathbf{d}_i$ from (1) into (5), and using (6),

$$\begin{aligned}\mathbf{s}_i - \mathbf{y}_i &= \mathbf{A}_P\mathbf{m}_i + \mathbf{H}_{P+1}\mathbf{m}_{i-1} + \cdots + \mathbf{H}_N\mathbf{m}_{i-N+P} \\ &\quad - \mathbf{H}_1\mathbf{m}_{i-1} - \cdots - \mathbf{H}_N\mathbf{m}_{i-N}\end{aligned} \quad (7)$$

Converting to z-transforms,

$$\begin{aligned}\mathbf{s}(z) - \mathbf{y}(z) &= \left[(\mathbf{A}_P + \mathbf{H}_{P+1}z^{-1} + \cdots + \mathbf{H}_N z^{-N+P})\right. \\ &\quad \left. - (\mathbf{H}_1 z^{-1} + \cdots + \mathbf{H}_N z^{-N})\right]\mathbf{m}(z).\end{aligned} \quad (8)$$

In conventional controller terms (Fig. 1), the controller transfer function $\mathbf{C}(z)$ is given as

$$\mathbf{m}(z) = \mathbf{C}(z)(\mathbf{s}(z) - \mathbf{y}(z))$$

Therefore, in conventional controller terms the transfer function of the FMC is

$$\mathbf{C}(z) = \Big[(\mathbf{A}_P + \mathbf{H}_{P+1}z^{-1} + \cdots + \mathbf{H}_N z^{-N+P}) - (\mathbf{H}_1 z^{-1} + \cdots + \mathbf{H}_N z^{-N})\Big]^{(-1)}. \quad (9)$$

where $\mathbf{A}^{(-1)}$ indicates the left inverse of a non-square matrix, $\mathbf{A}$, defined so that $\mathbf{A}^{(-1)}\mathbf{A} = \mathbf{I}_n$ ($\mathbf{A}$ is of dimension m x n). Of course, if A is square, then the left inverse is the usual matrix inverse. The output of the closed loop system is expressed as

$$\mathbf{y}(z) = \Big[\mathbf{I} + \mathbf{G}(z)\mathbf{C}(z)\Big]^{(-1)}\mathbf{G}(z)\mathbf{C}(z)\mathbf{s}(z). \quad (10)$$

One way to examine stability of the closed loop system is to examine the system poles. Obviously, not a trivial task for (10). However, if we convert the controller to the Internal Model configuration (Fig. 2), stability analysis is simpler. Garcia and Morari (1985) show that the output of the system in Fig. 2 is given by

$$\mathbf{y}(z) = \mathbf{G}(z)\Big[\mathbf{I} + \mathbf{G}_c(z)\Big(\mathbf{G}(z) - \mathbf{G}_m(z)\Big)\Big]^{(-1)}\mathbf{G}_c(z)\Big(\mathbf{s}(z) - \mathbf{y}(z)\Big) + \mathbf{d}(z). \quad (11)$$

Now, if exact process modeling is assumed, then $\mathbf{G}_m$ is set to $\mathbf{G}$ and (11) becomes

$$\mathbf{y}(z) = \mathbf{G}(z)\mathbf{G}_c(z)\Big(\mathbf{s}(z) - \mathbf{y}(z)\Big) + \mathbf{d}(z). \quad (12)$$

Therefore, the closed-loop system is stable if $\mathbf{G}_c(z)$ and $\mathbf{G}(z)$ are stable.

We will assume that the process $\mathbf{G}(z)$ is stable. Therefore, stability of the controller $\mathbf{G}_c(z)$ is sufficient for closed-loop stability. Consequently, for system stability, the controller poles (given by the roots of $\det(\mathbf{G}_c(z)) = 0$) must lie within the unit circle for stability.

For the multivariable FMC, $\mathbf{G}_c(z)$ is developed as follows. From Fig. 2,

$$\tilde{\mathbf{d}}(z) = \mathbf{y}(z) - \mathbf{G}_m(z)\mathbf{m}(z)$$

Therefore,

$$\mathbf{s}(z) - \mathbf{y}(z) = \mathbf{s}(z) - \tilde{\mathbf{d}}(z) - \mathbf{G}_m(z)\mathbf{m}(z) \quad (13)$$

Now, if exact process modeling is assumed, then $\mathbf{G}_m$ is set to $\mathbf{G}$ and (13) becomes

$$\mathbf{s}(z) - \mathbf{y}(z) = \mathbf{s}(z) - \tilde{\mathbf{d}}(z) - (\mathbf{H}_1 z^{-1} + \cdots + \mathbf{H}_N z^{-N})\mathbf{m}(z). \quad (14)$$

Equating (14) and (8),

$$\mathbf{s}(z) - \tilde{\mathbf{d}}(z) = \Big[\mathbf{A}_P + \mathbf{H}_{P+1}z^{-1} + \cdots + \mathbf{H}_N z^{-N+P}\Big]\mathbf{m}(z).$$

Now,

$$\mathbf{m}(z) = \mathbf{G}_c(z)\Big[\mathbf{s}(z) - \tilde{\mathbf{d}}(z)\Big]. \quad (15)$$

Therefore,

$$\mathbf{G}_c(z) = \Big[\mathbf{A}_P + \mathbf{H}_{P+1}z^{-1} + \cdots + \mathbf{H}_N z^{-N+P}\Big]^{(-1)}. \quad (16)$$

For system stability, the controller poles (given by the roots of $\det(\mathbf{G}_c(z)) = 0$ in equation (16)) must lie within the unit circle for stability. Unfortunately, the task of finding the minimum P for stability of the controller, and hence, stability of the closed loop system is not a trivial task even for a system with the same number of manipulated variables and system outputs. We have not yet been able to formulate stability theorems as for the SISO FMC (Otto, 1986). However, note that if P is picked equal to N (steady state control) stability is guaranteed if the modeling is exact.

The preceding development assumed that P, the control horizon, was the same for each controlled variable. However, each controlled variable may have a different value of P, depending on how tightly each one is to be controlled. To accomodate different values of P for each process output, we redefine $\mathbf{T}^P$ as

$$\mathbf{T}^P = \Big[T^{P_1}\ T^{P_2}\ \cdots\ T^{P_q}\Big]$$

where P_k is the control horizon for the kth process output. Similarly, $\mathbf{A}_P$ is redefined as

$$\mathbf{A}_P = \begin{bmatrix} a_{11}^{P_1} & a_{12}^{P_1} & \cdots & a_{1r}^{P_1} \\ a_{21}^{P_2} & a_{22}^{P_2} & \cdots & a_{2r}^{P_2} \\ \vdots & & & \vdots \\ a_{q1}^{P_q} & a_{q2}^{P_q} & \cdots & a_{qr}^{P_q} \end{bmatrix} \quad (17)$$

where $a_{jk}^{P_i}$ is the step-response model coefficient at horizon P_i of the model between the kth process input and the jth process output. The expression for the controller transfer function (16) gets similarly complex and will not be shown here. The mimimum value of each P_i can be calculated using stability analysis.

Filtering for Robustness

If modeling errors exist, stability cannot be guaranteed even for all $P_i = N$. Furthermore, even if one had exact modeling, a policy of setting $P_i = N$ may produce a controller which moves the manipulated variables too vigorously when large amounts of noise are present in the measurements. We need to modify the controller to produce robustness (tolerance to modeling errors) and noise rejection. Garcia and Morari (1985) have shown that if a filter $\mathbf{F}(z)$ is added to the controller input, the closed-loop system can be made stable for arbitrarily large modeling errors (other than the wrong sign on the model gains) by filtering heavily enough. With filtering, the calculation of the manipulated variable (15) becomes

$$\mathbf{m}(z) = \mathbf{G}_c(z)\mathbf{F}(z)\Big[\mathbf{s}(z) - \tilde{\mathbf{d}}(z)\Big]. \quad (18)$$

In Garcia and Morari (1985) the filter $\mathbf{F}(z)$ is diagonal and of the exponential type

$$\mathbf{F}(z) = \mathrm{diag}\left\{\frac{1-\alpha_i}{1-\alpha_i z^{-1}}\right\},\ 0 \le \alpha_i < 1,\ \ i = 1,\dots,q.$$

Similar to Otto (1986), we assume the filter $\mathbf{F}(z)$ is diagonal and of the form

$$\mathbf{F}(z) = \mathrm{diag}\left\{\frac{1}{1+f_i - f_i z^{-1}}\right\},\ \ f_i > 0,\ \ i = 1,\dots,q.$$

$$= \Big[\mathbf{I} + \Phi - \Phi z^{-1}\Big]^{-1}$$

where $\Phi = \mathrm{diag}\Big\{f_1\ f_2\ \cdots\ f_q\Big\}$. As in Otto (1986), f_i is picked as

$$f_i = \begin{cases} 0 & \text{for } P_i \le N \\ \dfrac{P_i - N}{3} & \text{for } P_i > N \end{cases}$$

When a particular P_i is larger than the length of the step response model, the ith controlled variable is filtered, and provides robustness to modeling error. Therefore, adjustment of the control horizon, P, can move the controller from high performance control to noise-rejecting, sluggish, robust control.

Feedforward Control

Feedforward control is easily accomodated by the MIMO FMC. Given a model of the effect of a vector of *measured* disturbances, **x**, on the process outputs of the same form as the discrete impulse model used for the process,

$$\mathbf{y}_i = \mathbf{H}'_1\mathbf{x}_{i-1} + \mathbf{H}'_2\mathbf{x}_{i-2} + \cdots + \mathbf{H}'_N\mathbf{x}_{i-N}$$

where the matrices $\mathbf{H}'_j$ are of appropriate dimension. Under the additional assumption that measured disturbances remain at their present value, the prediction vector at the jth sample in the future is

$$\mathbf{T}^j_i = \mathbf{A}_j\mathbf{m}_i + \mathbf{H}_{j+1}\mathbf{m}_{i-1} + \cdots + \mathbf{H}_N\mathbf{m}_{i-N+j} + \mathbf{d}_i + \mathbf{A}'_j\mathbf{x}_i + \mathbf{H}'_{j+1}\mathbf{x}_{i-1} + \cdots + \mathbf{H}'_N\mathbf{x}_{i-N+j}$$

$\mathbf{A}'_j = \sum_{k=1}^{j} \mathbf{H}'_k$ is the jth "matrix coefficient" of the measured disturbances step response. With this change, equation (3) becomes

$$\mathbf{T}^j_i = \mathbf{T}^{j+1}_{i-1} + \mathbf{A}_j\Delta\mathbf{m}_i + \mathbf{A}'_j\Delta\mathbf{x}_i + \Delta\mathbf{d}_i \tag{19}$$

Therefore, the only change to the control algorithm is to update the process variable predictions with the expected effect of the measured disturbances. No other change to the control algorithm is necessary.

Controller Tuning

The multivariable FMC has only two types of adjustments. The above development focused on the control horizons, P_i, which smoothly take the control action from

1. Extremely aggressive control by setting all P_i to their minimum values, to
2. Steady state control where the controller moves the manipulated variables only to statically compensate the loop, to
3. Extremely sluggish, noise rejecting, robust control which should be stable in all practical situations.

There is only one control horizon for each controlled variable. In addition to the control horizons, the controller has only one other tuning parameter, the control interval, Δt. The number of points in the model, N, is usually fixed. For ease of communicating with the engineer commisioning the loop and with operating personnel, the two types of parameters can be renamed:

1. the closed loop settling time of the controlled variable, $P_i{*}\Delta t$, and
2. the open loop settling time of the process, $N{*}\Delta t$.

In practice, the open loop settling time of the process is rarely changed. The closed loop settling time is approximate (except for all $P_i = N$ with exact modeling) but is useful for explaining the tradeoff between loop performance and robustness. Decreasing the closed loop settling time tends to make the system less tolerant to modeling errors. This feature has strong intuitive appeal and is expected to be well received by operating personnel.

III. FMC PERFORMANCE

The features and performance of the multivariable FMC are evaluated on a simulation of the Wood and Berry (1973) methanol/water distillation column (Fig. 3). This process model is used frequently in the literature for the comparison of multivariable control algortihms (Garcia and Morari, 1985). The model for the column is

$$\begin{bmatrix} y_1(s) \\ y_2(s) \end{bmatrix} = \begin{bmatrix} \dfrac{12.8e^{-s}}{16.7s+1} & \dfrac{-18.9e^{-3s}}{21.0s+1} \\ \dfrac{6.6e^{-7s}}{10.9s+1} & \dfrac{-19.4e^{-3s}}{14.4s+1} \end{bmatrix} \begin{bmatrix} m_1(s) \\ m_2(s) \end{bmatrix} + \begin{bmatrix} \dfrac{3.8e^{-8s}}{14.9s+1} \\ \dfrac{4.9e^{-7s}}{13.2s+1} \end{bmatrix} d(s) \tag{20}$$

where time is measured in minutes. The physical meaning of the variables and nominal operating conditions for the column are given in Table 1.

As is usual in the literature, an error-free system model is assumed for the simulations. The multivariable FMC was implemented with a sampling period of 1 minute. The transfer function model (20) was converted to matrices of impulse response models of 60 samples each. Using Jury's stability analysis (Jury, 1964) on the poles of $\mathbf{G}_c(z)$ in equation (16), the minimum control horizons are $P_{1min} = 2$ and $P_{2min} = 4$. The manipulated variable moves were calculated by equation (21). Note that the matrix inversion in (21) only needs to be performed when a control horizon, P_1 or P_2, is changed.

Servo behavior The system response to changes in the overhead composition setpoint to 97.0 for various values of control horizons is shown in Fig. 4. Note that for both control horizons at their minimum values, the system exhibits deadbeat response. However, the deadbeat response comes at the expense of strong input actions. The input actions are moderated by increasing both control horizons, but at the expense of some interaction at the bottoms composition output. Although not shown, the input actions can also be alleviated by increasing the control horizon for just the bottoms composition, presuming some degradation in control can be tolerated. However, the resulting interaction at the bottoms composition output is more severe than the response shown. The system responds in a similar manner to changes in the bottoms composition setpoint.

Regulatory behavior Responses to a disturbance change of 0.34 lb/min feed flow rate are shown in Fig. 5. As with the servo response, increasing the control horizon reduces the input actions but causes more deviation at the system outputs.

The regulatory and servo behavior of the multivariable FMC on the distillation column are similar to those of the IMC (Garcia and Morari, 1985).

IV. SUMMARY

The major advantages of the multivariable FMC include:

- Each controlled variable has only one tuning parameter, the control horizon, also called the closed loop settling time. The adjustment of these tuning parameters can move the controller from high performance control to noise-rejecting, sluggish, robust control.
- The mimimum value of each control horizon is calculated using stability analysis, providing the operator with an indication of the maximum performance available.
- Feedforward compensation is trivial.
- Computationally simpler than other model-based predictive digital controllers.
- Graphical display of the process models and process variable predictions have intuitive appeal to process operators.

Current work on the multivariable FMC is concentrating on modifications to handle process contraints and the identification of MIMO impulse response models.

APPENDIX

Ignoring any filtering and feedforward compensation, the change in the manipulated variable vector, $\Delta\mathbf{m}_i$ is calculated so that, if it were held constant, the Euclidean norm of the error between the prediction and setpoint at P sample intervals in the future is minimized. Given equations (3) and (4), the calculation of Δm_i can be expressed as the minimization problem:

$$\min_{\Delta m_i} V = \min_{\Delta m_i} \left[\mathbf{e}^P{}_i\right]^T \mathbf{e}^P{}_i.$$

where

$$\mathbf{e}^P{}_i = \mathbf{s}^P{}_i - \mathbf{T}^P{}_i$$
$$= \mathbf{s}^P{}_i - \left(\mathbf{T}^P{}_{i-1} + \mathbf{A}^P \Delta\mathbf{m}_i + \Delta\mathbf{d}_i\right)$$

where $\mathbf{A}^P$ is defined by (17).

The unconstrained solution to this problem is found by solving

$$\frac{\partial V}{\partial \Delta\mathbf{m}_i} = 0 \quad \text{and} \quad \frac{\partial^2 V}{\partial \Delta\mathbf{m}_i{}^2} > 0$$

which gives a change in the manipulated variable,

$$\Delta\mathbf{m}_i = \left(\mathbf{A_P}^T \mathbf{A_P}\right)^{-1} \mathbf{A_P}^T \left(\mathbf{s}^P{}_i - \mathbf{T}^P{}_{i-1} - \Delta\mathbf{d}_i\right) \qquad (21)$$
$$= \mathbf{A_P}^{(-1)} \left(\mathbf{s}^P{}_i - \mathbf{T}^P{}_{i-1} - \Delta\mathbf{d}_i\right)$$

If $\mathbf{A_P}$ is square, then,

$$\Delta\mathbf{m}_i = \mathbf{A_P}^{-1} \left(\mathbf{s}^P{}_i - \mathbf{T}^P{}_{i-1} - \Delta\mathbf{d}_i\right) \qquad (22)$$

REFERENCES

Cutler, C. R. and B. L. Ramaker (1980). Dynamic matrix control - a computer control algorithm. *Proc. JACC*, San Francisco, paper WP5-B.

Garcia, C. E. and M. Morari (1982). Internal model control. 1. A unifying review and some new results. *Ind. Eng. Chem. Process Des. Dev.*, **21**, 308-323.

Garcia, C. E. and M. Morari (1985). Internal model control. 2. Design procedures for multivariable systems. *Ind. Eng. Chem. Process Des. Dev.*, **24**, 472-484.

Jury, E. I. (1964). *Theory and Application of the Z-Transform Method.* Huntington, New York.

Marchetti, J. L., D. A. Mellichamp, and D. E. Seborg (1983). Predictive control based on discrete convolution models. *Ind. Eng. Chem. Process Des. Dev.*, **22**, 488-495.

Otto, R. E. (1986). Forward modeling controllers: a comprehensive SISO controller. 1986 AIChE Annual Meeting.

Richalet, J., A. Rault, J. L. Testud and J. Papon (1978). Model predictive heuristic control: applications to industrial processes. *Automatica*, **14**, 413-428.

Rouhani, R. and R. K. Mehra (1982). Model algorithmic control (MAC); basic theoretical properties. *Automatica*, **18**, 401-414.

Stoica, P. and T. Söderström (1982). Uniqueness of prediction error estimates of multivariable moving average models. *Automatica*, **18**, 617-620.

Wood, R. K. and M. W. Berry (1973). Terminal composition control of a binary distillation column. *Chemical Engineering Science*, **28**, 1707-1717.

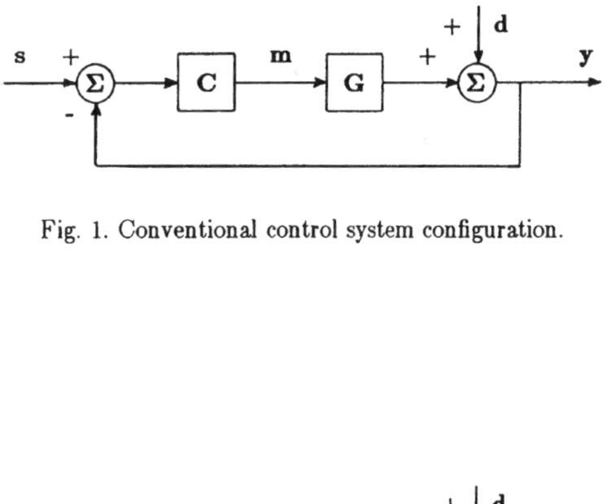

Fig. 1. Conventional control system configuration.

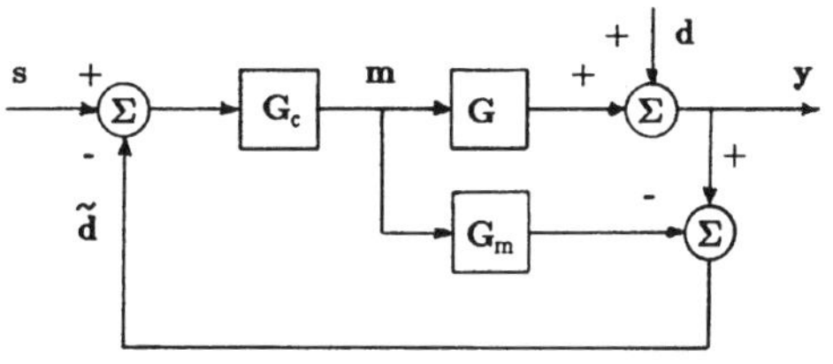

Fig. 2. Internal model control configuration.

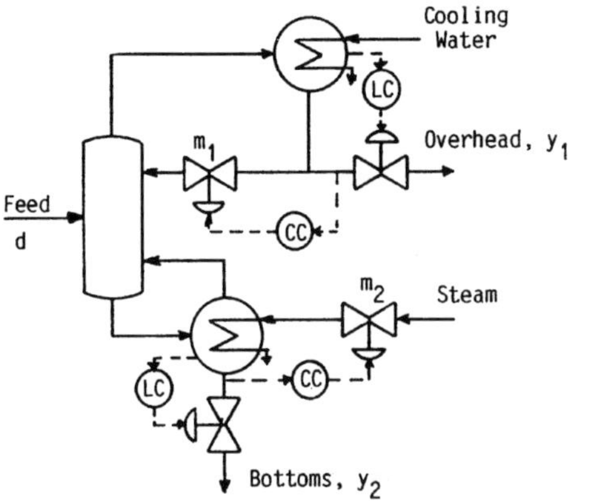

Fig. 3. Diagram of Wood and Berry column.

Table 1. Wood and Berry Column Variable Summary

Variable	Description	Operating steady state
y_1	overheads composition	96.25 mol % methanol
y_2	bottoms composition	0.50 mol % methanol
m_1	reflux flow rate	1.95 lb/min
m_2	steam flow rate	1.71 lb/min
d	feed flow rate	2.45 lb/min

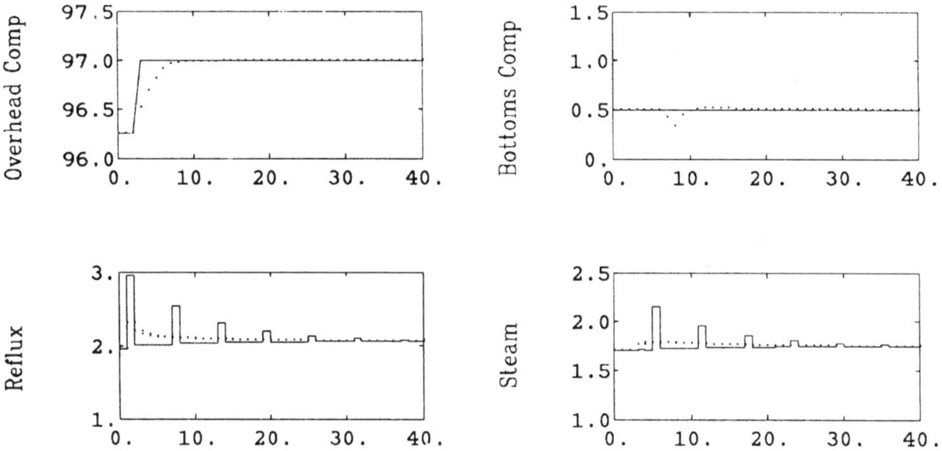

Fig. 4. Column response to a setpoint change in overhead composition: (—) $P_1 = 2$ and $P_2 = 4$, (···) $P_1 = 4$ and $P_2 = 6$ (time in minutes).

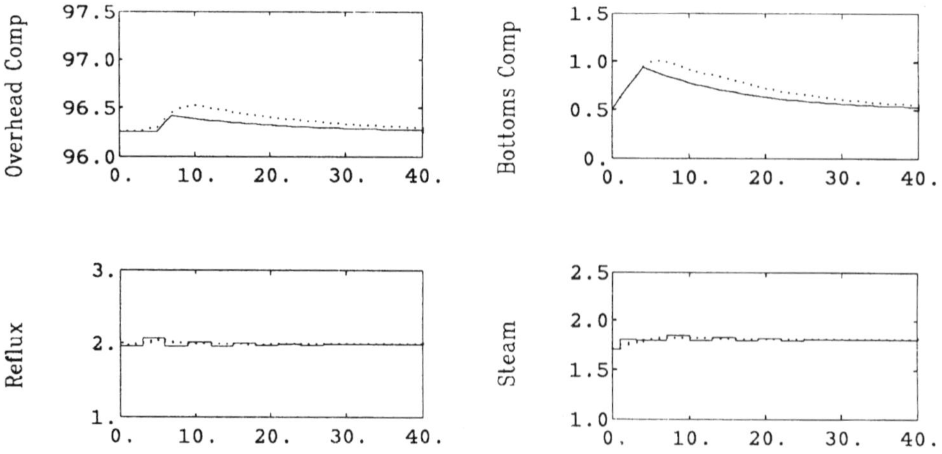

Fig. 5. Column response to a disturbance change of +0.34 lb/min: (—) $P_1 = 2$ and $P_2 = 4$, (···) $P_1 = 4$ and $P_2 = 6$ (time in minutes).

A STUDY ON ROBUST STABILITY OF MODEL PREDICTIVE CONTROL

I. Hashimoto*, M. Ohshima* and H. Ohno**

**Department of Chemical Engineering, Kyoto University, Yoshida Honmachi, Sakyo-ku, Kyoto 606, Japan*
***Department of Chemical Engineering, Kobe University, Rokkodai-machi 1–1, Nada-ku, Kobe 657, Japan*

ABSTRACT

The robust stability of Model Predictive Control (MPC) is analyzed without performing any numerical computations. Some robust stability theorems are newly derived. They assure that a stable MPC system is easily realized only by tuning a couple of control parameters even if some large plant-model mismatch exist. A guideline for the tuning the control parameters is also provided.

Keywords

Predictive Control, Robust Stability, Process Models, Model Algorithmic Control, Stability Criteria

INTRODUCTION

Model Predictive Control (MPC) was significantly developed as a heuristic digital control scheme in the 70's (Richalet et al,1978; Cutler et al, 1979, 1982) and is now drawing much attention in various industrial fields.

The reasons why this control scheme attracts such attention and is well received especially by process control engineers are: its algorithm is easy to understand, and it is easy to handle various constraints in determining the control action. Moreover the closed-loop system can be very robust, and so forth.

Theoretical studies on the properties of MPC began in the latter half of the 70's, bringing about quite a few important results. Various attempts such as introducing optimization techniques into the control algorithm, amalgamating adaptive control with this control scheme, etc. have been reported.

With regard to the robustness properties, some interesting papers were published. For example, Rouhani and Mehra (1982) derived the characteristic equation of Model Algorithmic Control (MAC) assuming one step prediction. By numerically examining the location of the characteristic roots, they analyzed the robustness of MAC, assuming that the precise mathematical expression of the actual process and its model are given a priori. Morari and Garcia (1982,1985) theoretically analyzed the properties of Dynamic Matrix Control (DMC) and derived the relationship between the stability of the closed-loop system and various tuning parameters. By studying MPC in the framework of Internal Model Control (IMC), they also obtained some theoretical results on the robust stability for the multi-input multi-output case.

In this paper, we present some of the new theoretical results concerning the robust stability of MAC. Without assuming the detailed characteristics of the actual process dynamics as well as the parameter values of its model, we thoroughly analyzed what role the coincidence horizon and the parameter α of the reference trajectory play in endowing sufficient robustness with MAC.

GENERAL DESCRIPTION OF THE MODEL PREDICTIVE CONTROL

The process is assumed to be stable and to be expressed by

$$y(t) = \frac{z^{-1} N(z)}{D(z)} u(t) \tag{1}$$

where N(z) and D(z) are both polynomials of backward shift operator z^{-1}. D(z) is moreover a stable polynomial of z^{-1} since the process is stable.

The model of this process is also assumed to be given by the following Moving Average type model.

$$\begin{aligned} y_M(t) &= z^{-1}\tilde{N}(z)\, u(t) \\ &= z^{-1}\{\tilde{h}_1+\tilde{h}_2 z^{-1}+\ldots+\tilde{h}_s z^{-s+1}\}\, u(t) \end{aligned} \tag{2}$$

The output at time(t+j), $(L \leq j \leq L+P-1)$ is predicted by using the following equation:

$$y_P(t+j) = y_M(t+j) + y(t) - y_M(t) \tag{3}$$

where $y_P(t+j)$ expresses the predicted value of the output y(t+j).

At each instant t, the set of M future inputs {u(t),u(t+1),..., u(t+M-1)} is determined in such a way that the predicted P output values over the coincidence horizon [t+L,t+L+P-1] are closest possible to the so-called reference trajectory $y_R(t+j)$, $(L \leq j \leq L+P-1)$ (see Fig.1).

The reference trajectory is a future path leading to an ultimate set point R that will be followed by the actual process output. It is usually given by the following first order response which starts at the output of the process at time t.

$$y_R(t+j) = \alpha^j y(t) + (1-\alpha^j)R(t+j) \tag{4}$$

At each instant t, the optimal inputs {u(t),u(t+1),..., u(t+M-1)} are determined by solving the following static optimization problem.

$$\min_{u(t),..u(t+M-1)} \sum_{j=L}^{L+P-1} (y_R(t+j) - y_P(t+j))^2 \qquad (5)$$

Only the first optimum input u(t) is actually applied to the process. The actual output value y(t+1) is observed. The reference trajectory of Eq.(4) is initialized on the observed value y(t+1), and then the optimization problem given by Eq.(5) is solved at time (t+1). This procedure is repeated at each sampling time. This is a brief description of the algorithm of a model predictive control.

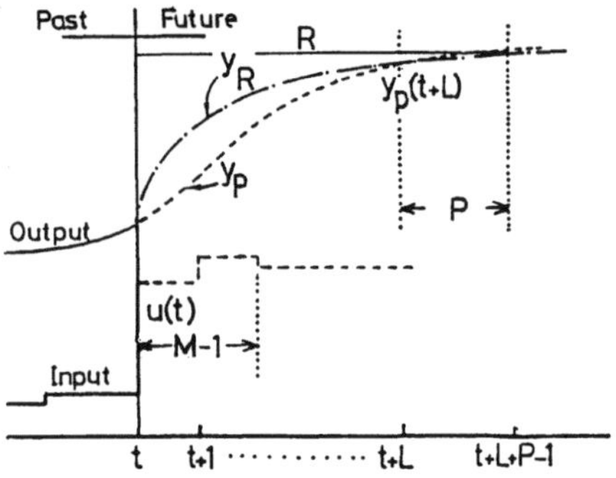

Fig.1 The basic procedure of MPC

ROBUST STABILITY OF SISO SYSTEMS

In this section, we limit our study to the case where M=1. In other words, we discuss only the case where the input at time t, u(t), is derived from Eq.(5) assuming that the input u(t) at time t is kept constant in the future, i.e. u(t)= u(t+1)=..= u(t+L+P-2). Only u(t) is applied to the process.

Based on this assumption, the prediction of output value, $y_P(t+j)$, is given by

$$y_P(t+j) = \sum_{i=1}^{j}\tilde{h}_i u(t) + \sum_{i=1}^{s-j}\tilde{h}_{i+j} u(t-i) + y(t) - y_M(t) \qquad (6)$$

The closed-loop transfer function becomes

$$y(t) = \frac{\{(1-\alpha^L)(\sum_{i=1}^{L}\tilde{h}_i)z^L + (1-\alpha^{L+1})(\sum_{i=1}^{L+1}\tilde{h}_i)z^{L+1} + \ldots + (1-\alpha^{L+P-1})(\sum_{i=1}^{L+P-1}\tilde{h}_i)z^{L+P-1}\}\, z^{-1}\frac{N}{D}}{\{(\sum_{i=1}^{L}\tilde{h}_i)(\tilde{N}_L - z^{-1}\tilde{N}) + \ldots + (\sum_{i=1}^{L+P-1}\tilde{h}_i)(\tilde{N}_{L+P-1} - z^{-1}\tilde{N})\} + z^{-1}\frac{N}{D}\{(1-\alpha^L)(\sum_{i=1}^{L}\tilde{h}_i) + .. + (1-\alpha^{L+P-1})(\sum_{i=1}^{L+P-1}\tilde{h}_i)\}} R(t) \qquad (7)$$

where

$$\tilde{N}_j(z) := (\sum_{i=1}^{j}\tilde{h}_i + \sum_{i=1}^{s-j}\tilde{h}_{i+j} z^{-i}) \qquad (8)$$

Therefore, the characteristic equation of the controlled system is given by

$$\{(1-z^{-1})\,[(\sum_{i=1}^{L}\tilde{h}_i)\bar{N}_L(z) + .. + (\sum_{i=1}^{L+P-1}\tilde{h}_i)\bar{N}_{L+P-1}(z)]\,D(z) + z^{-1}N(z)\,[(1-\alpha^L)(\sum_{i=1}^{L}\tilde{h}_i) + .. + (1-\alpha^{L+P-1})(\sum_{i=1}^{L+P-1}\tilde{h}_i)]\} = 0 \qquad (9)$$

where

$$\bar{N}_j(z) := (\tilde{N}_j(z) - z^{-1}\tilde{N}(z)) \,/\, (1-z^{-1}) \qquad (10)$$

(The derivation of this equation is shown in Appendix 1.) As for the condition under which a stable control system can be realized by adjusting only the parameter α, the following theorem can be derived. (The proof is given in Appendix 2)

[Theorem 1] When the process is stable and has a nonzero steady state gain, the closed-loop system can be stabilized and made offset-free by adjusting the parameter α, provided that the parameters L, P and the process model used can make the following polynomial stable.

$$N^*(z) = (\sum_{i=1}^{L}\tilde{h}_i)\bar{N}_L(z) + .. + (\sum_{i=1}^{L+P-1}\tilde{h}_i)\bar{N}_{L+P-1}(z) \qquad (11)$$

With regard to the way of adjusting the parameter α, the following corollary can also be derived.(The proof is given in Appendix 3)

[Corollary 1] It is assumed that the process is stable and has a nonzero steady state gain, and that L, P, and the model are chosen such that $N^*(z)$ becomes a stable polynomial. Then, no matter how large the mismatch between the process and its model is, the closed-loop system can be made stable and offset-free by adjusting α such that the following relationship is satisfied.

$$\beta \cdot \frac{N(1)}{D(1)} \cdot \frac{1}{N^*(1)} > 0 \qquad (12)$$

where

$$\beta = \{(1-\alpha^L)(\sum_{i=1}^{L}\tilde{h}_i) + (1-\alpha^{L+1})(\sum_{i=1}^{L+1}\tilde{h}_i) + .. + (1-\alpha^{L+P-1})(\sum_{i=1}^{L+P-1}\tilde{h}_i)\} \qquad (13)$$

The physical meaning of these theorems becomes clearer in the case where L=1 and P=1.

[Corollary 2] It is assumed that the process is stable and has a nonzero steady state gain, and that the coincidence horizon is chosen such that L=1 and P=1. Then, if the model consisting of a stable polynomial is used, the closed-loop system can be made stable and offset-free by adjusting only the parameter α. (The proof is given in Appendix 4.)

Furthermore, if we use a model whose steady state gain has the same sign as that of the process:

$$\mathrm{sign}(\tilde{N}(1)) = \mathrm{sign}(N(1)/D(1)) \qquad (14)$$

there exists a certain value $\alpha^* < 1$, and for an arbitrary α ($\alpha^* < \alpha < 1$), the closed loop becomes stable and offset-free.

On the contrary, if we use a model whose steady state gain has the sign opposite to that of the process, i.e.

$$\mathrm{sign}(\tilde{N}(1)) \neq \mathrm{sign}(N(1)/D(1)) \qquad (15)$$

there exists a certain value $\alpha^* > 1$, and for an arbitrary α ($\alpha^* > \alpha > 1$), the closed loop becomes stable and offset-free.

Let us apply this corollary to the case of a nonminimum phase process. We can make two different types of models by emulating the behavior of the process, 1) when it approaches the steady state (Fig.2) and 2) when it is in the rising period (Fig.3).

If these models, regardless of their types, are composed of a stable polynomial, the closed-loop system can be made stable and offset-free by adjusting only the parameter α.

When the model shown in Fig.2 is utilized, the closed-loop system can be stabilized by adjusting the parameter α in such a way that α decreases from 1. On the contrary, for the model shown in Fig.3, its closed loop can be stabilized by setting the parameter α larger than 1.

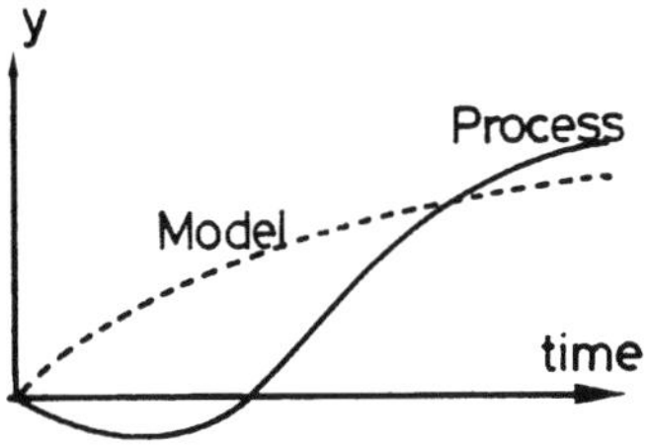

Fig.2 A model for a non-minimum phase process

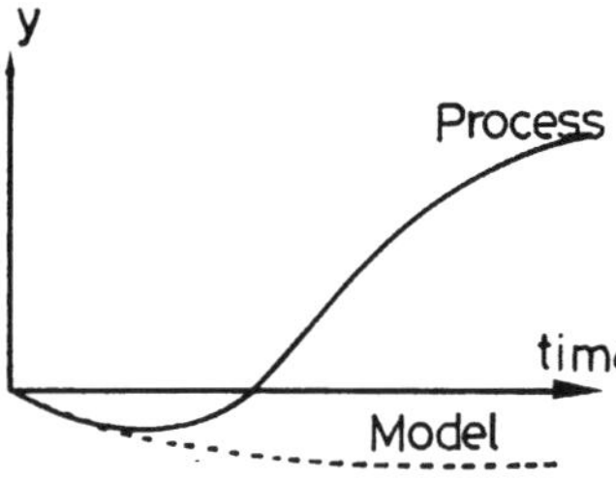

Fig.3 A model for a non-minimum phase process

ROBUST STABILITY AND COINCIDENCE HORIZON

In the foregoing section, it was clarified that a stable closed loop can be realized by adjusting only the parameter α, if $N^*(z)$ can be made a stable polynomial.

$N^*(z)$ is a function of the process model, $\tilde{N}(z)$, and the coincidence horizon (L,P). Therefore, it is possible to make $N^*(z)$ a stable polynomial by choosing a certain model.

In fact, Rouhani and Mehra (1982) proposed, for non-minimum phase systems, an approach using the model parameters as the tuning parameters of the control system with the coincidence horizon set such that L=1 and P=1. However, if we are to use the model parameters for tuning the performance of the controlled system in addition to stabilizing the system, the control algorithm becomes too complex. As a matter of course, tuning procedure of the control system also becomes complicated because the number of parameters to be adjusted increases.

In this section, we present a sufficient condition for $N^*(z)$ to be made a stable polynomial, without using the model parameters as the tuning parameters, but only by adjusting the coincidence horizon, i.e., on condition that the model once built through an identification stage is not changed throughout the design process of the control system.

[Corollary 3] It is assumed that the process is stable and has a nonzero steady state gain. If we use a moving average type model whose coefficients, $\{\tilde{h}_i\}$ (i=1,..,s), all have the same sign, $N^*(z)$ can be made a stable polynomial by setting L longer than the number of the coefficients, s. (The proof is given in Appendix 5.)

This corollary cannot be applied to non-minimum phase models, because they do not satisfy the premise that all the coefficients of the model $\{\tilde{h}_i\}$ have the same sign.

However, by using the reference trajectory newly defined by Eq.(16), we can derive a more general sufficient condition which is applicable to a greater variety of models including non-minimum phase models.

The new reference trajectory is defined by a first order response with the starting point at the predicted output value, $y_P(t+L-1)$, at time (t+L-1). i.e.

$$y_R(t+j)=\alpha^{j-L+1}y_P(t+L-1)+(1-\alpha^{j-L+1})\,R(t+j) \tag{16}$$

where it is assumed that $y_P(t)=y(t)$, when L=1.

This reference trajectory is thereafter called type 2. The one of Eq.(4) is called type 1.

By using the reference trajectory of type 2 and letting P=1, the characteristic equation of the closed-loop system is given by

$$\begin{aligned}(1-z^{-1})\bar{\bar{N}}_L(z)+(1-\alpha)\,(z^{-1}\tilde{N}(z)\\ -z^{-1}\tilde{N}(z)D(z)+\tilde{N}_{L-1}(z)D(z))=0\end{aligned} \tag{17}$$

where

$$\bar{\bar{N}}_L(z)=\tilde{h}_L+\tilde{h}_{L+1}z^{-1}+\ldots.+\tilde{h}_s z^{-s+L} \tag{18}$$

(The derivation of this equation is shown in Appendix 5.)

The necessary and sufficient condition for the closed-loop system to be made stable by adjusting the parameter α is, as is clear from Eq.(17), that $\bar{\bar{N}}_L(z)$ is a stable polynomial.

[Corollary 4] If L is chosen such that the following inequality is satisfied,

$$\tilde{h}_L \geqq \tilde{h}_{L+1} \geqq \ldots. \geqq \tilde{h}_s > 0 \quad \text{or} \quad (\tilde{h}_L \leqq \tilde{h}_{L+1} \leqq \ldots. \leqq \tilde{h}_s < 0) \tag{19}$$

then, $\bar{\bar{N}}_L(z)$ becomes a stable polynomial. (The proof is given in Appendix.7)

For a model which shows a stable response, there always exists the L that satisfies Eq.(19).

From this corollary, even when a non-minimum phase model shown in Fig.4 is utilized, it is now clear that a stable and offset-free control system can always be realized by adjusting the parameter α, if L is chosen so as to be greater than L^*.

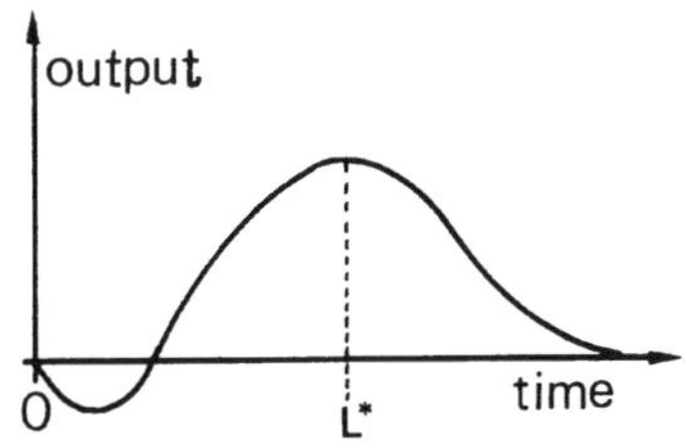

Fig.4 Coincidence horizon for a non-minimum phase process

ROBUST STABILITY OF MULTIVARIABLE SYSTEMS

In this section, we consider the multivariable case where the process is expressed by a linear n-input and n-output system as follows:

$$\mathbf{y}(t) = z^{-1}\mathbf{N}(z)\mathbf{D}(z)^{-1}\mathbf{u}(t) \tag{20}$$

where

$$\mathbf{y}(t)=[y^1(t), y^2(t), \ldots, y^n(t)]^T$$

$$\mathbf{u}(t)=[u^1(t), u^2(t), \ldots, u^n(t)]^T$$

$z^{-1}\mathbf{N}(z)\mathbf{D}(z)^{-1}$ is a right coprime factorization of the transfer function matrix. Here, the model is assumed to be given by

$$\mathbf{y}_M(t)= z^{-1}\tilde{\mathbf{N}}(z)\ \mathbf{u}(t) \tag{21}$$

where $\mathbf{y}_M(t)=[y_M^1(t), y_M^2(t), \ldots, y_M^n(t)]^T$ is the output vector of the model. $\tilde{\mathbf{N}}(z)$ is the matrix of polynomials of the backward shift operator z^{-1}. The (i,j) element of $\tilde{\mathbf{N}}(z)$ is given by

$$\tilde{N}^{ij}(z)=\{\tilde{h}_1^{ij}+\tilde{h}_2^{ij}z^{-1}+, \ldots, +\tilde{h}_{s^{ij}}^{ij}z^{-s^{ij}+1}\} \tag{22}$$

Prediction of the i-th output , $y_P^i(t)$, is also given by

$$y_P^i(t+\iota)=\sum_{j=1}^{n}\{\sum_{k=1}^{\iota}\tilde{h}_k^{ij}u^j(t)+\sum_{k=1}^{s^{ij}-\iota}\tilde{h}_{k+\iota}^{ij}u^j(t-k)\}+y^i(t)-y_M^i(t) \tag{23}$$

The above equation is obtained on the assumption that the input u(t) determined at time t will remain constant thereafter. In other words, the control horizon is set such that M=1 for all inputs. Furthermore, it is assumed that the reference trajectory is given by that of type 1, and that one step prediction is performed.

For the i-th output, the reference trajectory y_R^i is given by a first order response which starts at the observed output value at time t, $y^i(t)$, as follows:

$$y_R^i(t+L^i)=\alpha_i^{L^i}y^i(t)+(1-\alpha_i^{L^i})R^i(t+L^i) \tag{24}$$

Then, the input u(t) is determined such that the prediction of each output, $y_P^i(t+L^i)$, approaches as closely as possible to the reference trajectory, $y_R^i(t+L^i)$. The transfer function matrix between inputs and outputs is given by

$$\mathbf{y}(t)=z^{-1}\mathbf{N}(z)\mathbf{D}^{-1}(z)\ \{(1-z^{-1})\mathbf{N}^*(z)+z^{-1}(\mathbf{I}-\boldsymbol{\alpha})\ \mathbf{N}(z)\mathbf{D}^{-1}(z)\}^{-1}(\mathbf{I}-\boldsymbol{\alpha})\ \mathbf{R} \tag{25}$$

where

$$(1-z^{-1})\mathbf{N}^*(z) = \bar{\mathbf{N}}(z) - z^{-1}\tilde{\mathbf{N}}(z) \tag{26}$$

The (i,j) element of $\bar{\mathbf{N}}$ is

$$\bar{N}^{ij}(z) = \sum_{k=1}^{L^i}\tilde{h}_k^{ij}+\sum_{k=1}^{s^{ij}-L^i}\tilde{h}_{k+L^i}^{ij}z^{-k} \tag{27}$$

Therefore, the stability of the closed loop depends on whether or not all the roots of the following equation are within the unit circle on the complex plane.
i.e.

$$\det\mathbf{D}(z)\det\{(1-z^{-1})\mathbf{N}^*(z)+z^{-1}(\mathbf{I}-\boldsymbol{\alpha})\ \mathbf{N}(z)\mathbf{D}^{-1}(z)\}=0 \tag{28}$$

where

$$\mathbf{I}-\boldsymbol{\alpha} = \begin{bmatrix} 1-\alpha_1^{L^1} & & & 0 \\ & 1-\alpha_2^{L^2} & & \\ & & \ddots & \\ 0 & & & 1-\alpha_n^{L^n} \end{bmatrix}, \qquad \mathbf{R}= \begin{bmatrix} R^1(t+L^1) \\ R^2(t+L^2) \\ \vdots \\ R^n(t+L^n) \end{bmatrix}$$

The polynomial of Eq.(28) contains the adjustable parameters $\{\alpha_i\}$, as in the case of SISO. For the condition under which the controlled system can be stabilized by adjusting $\{\alpha_i\}$, the following theorem is derived. (The proof is given in Appendix. 9)

[Theorem 2] It is first assumed that the process is stable and that all of the outputs have nonzero steady state gains. If the model and L are set such that $\det\mathbf{N}^*(z)$ becomes a stable polynomial, then the closed-loop control can be stabilized and made offset-free by adjusting the parameter $\{\alpha_i\}$ so as to satisfy the following relationship:

$$\mathrm{Re}\{\lambda_k[(\mathbf{I}-\boldsymbol{\alpha})\ \mathbf{N}(1)\mathbf{D}^{-1}(1)\mathbf{N}^{*-1}(1)]\}>0 \tag{29}$$

where $\lambda_k[A]$ means the k-th eigenvalue of matrix A.

Garcia and Morari (1985) clarified the conditions that need to be imposed on the process and its model in order for MAC to be stabilized and made offset-free by adjusting only the parameter $\{\alpha_i\}$. In their study, they assume that one step prediction is performed and, moreover, that all the parameters $\{\alpha_i\}$ are identical, namely $\alpha_1=\alpha_2=, \ldots, =\alpha_n$.

Theorem 2 provides a more general condition including the results obtained by Morari and Garcia.

CONCLUSION

We analyzed the robust stability of MAC with respect to mismatch between an actual process and its model. For both SISO and MIMO cases, we derived the conditions under which a stable and offset-free closed-loop system can be realized by adjusting only the parameter α of the reference trajectory. For a stable SISO process, if we use a moving average type model, the stability of the closed loop can be ensured by adjusting the coincidence horizon and the parameter α, no matter how large the model mismatch is. For general MIMO cases, whether or not the closed-loop system can be stabilized depends on the magnitude of mismatch in the process gain.

APPENDIX

1. (Derivation of Eq.(9))
The solution u(t) of Eq.(5) satisfies the following normal equation, when M=1.

$$[\sum_{i=1}^{L}\tilde{h}_i, \sum_{i=1}^{L+1}\tilde{h}_i, \ldots, \sum_{i=1}^{L+P-1}\tilde{h}_i] \begin{bmatrix} e(L) \\ e(L+1) \\ \vdots \\ e(L+P-1) \end{bmatrix} = 0 \tag{1-1}$$

where

$$e(j):=y_R(t+j)-\{\sum_{i=1}^{j}\tilde{h}_i u(t)+\sum_{i=1}^{s-j}\tilde{h}_{i+j}u(t-i)+y(t)-y_M(t)\} \tag{1-2}$$

Using Eq.(4), we put Eq.(1-1) into the polynomial form:

$$\Bigg[\{(\sum_{i=1}^{L}\tilde{h}_i)(\tilde{N}_L - z^{-1}\tilde{N}) + \ldots + (\sum_{i=1}^{L+P-1}\tilde{h}_i)(\tilde{N}_{L+P-1} - z^{-1}\tilde{N})\}$$

$$+ z^{-1}\,\frac{N}{D}\,\{(1-\alpha^L)(\sum_{i=1}^{L}\tilde{h}_i) + \ldots + (1-\alpha^{L+P-1})(\sum_{i=1}^{L+P-1}\tilde{h}_i)\}\Bigg]\,u(t)$$

$$= \{(1-\alpha^L)(\sum_{i=1}^{L}\tilde{h}_i)z^L + \ldots + (1-\alpha_i^{L+P-1})(\sum_{i=1}^{L+P-1}\tilde{h})z^{L+P-1}\}\,R(t) \qquad (1\text{-}3)$$

From this equation, the closed-loop transfer function Eq.(7) is easily derived.

Since $\tilde{N}_j(1)=\tilde{N}(1)$, $\tilde{N}_j(z)-z^{-1}\tilde{N}(z)$ is divisible by $(1-z^{-1})$. Thus, $\bar{N}_j$ is a polynomial with a finite order. Substituting $(1-z^{-1})\bar{N}_j$ for $\tilde{N}_j-z^{-1}\tilde{N}(z)$ in Eq.(7), the denominator of Eq.(7) is given by Eq.(9).

2. (Proof of Theorem 1)

For the sake of brevity, we express the backward shift operator z^{-1} as d. Then, Eq.(9) becomes a polynomial of d as follows.

$$D(d)\,\{(1-d)N^*(d)D(d) + \beta dN\} = 0 \qquad (2\text{-}1)$$

If we can choose a proper β such that the polynomial of Eq.(2-1) has all its roots outside the unit circle on the complex plane, the closed loop becomes stable. From the premise that the process has a nonzero steady state gain, it holds that $N(1)\neq 0$. Therefore, Eq.(2-1) does not have such a root as d=1, when $\beta\neq 0$. It is also clear from the premises of the theorem that all the roots of both equations, $D(d)=0$ and $N^*(d)=0$, exist outside the unit circle. Therefore, we can draw a proper circle with its center at (1,0) such that all the roots of $N^*(d)D(d)=0$ do not exist within this circle (including its boundary).

Let us draw a closed curve C_d consisting of a circle with its center at (1,0) and the unit circle with its center at the origin, as shown in Fig.5. Then, in this closed curve, C_d, the polynomial, $(1-d)N^*(d)D(d)=0$, has a single real root, i.e., d=1.

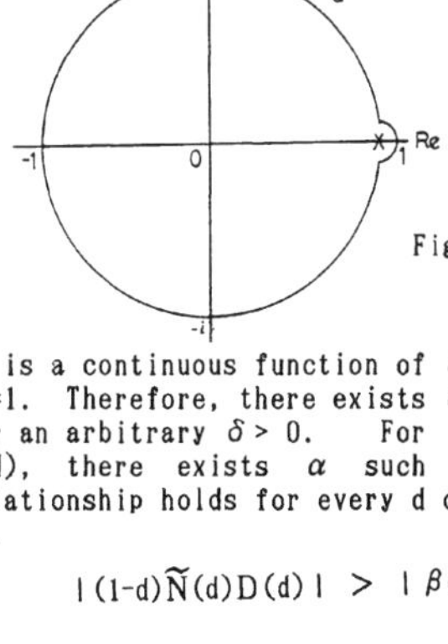

Fig.5 A single closed curve in d-domain

β is a continuous function of α such that $\beta=0$ for $\alpha=1$. Therefore, there exists α such that $|\beta|<\delta$ for an arbitrary $\delta>0$. For arbitrary N(d) and D(d), there exists α such that the following relationship holds for every d on the closed curve C_d.

$$|(1-d)\tilde{N}(d)D(d)| > |\beta dN(d)| \qquad (2\text{-}2)$$

Let β_0 be the value of β for some α such that Eq.(2-2) holds.

Rouche's theorem says that the following two equations,

$$(1-d)N^*(d)D(d)=0 \text{ and } (1-d)N^*(d)D(d)+\beta_0 dN(d)=0$$

have an equal number of roots inside C_d. Therefore, when $\beta=\beta_0$, Eq.(2-1) has a single real root inside C_d. Thus, the closed-loop system is stable if the root existing inside C_d is not within the unit circle. If the real root inside C_d is not larger than 1 for the present β_0, we can make this real root larger than 1, by putting $\beta=-\beta_0$.

Eq.(2-1) has a single root inside C_d for every β, $|\beta|\leqq|\beta_0|$ and especially for $\beta=0$, the root of Eq.(2-1) is 1.

The roots of a polynomial are continuous functions of its coefficients. Therefore, the root of Eq.(2-1) becomes larger than 1, when β continuously decreases from β_0 to $-\beta_0$.

It is also easy to check that the control system is offset-free for a step change of the set point.

Taking into account that $R(t)=(1-z^{-1})R$ (R:constant) and applying the final value theorem to Eq.(7), we can obtain the following relationship:

$$y(\infty)=R \qquad (2\text{-}3)$$

3. (Proof of Corollary 1)

Eq.(2-1) has the root, d=1, when $\beta=0$ (i.e., $\alpha=1$). Let us examine the locus of this root as β changes. Let x be the root existing inside C_d, and Δx and $\Delta\beta$ be the small perturbations from $\beta=0$ and x=1 respectively.

$$\text{i.e.} \quad x=1+\Delta x, \quad \beta=0+\Delta\beta \qquad (3\text{-}1)$$

From the characteristic equation, the following relationship holds.

$$-\Delta x\; N^*(1)D(1) + \Delta\beta N(1) = 0 \qquad (3\text{-}2)$$

If β is changed such that $\Delta\beta N(1)/(D(1)N^*(1))>0$, then, Δx is positive. Therefore, the root, d=1, becomes larger than 1 as β changes.

4. (Proof of corollary 2)

When L=1 and P=1, then $N^*(1)=\tilde{h}_1\tilde{N}(1)$ and $\beta=(1-\alpha^L)\tilde{h}_1$. The result is immediate from theorem 1 and corollary 1.

5. (Proof of corollary 3)

Taking into account that

$$\bar{N}_j(z) = \sum_{i=1}^{j}\tilde{h}_i + \sum_{i=2}^{j+1}\tilde{h}_i z^{-1} + \ldots + \sum_{i=s}^{j+s-1}\tilde{h}_i z^{-s+1} \qquad (5\text{-}1)$$

and Eq.(11), we obtain the coefficients of $N^*(z)$ as follows

$$\text{for the } z^0,\quad (\sum_{i=1}^{L}\tilde{h}_i)^2 + (\sum_{i=1}^{L+1}\tilde{h}_i)^2 + \ldots + (\sum_{i=1}^{L+P-1}\tilde{h}_i)^2 \qquad (5\text{-}2)$$

$$\text{for the } z^{-1},\quad (\sum_{i=1}^{L}\tilde{h}_i)(\sum_{i=2}^{L+1}\tilde{h}_i) + (\sum_{i=1}^{L+1}\tilde{h}_i)(\sum_{i=2}^{L+2}\tilde{h}_i) + \ldots + (\sum_{i=1}^{L+P-1}\tilde{h}_i)(\sum_{i=2}^{L+P}\tilde{h}_i)$$

$$\vdots$$

$$\text{for the } z^{-s+1},\quad (\sum_{i=1}^{L}\tilde{h}_i)(\sum_{i=s}^{L+s-1}\tilde{h}_i) + (\sum_{i=1}^{L+1}\tilde{h}_i)(\sum_{i=s}^{L+s}\tilde{h}_i) + \ldots + (\sum_{i=1}^{L+P-1}\tilde{h}_i)(\sum_{i=2}^{L+P+s-2}\tilde{h}_i)$$

The difference between the coefficient of z^{-k+1} and that of z^{-k} is

$$(\sum_{i=1}^{L}\tilde{h}_i)(\tilde{h}_k-\tilde{h}_{L+k})+(\sum_{i=1}^{L+1}\tilde{h}_i)(\tilde{h}_k-\tilde{h}_{L+k+1})+..+(\sum_{i=1}^{L+P-1}\tilde{h}_i)(\tilde{h}_k-\tilde{h}_{L+k+P-1}) \tag{5-3}$$

Let L be s. Then the difference becomes positive for all k (k=1,..s-1), since $\tilde{h}_j=0$ for all j>s.
By Kakeya's Theorem, $N^*(z)$ is guaranteed to be a stable polynomial.

6. (Derivation of Eq.(17))

The input u(t) is determined such that

$$y_R(t+L)=y_P(t+L) \tag{6-1}$$

Using Eq.(6) and Eq.(16), we can put Eq.(6-1) into the polynomial form:

$$\left[\{\tilde{N}_L(z)-z^{-1}\tilde{N}_{L-1}(z)\}+(1-\alpha)\{z^{-1}N(z)/D(z) -z^{-1}\tilde{N}(z)+\tilde{N}_{L-1}(z)\}\right] u(t)=(1-\alpha)z^L R(t) \tag{6-2}$$

Since $\tilde{N}_L(1)=\tilde{N}_{L-1}(1)$, $\tilde{N}_L(z)-z^{-1}\tilde{N}_{L-1}(z)$ is divisible by $1-z^{-1}$. Therefore, the quote $\bar{N}_L$ is a polynomial. Substitute $(1-z^{-1})\bar{\bar{N}}_L(z)$ for $\tilde{N}_L(z)-z^{-1}\tilde{N}_{L-1}(z)$ in Eq.(6-2). Then, Eq.(17) is immediately derived.
For the case that P is not 1, the characteristic equation can also be obtained in the way Eq.(9) was derived.

7. (Proof of corollary 4)

The result is immediate from Kakeya's theorem.
Note that this corollary can be easily extended to the case that P is not 1.

8. (Derivation of Eq.(25))

Inputs $u(t)=[u^1(t),\ldots,u^n(t)]$ are determined so as to satisfy

$$y_R^i(t+L^i)=\sum_{j=1}^{n}\{\sum_{k=1}^{L^i}\tilde{h}_k^{ij}u^j(t)+\sum_{k=1}^{s^{ij}-L^i}\tilde{h}_{k+L^i}^{ij}u^j(t-k)\}+y^i(t)-y_M^i(t) \tag{8-1}$$

(for i=1,..,n)

Considering Eq.(24) in the above equation, we put it into the polynomial form. Then, the following equation is obtained in a matrix form.

$$\{\bar{N}(z)-z^{-1}\tilde{N}(z)+(I-\alpha)(z^{-1}N(z)D(z)^{-1})\}\ u(t)=(I-\alpha)R \tag{8-2}$$

Since $\bar{N}^{ij}(1)=\tilde{N}^{ij}(1)$, each element of $\bar{N}(z)-z^{-1}\tilde{N}(z)$ is divisible by $(1-z^{-1})$. Thus, $N^*(z)$ is a polynomial matrix with a finite order.
Substituting $(1-z^{-1})N^*(z)$ for $\bar{N}(z)-z^{-1}\tilde{N}(z)$ in Eq.(8-2), we obtain Eq.(25).

9. (Proof of Theorem 2)

We again express z^{-1} as d. From the premise that the process is stable, detD(d) has all its roots outside the unit circle. Therefore, the stability condition is that all the roots of Eq.(9-1) exist outside the unit circle.

$$\det\{(1-d)N^*(d)D(d)+d\,(I-\alpha)\,N(d)\}=0 \tag{9-1}$$

Eq.(9-1) can be decomposed as follows.

$$(1-d)^n\det N^*(d)D(d)+\sum_{i=1}^{n}(1-\alpha_i^{L^i})d(1-d)^{n-1}f_i(d) +\sum_{\substack{i=1\\ i\neq j}}^{n}\sum_{j=1}^{n}(1-\alpha_i^{L^i})(1-\alpha_j^{L^j})d^2(1-d)^{n-2}f_{ij}(d)+,\ldots, +\prod_{i=1}^{n}(1-\alpha_i^{L^i})d^n\det N(d)=0 \tag{9-2}$$

All the terms except the first term vanish, when $\alpha_i=1$ for all i (i=1,...,n). Therefore, there exist certain $\{\alpha_i\}$ that can make the absolute value of the first term larger than that of the residual term of Eq.(9-2). Express these $\{\alpha_i\}$ as $\{\alpha_i^*\}$.
Rouche's theorem says that, for those $\{\alpha_i^*\}$, $(1-d)^n\det N^*(z)\det D(z)=0$ and Eq.(9-2) have an equal number of roots inside the single closed curve C_d. All the roots of both equations, $\det D(z)=0$ and $\det N^*(z)=0$ do not exist inside C_d because of the premise of this theorem. Then, both $(1-d)^n\det N^*(z)\det D(z)=0$ and Eq.(9-1) have n roots inside C_d for $\{\alpha_i^*\}$.

The closed-loop system is stable and offset-free if there exist $\alpha_i^*\neq 1$ (for i=1,..,n) such that all the roots inside C_d are not within the unit circle.
Let Δd and $\Delta\beta$ be small perturbations around d=1 and $\beta=0$ ($\alpha=1$). It is clear from Eq.(28) that the following equation holds for the small perturbations.

$$\det\{-\Delta d N^*(1)D(1)+\Delta\beta N(1)\}=0 \tag{9-3}$$

Δd corresponds to the eigenvalues of the following matrix,

$$\Delta\beta N(1)D^{-1}(1)\ N^{*-1}(1) \tag{9-4}$$

Therefore, the system is stable for $\Delta\beta$ such that real parts of all the eigenvalues of Eq.(9-3) are positive.

Rouche's Theorem

When the absolute value of the polynomial $\phi(d)$ is smaller than that of $\phi(d)$ on a simple closed curve C_d, i.e., $|\phi(d)|<|\phi(d)|$
Both polynomials, $\phi(d)$ and $\phi(d)+\phi(d)$ have an equal number of roots inside C_d.

Kakeya's Theorem

Consider a polynomial equation with real coefficients,

$$a_0z^n+a_1z^{n-1}+\ldots+a_n=0$$

If the coefficients satisfy the following relationship.

$$a_0\geqq a_1\geqq a_2\geqq\ldots\geqq a_n>0$$

(where not all the equalities hold at the same time.) then, the absolute value of every root is less than 1.

The proof of both theorems is found in Takagi(1960).

REFERENCE

Cutler,C.R. and B.L.Ramaker(1979). Dynamic matrix control a computer control algorithm, paper51b AIChE 86th National mtg.LA.

Cutler,C.R.(1982) Dynamic matrix control of imbalanced systems, ISA Transactions, vol.21,1,pp.11-5

Garcia,C.E. and M.Morari(1982). Internal model control.1 a unifying review and some new results. I&EC Process Design and Development, 21, pp.308-323

Garcia,C.E. and M.Morari(1985). Internal model control. 2. design procedure for multivariable systems, I&EC Process Design and Development, 24, pp.472-484

Rouhani,R and R.K.Mehra(1982). Model algorithmic control(MAC) basic theoretical properties, Automatica, vol.18, 4, pp.401-414.

Richalet,J.,A.Rault,J.L.Testud and J.Papon(1978). Model predictive control:application to industrial processes. Automatica,vol.14, pp.413-428

Takagi,T(1930). Lecture on algebra , Kyoritsu

PREDICTIVE CONTROL: A REVIEW AND SOME NEW STABILITY RESULTS

Dae Gyu Byun and Wook Hyun Kwon

Department of Control and Instrumentation Engineering, Seoul National University, Seoul, Korea

ABSTRACT

Various kinds of predictive control design methods such as MAC(Model Algorithmic Control), DMC(Dynamic Matrix control), PCA(Predictive Control Algorithm), EHAC(Extended Horizon Adaptive Control), EPSAC(Extended Prediction Self Adaptive Control), GPC(Generalized Predictive Control), and RHTC(Receding Horizon Tracking Controller) are reviewed and compared in this paper. In addition, some new stability results for EHAC and GPC are presented.

1. Introduction

Recently, several computer control techniques called predictive control have been introduced and their successful applications to industrial multivariable processes have been reported [9,10,11,15,47]. They are expected to be widely used in the future in process industries [56]. The main idea of the strategy for the predictive control is to predict the effects of potential control actions on the future values of the process output and to find the best control actions which minimize the squared sum of deviations of predicted outputs from command signals. For this type of problems the future command signals over a range up to the prespecified prediction horizon must be known in advance. In most process control systems it is possible to know the finite future command signals over the small horizon. This information about future command signals is believed to be useful to the improvement of the transient response.

The strategy for the predictive controls is summarized as follows:

1) at the present moment t, the future outputs of the plant, $y(t+1),y(t+2),\ldots,y(t+N)$, are predicted based on a given model.
2) the control vector composed of $u(t),u(t+1),\ldots,u(t+N-1)$ is computed in such a way that it minimizes the given receding horizon cost function. In most cases, this control vector aims at driving the predicted outputs close to future command signals.
3) the first element $u(t)$ of the control vector is actually applied to the real plant and the whole procedure is repeated at the next time $t+1$.

The followings are existing predictive controls using this strategy. They range from the earliest MAC to the latest RHTC.

(1) Model Algorithmic Control (MAC)
 ; Impulse response model [1,2,3,6,7]
 ; State space model [4,5]
(2) Dynamic Matrix Control (DMC)
 ; Step response model [14,16,17,20]
(3) Predictive Control Algorithm (PCA)
 ; Impulse response model [12,13]
(4) Extended Horizon Adaptive Control (EHAC)
 ; ARMAX model [21,22,23]
(5) Extended Prediction Self Adaptive Control (EPSAC) ; ARMAX model [36,37]
(6) Generalized Predictive Control (GPC)
 ; CARIMA model [24,25,26]
(7) Receding Horizon Tracking Control (RHTC)
 ; State space model [28,29,30,55]
 ; ARMAX model [53,54,57,58]

Output predictive algorithmic control [4] and model predictive control [5] are very similar to MAC. In this paper, we classify them within the boundary of MAC. Though the first assumed models are ARMAX models, control laws in [53,54,57,58] are similar to the receding horizon regulator of [29,30]. So we classify all of them within the boundary of RHTC. Model Predictive Control (MPC), whose name is the same as in [5], is proposed for finding the approximate inverse of the plant model in the study of internal model control (IMC) [40,41,42,43]. It is in the form of mixup of MAC and DMC and thus not listed above.

In this paper, self tuning feature will not be discussed. For predictive controls where the control and the self tuning feature are combined, only those predictive controls whose control parts can be explicitly separated from the identification part will be discussed. MUSMAR (MUltiStep Multivariable Adaptive Regulator) [38,39] use also the predictive strategy but it takes the form of implicit self tuning controllers. So it will not be discussed here.

In [33,34,35] an optimal tracking problem called preview control is discussed where the finite future command signals are assumed to be known a priori but the cost with complete horizon instead of receding horizon is minimized. Similar approach as the preview control is found also in [52]. The preview control, however, is excluded in this paper since it is a little deviated from mainstream of the predictive controls.

Each of the predictive controls (1) to (7) has a few differences distinguished from others in the form of :

(a) underlying plant models
(b) cost functions
(c) assumptions about future control actions
(d) assumptions about command signals
(e) prediction methods

We review the predictive controls centered around these different points. Previous reviews are found in [49,50,51], which cover only the predictive control laws based on weighting sequence models such as MAC, DMC, and PCA. In [48], MAC, DMC, EHAC, EPSAC, and GPC are reviewed but GPC is a central concern. This paper covers wider class of predictive controls and focuses on the unifying approach. We also discuss the stability properties and present some new stability results.

The review in this paper is not exhausted but intended to discuss differences of the concepts of each predictive control rather than to search and list all references.

2. Underlying Models

Predictive control is sometimes called model based control. This implies the essential role of the model in the predictive control design. The early predictive controls such as MAC [1,2,3,6,7] and PCA are based on impulse response models and DMC on step response models. An impulse response sequence $\{h_i\}$ is simply related to a step response sequence $\{s_i\}$ by

$$s_k = \sum_{i=1}^{k} h_i \qquad k=1,2,... \tag{1}$$

and

$$h_k = s_k - s_{k-1} \tag{2}$$

These weighting sequence models give some advantages: there is no need to assume a system structure and it is easy to find experimentally. It is indicated in [2,40] that impulse response models improve the robustness for its nonminimal representations. The weighting sequence models, however, has too many parameters requiring the large computation efforts for identification and can't represent open loop unstable plants.

When there is no prior knowledge about the process structure or the order of states of the process is too high for the state space model to be useful for control system design, it is reasonable that the parameterized input/output model should be used for easy identification. EHAC and EPSAC use ARMAX models :

$$A(q^{-1})y(t)=B(q^{-1})u(t-1)+C(q^{-1})e(t)+d(t) \tag{3}$$

where q^{-1} is the backward shift operator and the A,B,and C polynomials are defined by

$$A(q^{-1}) = 1 + \sum_{i=1}^{n} a_i q^{-i} \tag{4}$$

$$B(q^{-1}) = \sum_{i=0}^{m} b_i q^{-i} \tag{5}$$

$$C(q^{-1}) = \sum_{i=0}^{n} c_i q^{-i} , \tag{6}$$

e(t) is an uncorrelated zero-mean random sequence, and d(t) is a load disturbance variable. It is shown in [59] that the ARMAX model is not satisfactory for dealing with offset problem, especially when the load disturbance d(t) is rapidly varying.

GPC uses CARIMA models which are known to be useful to the elimination of offset [60]. The CARIMA model is given by :

$$A(q^{-1})y(t) = B(q^{-1})u(t) + n(t) \tag{7}$$

where

$$n(t) = C(q^{-1})e(t)/\Delta \tag{8}$$

Although EPSAC [36,37] is based on the ARMAX model, the prediction equation used in [37] for EPSAC is similar to one obtained from the CARIMA model. For such formulation appropriate assumptions about noise or disturbance are needed.

If the internal structure of the process is known it is recommended to use a state space model. It is well suited to analyze the internal properties of the process and can be most easily transformed into other type of model. RHTC [28,29,30] is based on a deterministic state space model. It should be noted, however, that when the states are not accessible a state estimator has to be employed. For self-tuning scheme the state space approach requires too much computation efforts for solving the Riccati equation.

For some predictive controls, the first assumed model is transformed into other model form and then the control law is obtained based on the transformed model. In [4] a system is first given by a state space model but it is transformed into an impulse response model on which MAC is based. In [53,54,57,58] ARMAX models are first assumed but the control laws are obtained in the transformed state space models on which RHTC is based. Similarly, some stability results are obtained in the transformed model. In [24,26] a state space model, transformed from a CARIMA model, is used for analysis of stability properties of GPC.

3. Cost, Control Action, and Command Signal

The predictive control laws can be obtained from minimizing the following general cost function J :

$$\min_{u(t),u(t+1),...,u(t+M-1)} J = \{\sum_{j=d}^{N} \| y(t+j)-y_r(t+j) \|_{Q(j)} + \sum_{j=1}^{N} \| w(q^{-1})u(t+j-1) \|_{R(j)} \} \tag{9}$$

where $\| x \|_q^2$ means x'qx and user selectable parameters are as follows:

N = maximum prediction horizon
d = minimum prediction horizon
Q(j),R(j) = weighting matrices
$w(q^{-1})$ = costing function for u(t)
M = control horizon
y_r = reference trajectory

(1) Prediction Horizons

For the predictive controls to be meaningful the prediction horizon N must encompass the significant part of the process step response. Generally N can be chosen to cover the deadtime and the nonminimum phase response (NMP) part approaching to the risetime of the plant [17,25]. It is indicated in [4,40] that N greater than twice the system order has no effects on the system performance. N is deeply related to the stability properties, some of which are discussed in section 5.

Minimum prediction horizon d is normally taken to be unity for most predictive controls, but if the deadtime k of a plant is known a priori it can be k or more. GPC [48] provides the control law for d which is not equal to unity. In DMC [14] there arises singularity problem if there is the deadtime. By introducing the minimum prediction horizon this problem can be overcome since the elements of the step response sequence $\{s_i\}$ before the deadtime, which are all zeros, do not appear in the dynamic matrix.

(2) Weighting Matrices and Costing Function

Q(j) and R(j) are weighting matrices for the output error and the control input. In the first derivations of MAC, DMC, and EPSAC only output error terms are included in the cost function. So R(j) can be thought to be zero and Q(j) to be unity for them.

In the earliest predictive control MAC, where there is no limitations on control actions, the cost J becomes $(y(t+1)-y_r(t+1))^2$. This brings about instability for NMP plants.

DMC, however, introduces the idea of control horizon for the first time [14], hence the problem becomes different from MAC. Later R(j)=r with w(q-1)= $\Delta = 1-q^{-1}$ is introduced in [17,19,20] to deal with the matrix singularity problem of the original DMC solution, where r is called the move suppression factor.

PCA, similar to MAC, assumes R(j)=r and Q(j)=1. In MPC [40,42] R(j) and Q(j) are freely chosen.

EPSAC assumes two possibilities for control input as follows.

$$1.\ u(t) = u(t+1) = ... = u(t+N-1) \tag{10}$$

$$2.\ u(t+1) = u(t+2) = ... = u(t+N-1) = 0 \tag{11}$$

It should be noted that the infinite weighting on output error $(y(t+j)-y_r(t+j))$ is equivalent to imposing the constraint:

$$y(t+j)=y_r(t+j).$$

EHAC in [21] is based on the following cost (12)-(13), which is equivalent to putting $Q(j)=0$ for $0<j<N$, $Q(N)=\infty$, and $R(j)=1$.

$$y(t+N)=y_r(t+N). \qquad (12)$$

$$J = \sum_{j=1}^{N} u^2(t+j-1) \qquad (13)$$

This is interpreted to be the extension of deadbeat controllers allowing the process longer time than deadtime to reach the objective point with minimum control efforts. Later, EHAC in [23] considers two other possibilities for control inputs like the above (10) and (11).

In GPC, Q(j) is equal to unity and R(j) is freely chosen. The cost of GPC is thought to be rather general compared to other i/o model based predictive controls. It should be noticed that it suffices to assume $Q(j)=1$ when SISO systems are given. In RHTC, Q(j) and R(j) are freely chosen.

The costing function w(q-1) is generally chosen to be unity or the differencing operator $\Delta=1-q^{-1}$ where q is the shift operator. It has been known that the control increment Δu in the cost J gives integral action. For predictive controls such as MAC, DMC, and EPSAC, $w(q^{-1})$ can not be defined since there is no control input weighting in the cost. Later, $w(q^{-1})=\Delta$ is included for DMC [17,20].

GPC assumes $w(q^{-1})=\Delta$. RHTC provides two cases of $w(q^{-1})=1$ and $w(q^{-1})=\Delta$. PCA employs $w(q^{-1})=q^{j-1}-q^{-1}$, which is proposed for dealing with such problems in MAC as instability of NMP plants or singularity in case of deadtime [50]. In EHAC [21] $w(q^{-1})$ is equal to unity.

It is shown that the zero offset is guaranteed for constant command inputs in GPC [25] and RHTC with $w(q^{-1})=\Delta$ [28]. Even when Δu is not included in the cost a control input can be obtained in the form of control increment Δu by using Δu in the prediction equation. By this way, in DMC [14], EHAC [23], and EPSAC [37] the control laws are obtaind in the form of control increment without including Δu in the cost J.

(3) Control Horizon

The control horizon M means that the future M control inputs from the current time are free but after M intervals the change of controls is not allowed,

i.e. $\Delta u(t+j-1)=0$ for $j > M$.

This idea was first introduced in DMC [14,16] and thereafter have been used for EHAC [23], EPSAC [36,37], MPC [40,42], and GPC [24,25]. The control horizon has the effect that the control input becomes active for large M and sluggish for small M. This effect may be obtained also by adjusting the weighting matrix R(j). The control horizon M has such effect as to put infinite weighting on the control increment $\Delta u(t+j-1)$ for $j>M$. The control horizon, in addition, renders the reduction in the computation of the control law. The control horizon M in MAC, PCA, and RHTC is equal to the prediction horizon N.

Effects of the control horizon are well explained in GPC [25], where the case of $M=1$ is recommended for typical use. EHAC [23] and EPSAC [36,37] utilize only the case of $M=1$. In [17] effects of M for DMC are discussed concentrated on the case of $M=1$. Some stability properties related to M are discussed in section 5.

(4) Reference Trajectory

Most predictive controls assume that the future command signals over the prediction horizon are known a priori. The future command signal $y_r(t+j)$ is a constant set point in many process control systems, and there are many other systems whose command signals are varying but can be known on a finite receding horizon in advance.

In MAC a simple first order exponential trajectory is used as the future command signal:

$$y_r(t+j) = \alpha^j\, y(t) + (1-\alpha^j)\, C \qquad j=1,2,\cdots, \quad |\alpha| < 1 \qquad (14)$$

where C is a constant set point and y(t) is an actual plant output of the current time. The effect of α on the closed loop stability is discussed in [3,7]. Similar discussion is made in the framework of the internal model control and relatively good explanations are given [40].

There are no particular assumptions for the command signal in other predictive controls. The command signal like (14) can be used also for other predictive controls if necessary.

Following table summarizes the differences between the predictive controls. As discussed above, some predictive controls are extended to include features which are not in the original derivation. Table 1. are based on contents appeared first in the literature.

Table 1. Summary of the Predictive controls

	MAC	DMC	PCA	EHAC	EPSAC	GPC	RHTC
plant model model	impul -se	step	impul -se	ARMAX	ARMAX	CARIMA	state space
prediction horizon N	N	N	N	N	N	N	N
minimum horizon d	1	1	1	1	1	d	1
costing function w	X	X	qj-1 -q-1	1	X	1-q-1	1, 1-q-1
error weighting Q	1	1	Q	0 Q(N)=∞	1	Q	Q
control weighting R	0	0	R	1	0	R	R
control horizon M	N	M	N	N	1	M	N

X : not included in the cost

The difference in the prediction methods is discussed in next section.

4. Controller Structure

The full description of control algorithms of the predictive controls is not presented here, but only the structure of the closed loop systems is focused on.

In case of i/o model based predictive controls it is general approach to use the predicted output y(t+j) instead of y(t+j) in the cost J. The predicted output y(t+j) is given by

$$y(t+j) = y_p(t+j) + y_f(t+j)$$

where $y_p(t+j)$ is a quantity depending on past control inputs and measured outputs and $y_f(t+j)$ depending on future control actions to be determined. Minimizing this cost, the predictor $y_p(t+j)$ is explicitly included in the control input.

In RHTC which is based on state space models [28,55], the predicted output y(t+j) is not used in the the cost J whereas the control law is obtained from the well known methods for linear optimal control problems. In this case a state feedback control law is obtained and the solution of the Riccati equation is required.

According to selected design parameters of the cost J and to minimization methods, algorithms of the predictive controls takes a little different form, but all of them can be cast into the structure like Fig.1.

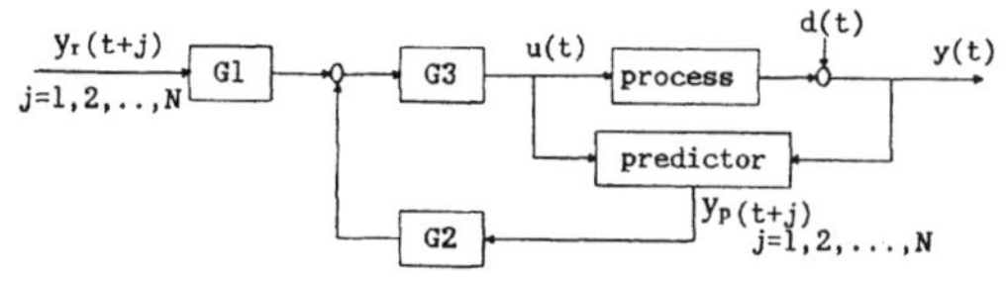

Fig.1 Structure of the predictive controls

The future command signals, $y_r(t+1)$, $y_r(t+2)$, ..., $y_r(t+N)$, must be known at each instant of time t to construct the control input for all the predictive controls except EHAC. In the case of EHAC only $y_r(t+N)$ is used since EHAC has no weighting on the output error $y(t+j)-y_r(t+j)$ for $0<j\leq N-1$. The block G_1 appears due to the reference $y_r(t+j)$ and is thought to function as a prefiltering which is believed to be useful for the transient response.

For all i/o model based predictive controls except EHAC, the predictor in Fig. 1 should provide the vectored predicted output [$y_p(t+1)$; $y_p(t+2)$; ... ; $y_p(t+N)$]. In EHAC only $y_p(t+N)$ is needed. In RHTC, however, the predictor is not introduced so far. Since RHTC is in the form of state feedback control, a state observer should be used in the place of the predictor when the states are not directly accessible.

For all i/o model based predictive controls, the block G_1 is equal to the block G_2 and the block G_3 is equal to unity in Fig.1. This type of structure is known to be sometimes unable to satisfy both the command following and disturbance rejection [45]. This fact coincides with the observation that predictive controls give less improvement for load rejection than they do for command following [19]. This is partly because simple noise or disturbance models are used in the predictor design for easy derivation of the control algorithm. It should be pointed out that more correct disturbance models have to be employed in the predictor design for better load response. In [19] a load estimator is used to improve the load response of DMC. In the case of RHTC the blocks G_1,G_2, and G_3 have different gains but the performance with respect to the command following and disturbance rejection is not known.

It is indicated in [43] that a major reason for the success of predictive controls in industry is in the constraints handling capability within the predictive control algorithms. Several results are reported on the constraints handling problem in [8,18,27,44]. But if the constraint handling algorithms are introduced the controller structure will be no more the form of Fig.1 whereas some computer programming methods will replace the mathematical formula.

5. Stability Properties

There have been some approaches to the analysis of stability properties of i/o model based predictive control laws. It is shown in [3,7] that MAC is unstable for NMP systems and that some curious method should be used for maintaining stability which is seen to be unrealistic. The stability property of DMC is dealt with in [17] and shown that it is a complex function of design parameters. Some stability properties for MPC are presented in [40], among which some results on practical settings of N and M are included. All the above results, however, are based on truncated weighting sequence models, which cannot represent unstable plants.

In [24,26] some stability results are presented for GPC, most of which are about the limiting case of N. But the results are not fruitful since the limiting case of N will hardly be used in practical situations. Some stability results for RHTC are provided in [28]. They are focused on how the terminal weighting matrix Q(N) can affect the closed loop stability. Similar results for continuous time regulators are found in [31]. The results for GPC and RHTC can be used for other predictive controls since the used models, CARIMA and state space, are in the general form which can be easily transformed into other model forms.

It is possible to analyze stability properties of one control law in the transformed model. The results in [24,26] for GPC are obtained in the transformed state space model although the first assumed model is a CARIMA model. We observe that the parameters appearing in the original algorithm can be represented in terms of parameters of the transformed model. Similar observation is used in [5] for the stability analysis of MAC. We present some new stability results for EHAC and GPC using these ideas.

Since the disturbance can be ignored as far as the stability properties are concerned, we transform the model (3) without the disturbance term into the observable canonical state space model $\{A_e,B_e,C_e\}$. We first consider the stability of EHAC in the transformed state space model with the cost function (13).

Theorem 1:
If the system $\{A_e,B_e,C_e\}$ is completely controllable and A_e is invertible , then EHAC minimizing the cost (13) is a stabilizing control for any $N\geq n+1$ where n is the system order.

Proof :
If we define $x_r(t+N)=[y_r(t+N)\ 0\ \dots\ 0]'$, then the constraint (11) becomes equal to

$$x(t+N)=x_r(t+N) \tag{15}$$

This is possible because

$$y(t+N)=Cx(t+N)=Cx_r(t+N)=y_r(t+N)$$

where $C=[1\ 0\ \dots\ 0]$ in the observable canonical state space model. Minimizing the cost (13) with constraint (15) based on the transformed state space model gives the following control law

$$u(t)=B_e'(A_e')^{N-1}W^{-1}(N-1)\{x_r(t+N)-A_e^N x(t)\} \tag{16}$$

where

$$W(N)=\sum_{j=0}^{N} A_e^j B_e B_e'(A_e')^j$$

The control law u(t) is a stabilizing control law by the result of [55]. Q.E.D.

Similarly, we transform the CARIMA model without disturbance term into the incremental state space model.

$$\begin{aligned} x(t+1) &= Ax(t)+B\Delta u(t) \\ y(t) &= Cx(t) \end{aligned} \tag{17}$$

If we define the augmented polynomial $A_a(q^{-1})$ to be

$$A_a(q^{-1})=A(q^{-1})\Delta=1+\alpha_1 q^{-1}+\alpha_2 q^{-2}+\dots+\alpha_{n+1}q^{-n-1} \tag{18}$$

then the parameters A,B, and C in observable canonical form are given as

$$A=\begin{vmatrix} -\alpha_1 & 1 & 0 & \dots & 0 \\ -\alpha_2 & 0 & 1 & \dots & 0 \\ \cdot & & 0 & 1 & \cdot \\ \cdot & & & \cdot & \cdot \\ \cdot & & & & \cdot\,0 \\ & & & & 1 \\ -\alpha_{n+1} & \dots & & & 0 \end{vmatrix}$$

$$B=[b_0\ b_1\ \dots\ b_m\ \dots\ 0]'$$

$$C=[1\ 0\ \dots\dots\dots\ 0] \tag{19}$$

and in controllable canonical form given as

$$A=\begin{bmatrix} -\alpha_1 & -\alpha_2 & \cdots & -\alpha_{n+1} \\ 1 & 0 & \cdots & 0 \\ 0 & 1 & & \cdot \\ \cdot & & \cdot\ \cdot & \cdot \\ \cdot & & \cdot\ \cdot & \cdot \\ \cdot & & \cdot\ \cdot & \cdot \\ 0 & \cdots & 1 & 0 \end{bmatrix} \qquad (20)$$

$$B=[1\ 0\ \cdots\ 0]'$$

$$C=[b_0\ b_1\ \ldots\ b_m\ \ldots\ 0]$$

It is shown in [25] that the control algorithm of GPC is given by :

$$\Delta U(t) = (F'F + rI)^{-1}F'(Y_r(t) - Y_p(t)) \qquad (21)$$

where

$$\Delta U(t) = [\Delta u(t), \Delta u(t+1), \ldots, \Delta u(t+N-1)]'$$

$$F = \begin{bmatrix} f_0 & & 0 \\ f_1 & f_0 & \\ \cdot & & \cdot \\ \cdot & & f_0 \\ \cdot & & \cdot \\ \cdot & & \cdot \\ f_{N-1} & \cdots & f_{N-M} \end{bmatrix} \qquad (22)$$

$$Y_r(t) = [y_r(t+1), y_r(t+2), \ldots, y_r(t+N)]'$$

$$Y_p(t) = [y_p(t+1), y_p(t+2), \ldots, y_p(t+N)]'$$

In eq.(22) f_i is known to be a parameter of the step response, so we can express f_i as :

$$f_i = CA^iB$$

Now, it is possible to express the control law (21) in terms of the parameters of the state space model (17):
i.e.

$$F = \begin{bmatrix} CB & & 0 \\ CAB & CB & \\ \cdot & \cdot & \\ \cdot & & \cdot \\ \cdot & & CB \\ \cdot & & \cdot \\ CA^{N-1}B & \cdots & CA^{N-M}B \end{bmatrix} \qquad (23)$$

$$Y_p(t) = \begin{bmatrix} y_p(t+1) \\ y_p(t+2) \\ \cdot \\ \cdot \\ \cdot \\ y_p(t+N) \end{bmatrix} = \begin{bmatrix} CA \\ CA^2 \\ \cdot \\ \cdot \\ \cdot \\ CA^N \end{bmatrix} x(t) = L_2x(t) \qquad (24)$$

Using the equations (23) and (24) we can obtain the following two stability results for GPC.

Theorem 2:

Assume $M=1$, $d=1$, $r=0$, and N to be any positive constant. The control law (21) of GPC is a stabilizing control law if and only if all the roots of the polynomial (25)

$$H(z)=Kz^{n+1}+(\alpha_1K+P_1)z^n+\ldots+(\alpha_{n+1}K+P_{n+1}) \qquad (25)$$

where

$$K = (CB)^2+(CAB)^2+\ldots+(CA^{N-1}B)^2$$
$$[P_1,P_2,\ldots,P_{n+1}]=[CBCA+CABCA^2+\ldots +CA^{N-1}BCA^N]$$

lie within the unit circle.

Proof:

For the assumed parameter setting, the control law $\Delta u(t)$ in the vector $\Delta U(t)$ of (21) is written as :

$$\Delta u(t) = -L_1L_2x(t) + L_1Y_r$$

where

$$L_1=\frac{1}{(CB)^2+(CAB)^2+\ldots+(CA^{N-1}B)^2}[CB\ \ CAB\ \ldots\ CA^{N-1}B]$$

Then the closed loop matrix A_c is given by $A_c=A-BL_1L_2$. If we use the matrices defined in (20), we obtain the following A_c:

$$\begin{bmatrix} -(\alpha_1+P_1/K) & -(\alpha_2+P_2/K) & \cdots & -(\alpha_{n+1}+P_{n+1}/K) \\ 1 & 0 & \cdots & 0 \\ & 1 & 0 & \cdot \\ & & \cdot & \cdot \\ & 0 & \cdot & \cdot \\ & & 1 & 0 \end{bmatrix}$$

whose characteristic polynomial is (25). Q.E.D.

It is pointed out in [25] that a value of $M=1$ is adequate for typical industrial plant models. Theorem 2 provides an stability result for this practical case.

Theorem 3:

Assume $M=N$, $d=1$, $r=0$, and N to be any positive constant. The control law (21) of GPC is a stabilizing control law if and only if all the roots of the polynomial (26)

$$H(z) = b_0z^m + b_1z^{m-1} + \ldots + b_m \qquad (26)$$

lie within the unit circle.

Proof:

For the assumed parameter setting, the control law (21) is written as :

$$\Delta U(t) = - F^{-1}L_2x(t) + F^{-1}Y_r(t)$$

Since we need only the first element $\Delta u(t)$ of $\Delta U(t)$, after some manipulations we obtain :

$$\Delta u(t)=-[-\alpha_1/b_0;1/b_0;0;\ldots;0]x(t) + 1/b_0\, y_r(t+1)$$

Then using the matrices defined in (19) we can obtain the following characteristic polynomial of the closed loop system :

$$H_c(z) = z(b_0z^m + b_1z^{m-1} + \ldots + b_m)$$

which gives (26). Q.E.D

Theorem 3 says that the control law (21) is unstable for NMP systems under the assumed parameter setting. This coincides with existing results for MAC in [3] and MPC in [40]. This fact is, however, an expected result since the assumed parameter setting is equivalent to parameter setting ; $M=N=d=1$ and $r=0$, in which case the control law (21) becomes a one step ahead control law.

6. Conclusions

This paper shows that many seemingly different controls which have been developed by so many different investigators and so many different approaches can be categorized as one group of control design methods called predictive control. Though the predictive control laws have been less known compared to other design methods, listed references, not complete, say that there are considerable research results related to the predictive control, most of which are developed independently. They are discussed in this paper around the following five items ; (a) underlying plant models (b) cost functions (c) assumptions about future control actions (d) assumptions about command signals (e) prediction methods.

Theoretical developments for predictive controls, however, are not complete and much are left to be done. Even stability properties are not known completely. This paper presents some new stability results.

Since all the predictive controls adopt the receding horizon strategy they are thought to share many common properties. We can see in this paper that the predictive controls have much similar structure with each other. Stability properties for one predictive control can be applied to other predictive controls in many cases. Peculiar properties in one predictive control may be attributable to the different points indicated in this paper but they are minor and can be realized in other predictive controls. Related to robustness properties there are few results available. Very recently, studies for robust model predictive control design have been reported [43,46].

Conclusively, it is necessary that unifying efforts should be attempted and further theoretical developments should be made.

REFERENCES

[1] J. Richalet, A. Rault, J.L. Testud, and J. Papon, "Model Predictive Heuristic Control : Applications to Industrial processes," Automatica, Vol.14, pp. 413-428, 1978.

[2] R.K. Mehra, R. Rouhani, A. Rault, and J.G. Reid, "Model Algorithmic Control : Theoretical Results on Robustness," Proc. JACC, pp.387-392, 1979.

[3] R. Rouhani and R.K. Mehra, "Model Algorithmic Control(MAC) : Basic Theoretical Properties," Automatica, Vol.18, no.4, pp.401-414, 1982.

[4] J.G. Reid, D.E. Chaffin, and J.T.Silverthorn, "Output Predictive Algorithmic Control : Precision Tracking with Application to Terrain Following," J.Guidance and Control, Vol.4, no.5, pp.502-509, Sept.-Oct., 1981.

[5] J.G. Reid and S. Mahmood, "A State Space Analysis of The Stability and Robustness Properties of Model Predictive Control (MPC)," Proc. ACC, WP1, pp.335-338, 1986.

[6] R.K. Mehra, R. Rouhani, and L. Praly, "New Theoretical Development in Multivariable Predictive Algorithmic Control," Proc. JACC, FA9-B, 1980.

[7] R.K. Mehra and R.Rouhani, "Theoretical Considerations on Model Algorithmic Control for Non-minimum Phase Systems," Proc.JACC, FA8-B, 1980.

[8] D.L. Little and T.F. Edgar, "Predictive Control Using Constrained Optimal Control," Proc.ACC, FA1, pp.1365-1371, 1986.

[9] F.Lebourgeois, "IDCOM Application and Experiences on a PVC Production Plant," Proc. JACC, San Francisco, FA9-C, 1980.

[10] J.C.Engrand, "Applications of Multivariable Control in a Refinery and Implementation on a Dedicated Minicomputer," Proc. JACC, San Francisco, FA9-D, 1980.

[11] M. Lecrique, A. Rault, M.Tessier, and J.L. Testud, "Multivariable Regulation of A Thermal Power Plant Steam Generator," IFAC World Congress, Helsinki, 1978

[12] P.M. Bruijn, H.B. Verbruggen, "Model Algorithmic Control Using Impulse Response Models," Journal A., Vol.25, no.2, pp.69-74, 1984.

[13] P.M. Bruijn, L.J. Bootsma, and H.B. Verbruggen, "Predictive Control Using Impulse Response Models," IFAC Symposium on Digital Computer Applications to Process Control, Dusseldorf, 1980.

[14] C.R. Cutler and B.L. Ramaker, "Dynamic Matrix Control - A Computer Control Algorithm," Proc. JACC, San Francisco, WP5-B, 1980.

[15] R.H. Luecke, J.C. Lewis, H.Y. Lin, and W.K. Yoon, "Dynamic Matrix Control of A Batch Distillation Column," Proc. ACC, WA7, pp.209-213, 1985.

[16] C.R. Cutler, "Dynamic Matrix Control of Imbalanced Systems," ISA Trans., Vol.21, no.1, pp.1-6, 1982.

[17] P.R. Maurach, D.A. Mellichamp, and D.E. Seborg, "Predictive Controller Design for SISO Systems," Proc. ACC, FP4, pp.1546-1552, 1985.

[18] A.M. Morshedi, C.R.Cutler, and T.A. Skrovanek, "Optimal Solution of Dynamic Matrix Control with Linear Programming Techniques (LDMC)," Proc. ACC, WA7, pp.199-208, 1985.

[19] Pu Yuan and D.E. Seborg, "Predictive Control Using an Observer for Load Estimation," Proc.ACC, TA1, pp.669-675, 1986.

[20] P.R. Maurath, D.E. Seborg, and D.A. Mellichamp, "Predictive Controller Design by Principal Components Analysis," Proc. ACC, TP7, pp.1059-1065, 1985.

[21] B.E. Ydstie, "Extended Horizon Adaptive Control," Proc. IFAC World Congress, Budapest, Hungary, 1984.

[22] B.E. Ydstie and L.K. Liu, "Single and Multivariable Control with Extended Prediction Horizons," Proc. ACC, FA4, pp.1303-1308, 1984.

[23] B.E.Ydstie, L.S.Kershenbaum, and R.W.H.Sargent, "Theory and Application of an Extended Horizon Self-Tuning Controller," AICHE J, Vol.31, no.11, pp.1771-1780, NOV. 1985.

[24] C. Mohtadi and D.W. Clarke, "Generalized Predictive Control, LQ, or Pole-placement : A Unified Approach," Proc. 25th Conf. on Decision and Control, FA-4, pp.1536-1541, Dec. 1986.

[25] D.W. Clarke, C.Mohtadi, and P.S.Tuffs, "Generalized Predictive Control - Part 1. Basic Algorithm," Automatica, Vol.23, no.2, pp.137-148, 1987.

[26] D.W. Clarke, C.Mohtadi, and P.S.Tuffs, "Generalized Predictive Control - Part 2. Extensions and Interpretations," Automatica, Vol.23, no.2, pp.149-160, 1987.

[27] J.M.Dion, L.Dugard, and N.M.Tri, "Multivariable Adaptive Control with Input-Output Constraints," Proc.CDC, TP1, pp.1233-1238, 1987.

[28] W.H. Kwon, D.G. Byun, and O.K. Kwon, "Receding Horizon Tracking Control as a Predictive Control and Its Stability Results," to appear in 1988 ACC, Atlanta, U.S.A.

[29] W.H. Kwon and A.E. Pearson, "A Modified Quadratic Cost Problem and Feedback Stabilization of a Linear System," IEEE Trans. AC., Vol.22, pp.838-842, Oct. 1977.

[30] W.H. Kwon and A.E. Pearson, "On Feedback Stabilization of Time Varying Discrete Time Systems," IEEE Trans. AC., Vol.23, pp.479-481, Jun. 1978.

[31] W.H. Kwon, A.M. Bruckstein, and T.Kailath, "Stabilizing State Feedback Design via The Moving Horizon Method," Int.J.Control, Vol.37, no.3, pp.631-643, 1983.

[32] D.L.Kleinman, "Stabilizing a Discrete, Constant, Linear System with Application to Iterative Methods for Solving the Riccati Equation," IEEE Trans. AC, pp.252-254, Jun. 1974.

[33] M. Tomizuka and D.E. Whitney, "Optimal Discrete Finite Preview Problems (Why and How is Future Information Important ?)," J. of Dynamic System, Measurement, and Control, pp.310-325, Dec. 1975

[34] T. Katayama, T. Ohki, T. Inoue, and T. Kato, "Design of an Optimal Controller for a Discrete-Time System Subject to Previewable Demand," Int.J.Control, Vol.41, no.3, pp.677-699, 1985.
[35] M. Tomizuka, "Optimal Continuous Finite Preview Problem," IEEE Trans. AC., pp.362-365, Jun. 1975.
[36] R.M.C. De Keyser and A.V. Cauwenberghe, "Towards Robust Adaptive Control with Extended Predictive Control," Proc. CDC, FA4, pp.1548-1549, Dec. 1986.
[37] A.R.V. Cauwenberghe and R.M.C. De Keyser, "Self-Adaptive Long Range Predictive Control," Proc. ACC, TP10, pp.1155-1160, 1985.
[38] C. Greco, G.Menga, E. Mosca, and G. Zappa, "Performance Improvement of Self-Tuning Controllers by Multistep Horizons : The MUSMAR Approach," Automatica, Vol.20, no.5, pp.681-699, 1984.
[39] E. Mosca, G.Zappa, and C.Manfredi, "Multistep Horizon Self-Tuning Controllers : The MUSMAR Approach," IFAC 9th World Congress, 14.4/F-2, pp.155-159, Budapest, Hungary, 1984.
[40] C.E.Garcia and M.Morari, "Internal Model Control 1. A Unifying Review and Some New Results," Ind.Eng.Chem.Proc.Des.Dev., Vol.21, pp.308-323, 1982.
[41] C.E.Garcia and M.Morari, "Internal Model Control 2. Design Procedure for Multivariable Systems," Ind.Eng.Chem.Proc.Des.Dev., Vol.24, pp.472-484, 1985.
[42] C.E.Garcia and M.Morari, "Internal Model Control 3 Multivariable Control Law Computation and Tuning Guidelines," Ind.Eng.Chem.Proc.Des.Dev., Vol.24, pp.484-494, 1985.
[43] D.M.Prett and C.E.Garcia, "Design of Robust Process Controllers," IFAC 10th Congress on AC, pp.291-296, 1987.
[44] N.L.Ricker, "Use of Quadratic Programming for Constrained Internal Model Control," Ind.Eng.Chem.Proc.Des.Dev., Vol.24, pp.925-936, 1985.
[45] E.Zafiriou and M.Morari, "Setpoint Tracking vs. Disturbance Rejection for Stable and Unstable Processes," Proc.ACC, WP8, pp.649-651, 1987.
[46] P.J.Campo and M.Morari, "Robust Model Predictive Control ," Proc.ACC, TA8, pp.1021-1026, 1987.
[47] D.Marques and M.Morari, "Model Predictive Control of Gas Pipeline Networks," Proc.ACC, WP1, pp.349-354, 1986.
[48] D.W.Clarke and C.Mohtadi, "Properties of Generalized Predictive Control," IFAC 10th Congress on AC, Vol.10, 1987.
[49] P.M. Bruijn, H.B. Verbruggen, and O.V. Appeldoorn, "Predictive Control : A Comparison and Simple Implementation," IFAC Low Cost Automation, pp.271-276, Spain, 1986.
[50] D.W. Clarke and L.Zhang, "Long-Range Predictive Control Using Weighting-Sequence Models," IEE Proc., Vol.34, Pt.D, no.3, pp.187-195, May 1987.
[51] G.D. Martin, "Long-Range Predictive Control," AICHE J., Vol.27, no.5, pp.748-753, Sep. 1981.
[52] C.Samson, "An Adaptive LQ Controller for Non-Minimum-Phase Systems," I.J.Control, Vol.35, no.1, pp.1-28, 1982.
[53] K.P.Lam, "Design of Stochastic Discrete Time Linear Optimal Regulators Part 1. Relationship Between Control Laws Based on a Time Series Approach and a State Space Approach," Int.J.Systems Sci., Vol.13, no.9, pp.979-1000, 1982.
[54] K.P.Lam, "Design of Stochastic Discrete Time Linear Optimal Regulators Part 1. Extension and Computational Procedures," Int.J.Systems Sci., Vol.13, no.9, pp.1001-1011, 1982.
[55] J. Rodellar, J.M. Martin-Sanches, and A.H. Barbat, "Driver Block Design and Applications to Structural Control," Proc. CDC, FA3, pp.1529-1535, Dec. 1986.
[56] T.F.Edgar, "Current Problems in Process Control," IEEE Control Systems Magazine, Vol.7,no.2, Apr.1987.
[57] D.W.Clarke, P.P.Kanjilal, and C.Mohatadi, "A Generalized LQG Approach to Self-Tuning Control Part 1. Aspects of Design," I.J.Control, Vol.41,no.6, pp.1509-1523, 1985.
[58] D.W.Clarke, P.P.Kanjilal, and C.Mohatadi, "A Generalized LQG Approach to Self-Tuning Control Part 1. Implementation and Simulation," I.J.Control, Vol.41,no.6, pp.1525-1544, 1985.
[59] D.W.Clarke, A.J.F.Hodgson, P.S.Tuffs, "Offset Problem and k-Incremental Predictors in Self-Tuning Control," IEE Proc., Vol.130, Pt.D, no.5, Sept. 1983.
[60] P.S.Tuffs and D.W.Clarke, "Self-Tuning Control of Offset: a Unified Approach," IEE Proc., Vol.132, Pt.D, no.3, May 1985.

ROBUSTNESS AND TUNING OF ON-LINE OPTIMIZING CONTROL ALGORITHMS

E. Zafiriou

Chemical and Nuclear Engineering Department, and Systems Research Center, University of Maryland, College Park, MD 20742, USA

Abstract

A significant number of Model Based Process Control algorithms solve on-line an appropriate optimization problem and do so at every sampling point. The major attraction of such algorithms, like the Quadratic Dynamic Matrix Control (QDMC), lies in the fact that they can handle static nonlinearities in the form of hard constraints on the inputs (manipulated variables) of a process. The presence of such constraints as well as additional performance or safety induced hard constraints on certain outputs or states of the process, result in an on-line optimization problem that produces a nonlinear controller, even when the plant and model dynamics are assumed linear. This paper provides a theoretical framework within which the stability and performance properties of such algorithms can be studied.

1 Introduction

The problem of input saturation is of extreme importance for process control applications, because of its presence in almost every chemical system, even when the process dynamics can be assumed linear. In addition to the input constraints, safety and certain performance specifications also require the presence of hard constraints on some output and state variables. The urgency of rigorous theoretical work in this area has been repeatedly pointed out by the industry (e.g., [1]). An approach that has been tried in the chemical industry during the past few years is to on-line solve an appropriate optimization problem and to do so at every sampling point. The repeated application of such methods (e.g., Quadratic Dynamic Matrix Control (QDMC) [2] on industrial problems with considerable success indicate that sufficient degrees of freedom exist in these formulations. A drawback that has prohibited their widespread use is the fact that no exact tuning procedure for the optimization parameters exist and such tuning often has to be carried out on-line by experienced designers.

The presence of hard constraints in the on-line optimization problem produces a nonlinear controller even when the plant and model dynamics are assumed linear. The fact that the overall control system (plant + controller) is nonlinear makes the study of its properties quite involved, especially since no analytic expression is available for the controller. The problems are compounded when robustness with respect to model-plant mismatch is also considered, because no straighforward extension of the results of the Robust Linear Control Theory to this particular problem exists, even though the plant and model dynamics are assumed linear. Some efforts have been made recently [1,3] to achieve robustness by modifying the "min" optimization problem that is solved on-line to a "min max" problem that minimizes the objective function over all possible plants. One of the problems of this approach is that either the computations for solving the optimization problem are too time consuming to be carried out on-line at every sample point or to simplify the computations one has to use simplistic model uncertainty descriptions that are unrealistic. Another, potentially serious problem is the fact that these methods inherently assume that by solving the "min max" problem to obtain a sequence of future inputs (manipulated variables) and then implementing the first one and repeating the computation at the next sample point, one is guaranteed robust stability and performance, provided that a sufficiently long horizon is used in the objective function. However, feedback from an uncertain plant exists in reality and it is not taken into account in the formulation of the optimization problem, which is an open-loop minimization of the objective function over all possible plants. This fact can conceivably lead to performance deterioration and instability. Note that the situation is quite different from studying (and guaranteeing) a stabilizing control algorithm when no model error is present, in which case the assumption is reasonable, although not proven for the general case.

The problems discussed just above, cannot possibly be satisfactorily addressed without considering the problem in its proper nonlinear framework. It is the author's opinion that instead of augmenting the objective functions to add robustness, an action that dramatically increases the computational load and at the same time produces no rigorous robustness guarantees, one should study the problem in its nonlinear nature, obtain conditions that guarantee nominal and robust stability and performance and tune the parameters of the original optimization problems (e.g., QDMC) to satisfy them.

2 The On-Line Optimization Problem

Although control algorithms of the type described in Section 1 have been applied to systems with nonlinear dynamic models (QDMC [4]), it is usually assumed that the dynamics are linear, the nonlinearity of the problem arising from the hard constraints. The properties of the controller are independent of the type of model description used for the plant (see, e.g., [5]). The impulse response description is a convenient one:

$$y(k+1) = H_1 u(k) + H_2 u(k-1) + \ldots + H_N u(k-N+1) \quad (1)$$

where y is the output vector, u is the input vector and N is an integer sufficiently large for the effect of inputs more than N sample points in the past on y to be negligible.

The QDMC-type algorithms [2,6,7,5] use a quadratic objective function that includes the square of the weighted norm of the predicted error (setpoint - predicted output) over a finite horizon in the future as well as penalty terms on u or Δu. The minimization of the objective function is carried out over the values of $u(\bar{k})$, $u(\bar{k}+1)$, ..., $u(\bar{k}+M-1)$, where $\bar{k}$ is the current sample point and M a specified parameter. The minimization is subject to possible hard constraints on the inputs u, their rate of change Δu, the outputs y and other process variables usually referred to as associated variables. More details on the formulation of the optimization problem can be found in the cited references. After the problem is solved on-line at $\bar{k}$, only the optimal value for the first input vector $u(\bar{k})$ is implemented and the problem is solved again at $\bar{k}+1$. The optimal $u(\bar{k})$ depends on the tuning parameters of the optimization problem, the current output measurement $y(\bar{k})$ and the past inputs $u(\bar{k}-1)$,..., $u(\bar{k}-N)$ that are involved in the model output prediction. Let f describe the result of the optimization:

$$u(k) = f(y(k), u(k-1), \ldots, u(k-N)) \tag{2}$$

The optimization problem of the QDMC-type algorithms can be written as a standard Quadratic Programming problem [2]:

$$\min_v q(v) = \frac{1}{2} v^T G v + g^T v \tag{3}$$

subject to

$$A^T v = b \tag{4}$$

where

$$v = [\ u(\bar{k})\ \ \ldots\ \ u(\bar{k}+M-1)\ ]^T \tag{5}$$

and the matrices G, A, and vectors g, b are functions of the tuning parameters (weights, horizon, M). The vectors g, b are also linear functions of $y(\bar{k})$, $u(\bar{k}-1)$,..., $u(\bar{k}-N)$. For the optimal solution v^* we have [8]:

$$\begin{bmatrix} G & -\hat{A} \\ -\hat{A}^T & 0 \end{bmatrix} \begin{bmatrix} v^* \\ \lambda^* \end{bmatrix} = - \begin{bmatrix} g \\ b \end{bmatrix} \tag{6}$$

where $\hat{A}$ consists of the rows of A that correspond to the constraints that are active at the optimum and λ^* is the vector of the Lagrange multipliers. The optimal $u(\bar{k})$, descibed by (2), corresponds to the first m elements of the v^* that satisfies (6), where m is the dimension of u.

3 Formulation of the Problem as a Contraction Mapping

The framework selected for the study of the properties of the overall nonlinear system is that of the Operator Control Theory [9]. In this approach, the stability and performance of the nonlinear system can be studied by applying the contraction mapping principle on the operator F that maps the "state" of the system (plant + controller) at sample point k to that at sample point $k+1$. The fact that the plant dynamics are assumed linear allows us to obtain results and carry out computations that are not yet feasible in the general case. We can define as the "state" of the system at sample point k the following vector

$$x(k) = \begin{bmatrix} x_1(k) \\ \vdots \\ x_N(k) \end{bmatrix} \tag{7}$$

where

$$\begin{aligned} x_1(k+1) \overset{\text{def}}{=} u(k) \quad &= f(y(k), u(k-1), \ldots, u(k-N)) \\ &= f(H_1 u(k-1) + \ldots + H_N u(k-N), u(k-1), \ldots, u(k-N)) \\ &\overset{\text{def}}{=} \Psi(u(k-1), \ldots, u(k-N)) \\ &= \Psi(x(k)) \\ x_2(k+1) \overset{\text{def}}{=} u(k-1) \quad &= x_1(k) \\ \vdots \qquad\qquad &\quad \vdots \\ x_N(k+1) \overset{\text{def}}{=} u(k-N+1) \quad &= x_{N-1}(k) \end{aligned} \tag{8}$$

The "state" vector $x(k)$ is defined so that knowledge of it allows the computation of $x(k+1)$ by applying the plant and controller equations on it. Indeed the operator F that maps $x(k)$ to $x(k+1)$ is given by

$$x(k+1) = F(x(k)) = \begin{bmatrix} \Psi(x(k)) \\ x_1(k) \\ \vdots \\ x_{N-1}(k) \end{bmatrix} \tag{9}$$

Note, however, that although f is known, since it describes the on-line optimizing control algorithm and it involves only the process model, Ψ is not exactly known, because it involves the "true" plant impulse response coefficients H_1,..., H_N.

Convergence of the successive substitution $x(k+1) = F(x(k))$ to the unique fixed point of the contraction implies stability of the overall nonlinear system; fast convergence implies good performance. The use of the contraction mapping principle allows the development of conditions for robust stability and performance in terms of some induced matrix norm of the derivative F' of the above operator F.

4 Stability Conditions

We shall now proceed to obtain stability conditions for the overall nonlinear system by obtaining conditions under which the mapping described by F is a contraction. The terms stability and instability of the control system are used in the global sense over the domain of F under consideration.

Let us first examine the differentiability of F. From (9) it follows that this is equivalent to differentiability of $\Psi(x)$ and from (8) to differentiability of f. Let us assume that for some point x in the domain of F, an infinitesimal change in x (which results in a change of g, b in (3), (4)) does not change $\hat{A}$, i.e., the set of active constraints at the optimum does not change (note that A is independent of x). Then from (6) it follows that the derivative of Ψ exists and it has a constant value in a neighbourhood of that x.

Let J_i be a set of indices for the active constraints of (3) and J_1,..., J_n correspond to all possible active sets of constraints when all xs in the domain of F are considered. Every such J_i corresponds to an $\hat{A}_i$. Then from the above discussion, it is evident that for all xs that correspond to the same J_i and for which an infinitecimal change in their value does not change the set of active constraints, the derivative of Ψ and therefore of F exist and it has the same value that depends on the particular set J_i:

$$F'_{J_i} = \begin{bmatrix} (\nabla_{x_1}\Psi)_{J_i} & (\nabla_{x_2}\Psi)_{J_i} & \cdots & (\nabla_{x_{N-1}}\Psi)_{J_i} & (\nabla_{x_N}\Psi)_{J_i} \\ I & 0 & \cdots & 0 & 0 \\ 0 & I & \cdots & 0 & 0 \\ \vdots & \vdots & \ddots & \vdots & \vdots \\ 0 & 0 & \cdots & I & 0 \end{bmatrix} \tag{10}$$

where from (8) it follows that

$$(\nabla_{x_j}\Psi)_{J_i} = (\nabla_{x_j} f)_{J_i} + (\nabla_y f)_{J_i} H_j \tag{11}$$

It is clear from the above discussion that $F(x)$ is quasi-linear and that it is differentiable everywhere except the points where an infinitecimal change will change the set of active constraints at the optimum of (3). It follows then that for F to be a contraction, it is necessary that

$$||F'_{J_i}|| \leq \theta < 1, \quad i = 1, \ldots, n \tag{12}$$

where $||.||$ is any consistent matrix norm[1], the same for all i. The above condition however, can be shown to be sufficient as well. Consider two points x^a, x^b and let the straight path connecting them in the domain of F be broken into the succesive segments $x^a \rightarrow x^1$, $x^1 \rightarrow x^2$,..., $x^l \rightarrow x^b$, the points of each of which correspond to the same J_i: J_{k_0}, J_{k_1},..., J_{k_l}, respectively. Then

$$\begin{aligned} ||F(x^a) \; &- \; F(x^b)|| \\ &= ||(F(x^a) - F(x^1)) + (F(x^1) - F(x^2)) \\ &\quad + \ldots + (F(x^l) - F(x^b))|| \\ &= ||F'_{J_{k_0}}(x^a - x^1) + F'_{J_{k_1}}(x^1 - x^2) \\ &\quad + \ldots + F'_{J_{k_l}}(x^l - x^b)|| \\ &= ||(a_0 F'_{J_{k_0}} + a_1 F'_{J_{k_1}} + \ldots + a_l F'_{J_{k_l}})(x^a - x^b)|| \\ &\leq (a_0 + a_1 + \ldots + a_l)\theta ||x^a - x^b|| \\ &= \theta ||x^a - x^b|| \end{aligned} \tag{13}$$

where a_j is the relative length of the respective segment as compared to $x^a \rightarrow x^b$. From (13) it follows that F is a contraction. The fact that there is only a finite number of J_is allows us to drop the θ from (12) to obtain:

Theorem 1 *F is a contraction if and only if there exists a consistent matrix norm $||.||$, for which*

$$||F'_{J_i}|| < 1, \quad i = 1, \ldots, n \tag{14}$$

The practical use of (14) is limited by the fact that finding an appropriate consistent norm is not a trivial task. The following two subsections provide conditions which are more readily computable. The third subsection formulates the respective robustness conditions.

4.1 Sufficient Condition

By selecting one particular consistent matrix norm and stating (14) for that norm, one can get a sufficient only condition.

Let us select the following norm, which can be shown to be a consistent one on $\mathbf{R}^{mN \times mN}$[10], where m is the plant dimension:

$$||A|| = ||DAD^{-1}||_\infty \tag{15}$$

where

$$||B||_\infty = \max_i \sum_{j=1}^{N} |b_{ij}| \tag{16}$$

$$D = \text{diag}(I, \eta I, \eta^2 I, \ldots, \eta^{N-1} I) \tag{17}$$

$$0 < \eta < 1 \tag{18}$$

Then

$$DF'_{J_i}D^{-1} = \left[\begin{array}{ccccc} (\nabla_{x_1}\Psi)_{J_i} & (\nabla_{x_2}\Psi)_{J_i}\eta^{-1} & \ldots & (\nabla_{x_{N-1}}\Psi)_{J_i}\eta^{-(N-2)} & (\nabla_{x_N}\Psi)_{J_i}\eta^{-(N-1)} \\ \eta I & 0 & \ldots & 0 & 0 \\ 0 & \eta I & \ldots & 0 & 0 \\ \vdots & \vdots & \ddots & \vdots & \vdots \\ 0 & 0 & \ldots & \eta I & 0 \end{array}\right] \tag{19}$$

From (15)–(19) we get

$$\begin{aligned} & ||DF'_{J_i}D^{-1}||_\infty < 1 \\ \Leftrightarrow \; & || \, (\nabla_{x_1}\Psi)_{J_i} \;\; (\nabla_{x_2}\Psi)_{J_i}\eta^{-1} \; \ldots \; (\nabla_{x_N}\Psi)_{J_i}\eta^{-(N-1)} \, ||_\infty \\ & < 1 \\ \Leftarrow \; & || \, (\nabla_{x_1}\Psi)_{J_i} \;\; (\nabla_{x_2}\Psi)_{J_i} \; \ldots \; (\nabla_{x_N}\Psi)_{J_i} \, ||_\infty < \eta^{N-1} \end{aligned} \tag{20}$$

Since any η in $(0, 1)$ will do and there is only a finite number of J_is, from (20) we can obtain:

Theorem 2 *The control system is asymptotically stable if*

$$|| \, (\nabla_{x_1}\Psi)_{J_i} \;\; (\nabla_{x_2}\Psi)_{J_i} \; \ldots \; (\nabla_{x_N}\Psi)_{J_i} \, ||_\infty < 1, \; i = 1, ..., n \tag{21}$$

Note that for single-input single-output plants (21) becomes

$$\sum_{j=1}^{N} \left|\frac{\partial \Psi_{J_i}}{\partial x_j}\right| < 1, \quad i = 1, \ldots, n \tag{22}$$

which for the unconstrained case is simply a sufficient condition for the closed-loop poles to lie inside the Unit Circle.

4.2 Instability Conditions

For every consistent matrix norm we have

$$\rho(A) \leq ||A|| \tag{23}$$

where $\rho(A)$ is the spectral radius of A, defined as $\rho(A) = \max_j |\lambda_j(A)|$, $\lambda_j(A)$, being the eigenvalues of A. Then from (14) and (23) we get

Theorem 3 *F is a contraction only if*

$$\rho(F'_{J_i}) < 1, \quad i = 1, \ldots, n \tag{24}$$

Note that if the optimization (3) is not subject to (4), then $n = 1$ and (24) becomes sufficient as well, because, given a matrix one can always find a consistent norm arbitrarily close to its spectral radius [10]. The reason that (24) is not sufficient in general is that such a consistent norm is in general a different one for two different matrices (different J_is), while (14) requires the same norm for all i. In the case of $n = 1$, (24) translates to the requirement that the closed-loop poles of the system are located inside the Unit Circle.

If (24) is not true, then F is not a contraction. This however does not necessarily imply that the control system is unstable. The following theorem provides a condition that is sufficient for instability.

Theorem 4 *The control system is unstable if*

$$\rho(F'_{J_i}) > 1, \quad i = 1, \ldots, n \tag{25}$$

The proof follows the argument that if a stable local equilibrium point existed, then for the J_i corresponding to that point we would have $\rho(F'_{J_i}) < 1$.

[1] A consistent matrix norm has the property $||AB|| \leq ||A|| \, ||B||$.

Theorem 4 can be used to predict instability of the overall nonlinear system. Theorem 3 on the other hand does not seem at a first glance to be of much use, since violation of (24) does not necessarily imply instability. From a practical point of view, however, violation of that condition for some i, should be taken as a very serious warning that the control system parameters should be modified. The reason is that when in the region of the domain of F that corresponds to that i, the system will behave as a virtually unstable system, the only hope for stability being to move to a region with $\rho(F'_{J_i}) < 1$. It might be the case that for a particular system in question this will always happen, making this system a stable one. But even in this case, a temporary unstable-like behavior might occur, thus making the control algorithm practically unacceptable. The example in Section 5 demonstrates a situation where violation of (24) is enough to produce an unstable system although (25) is not satisfied.

4.3 Robustness Conditions

From (11) we see that F'_{J_i} depends on the impulse response coefficient matrices $H_1,\ldots, H_N$ of the actual plant. These matrices are never known exactly and so in order to guarantee stability for the actual plant, one has to compute the conditions of Sections 4.2, 4.1, not just for the model, but for all possible plants. To do so, one needs to have some information on the possible modeling error associated with the H_is. Let $\mathcal{H}$ be the set of possible values for these coefficients. Then we can write the following conditions:

Theorem 5 *The control system is asymptotically stable for all plants with coefficients in $\mathcal{H}$ if*

$$\sup_{\mathcal{H}} \| \; (\nabla_{x_1}\Psi)_{J_i} \;\; (\nabla_{x_2}\Psi)_{J_i} \;\; \cdots \;\; (\nabla_{x_N}\Psi)_{J_i} \; \|_\infty < 1, \quad i = 1,\ldots,n \tag{26}$$

Theorem 6 *F is a contraction for all plants with coefficients in $\mathcal{H}$ only if*

$$\sup_{\mathcal{H}} \rho(F'_{J_i}) < 1, \quad i = 1,\ldots,n \tag{27}$$

In order to carry out the maximizations over $\mathcal{H}$ described by (27), (26), one needs to parametrize the "uncertain" $H_1,\ldots, H_N$, in terms of a fewer "uncertain" parameters. For example, in the simple case where the linear plant dynamics are described by the transfer function $\frac{Ke^{-sd}}{\tau s+1}$, where K, d, τ, are within some ranges, we can write $H_1,\ldots, H_N$, as functions of K, d, τ, and compute $\sup_{\mathcal{H}}$ as $\sup_{K,d,\tau}$. However, the situation is usually more complex, a fact that makes the efficient parametrization of the modeling error in $H_1,\ldots, H_N$, a very important research topic.

5 Illustration

Let us consider a system with the following transfer function:

$$P(s) = \begin{bmatrix} \frac{1}{s+1} & 0 \\ \frac{-2e^{-5s}}{s+1} & \frac{-s+2}{(s+2)(s+1)} \end{bmatrix} \tag{28}$$

A sampling time $T = 0.5$ is used and the following objective function is minimized on-line:

$$\min_{u(\bar{k}),\ldots,u(\bar{k}+M-1)} \sum_{l=1}^{P} \left[e(\bar{k}+l)^T\Gamma^2 e(\bar{k}+l) + u(\bar{k}+l-1)^T B^2 u(\bar{k}+l-1) \right] \tag{29}$$

where $\bar{k}$ is the current sample point, e is the predicted difference between the setpoints and the plant outputs and Γ, B, are weights.

$$\Gamma = \begin{bmatrix} 1 & 0 \\ 0 & 0.5 \end{bmatrix} \tag{30}$$

is selected signifying that the first output is more important than the second.

Let us first consider the unconstrained problem. First we select $P = M = 2$, which is a selection that is expected [6,7] to produce an unstable control system if $B = 0$. The reason is the right-half plane (RHP) zero of $P(s)$. Indeed, one can easily check that for these values of the tuning parameters,

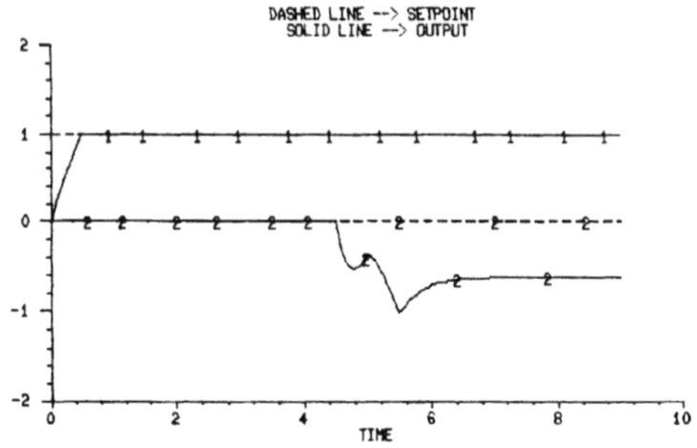

Figure 1: Unconstrained minimization.

we have $\rho(F'_{J_1}) > 1$, where J_1 corresponds to the case where no constraints are active at the optimum. Hence the necessary condition (24) predicts the instability. From theory [7] we know that by making B sufficiently large, we can stabilize the system. Indeed by making

$$B = \begin{bmatrix} 0 & 0 \\ 0 & 0.1 \end{bmatrix} \tag{31}$$

the system is stabilized ($\rho(F'_{J_1}) < 1$, which is sufficient for $n = 1$). The fact that the RHP zero is pinned to the second plant output, made it unnecessary to increase the 11 element of B. The response to a unit step change in setpoint 1 is shown in Fig. 1. The steady-state offset in output 2 is expected from theory and can be avoided by modifying the control algorithm, but we will not do so to avoid the unnecessary complication of the example.

Let us now assume that after looking at the response, the designer decides that a slight tightening of the specifications is in order, namely the addition in the optimization problem

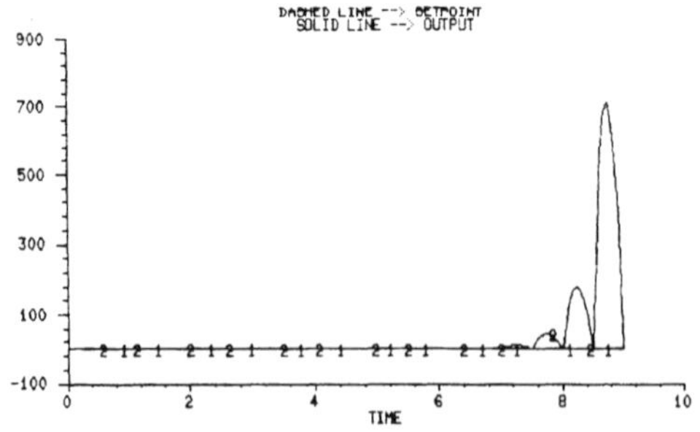

Figure 2: Minimization subject to lower bound constraint on output 2.

of a lower bound on output 2 at the value -0.9. Since output 2 only slightly violated this bound when the unconstrained algorithm was used, one might think that the response for the constrained algorithm should be almost the same as that in Fig. 1. This is not so, however. The response for the same setpoint change is shown in Fig. 2. The system is unstable. An instability warning was issued by the necessary condition for F to be a contraction (24), since $\rho(F'_{J_2}) > 1$, where J_2 corresponds to the case where the low constraint on output 2 is active at the optimum. Indeed by looking at a close-up of Fig. 2 in Fig. 3, we see that the system went unstable as soon as output 2 reached the low bound to which the on-line minimization was subject. The constraint remained active at the subsequent sample points and the system never stabilized.

A question that one may ask at this point is whether the

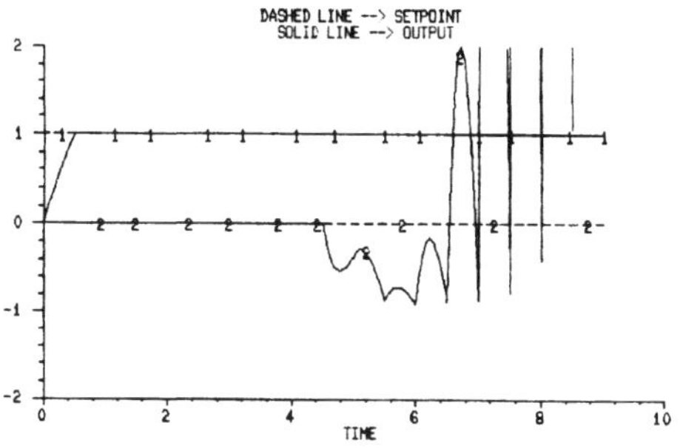

Figure 3: Close-up of Fig. 2.

use of a

$$B = \begin{bmatrix} 0 & 0 \\ 0 & \beta \end{bmatrix} \tag{32}$$

with a β larger than the previously used value of 0.1, will stabilize the system. We know that this would be the case for the unconstrained problem; however, for the constrained case that does not happen. By examining the analytic expression for F'_{J_2} one sees that β does not even appear in it and can therefore in no way influence the stability of the system when the constraint becomes active. When the constraint is reached, the algorithm puts as its higher priority keeping output 2 above the lower bound and to do so it inverts the 22 element of $P(s)$ and causes instability.

6 Conclusions

The main goal of this paper was to provide a theoretical framework for the study of the properties of control algorithms that are based on the on-line minimization of some objective function, subject to certain hard constraints. The selected framework seems to be quite promising since it allowed the derivation of necessary and/or sufficient conditions for nominal and robust stability of the overall nonlinear system. The simple example that was used demonstrated in a clear way that one cannot afford to neglect the nonlinear phenomena caused by the hard constraints to which the on-line optimization is subject. This example also indicates that inclusion of hard constraints on the plant outputs in the specifications can cause serious problems.

Future work in this framework should address the issue of performance, possibly by minimizing the contraction constant. Also the sufficient condition (21) should be used to obtain tuning guidelines for the optimization parameters P, M, Γ, B, so that stability is achieved for general or special cases.

Acknowledgements

Support from Shell Development Co. is gratefully acknowledged. The author wishes to thank Dr. C. E. Garcia for several useful discussions. The control software package CONSYD, developed at Caltech (Dr. M. Morari' group) and the University of Wisconsin (Dr. W. H. Ray's group), was used in the simulations.

References

[1] C. E. Garcia and D. M. Prett, "Advances in Industrial Model Predictive Control", Chemical Process Control Conf. III, Asilomar CA, 1986.

[2] C. E. Garcia, A. M. Morshedi and T. J. Fitzpatric, "Quadratic Programming Solution of Dynamic Matrix Control (QDMC)", Amer. Control Conf., San Diego CA, 1984.

[3] P. J. Campo and M. Morari, "Robust Model Predictive Control", Proc. Amer. Control Conf., p.1021, Minneapolis MN, 1987.

[4] C. E. Garcia, "Quadratic Dynamic Matrix Control of Nonlinear Processes: An Application to a Batch Reaction Process", AIChE Ann. Mtg., San Francisco CA, 1984.

[5] M. Morari, C. E. Garcia and D. M. Prett,"Model Predictive Control: Theory and Practice", IFAC Workshop on Model-Based Process Control, Atlanta GA, 1988.

[6] C. E. Garcia and M. Morari, "Internal Model Control. 1. A Unifying Review and Some New Results", Ind. Eng. Chem. Process Des. Dev., **21**, 308-323, 1982.

[7] C. E. Garcia and M. Morari, "Internal Model Control. 3. Multivariable Control Law Computation and Tuning Guidelines", Ind. Eng. Chem. Process Des. Dev., **24**, 484-494, 1985.

[8] R. Fletcher, *Practical Methods of Optimization; vol. 2: Constrained Optimization*, John Wiley and Sons, 1981.

[9] C. G. Economou, *An Operator Theory Approach to Nonlinear Controller Design*, Ph.D. Thesis, Caltech, 1985.

[10] G. C. Stewart, *Introduction to Matrix Computations*, NewYork: Academic Press, 1973.

APPLICATION OF SINGULAR VALUE METHODS FOR IDENTIFICATION AND MODEL BASED CONTROL OF DISTRIBUTED PARAMETER SYSTEMS

D. H. Gay and W. H. Ray

Department of Chemical Engineering, University of Wisconsin, Madison, WI 53706, USA

Abstract—In this paper the basic properties of the Singular Value Decomposition (SVD) for integral equation models of Distributed Parameter Systems (DPS) are presented in the context of process identification and model-based control. In addition, new methods of analysis and computation are described in which SVD-based techniques are used to provide a solution for the DPS identification and model-based control problem. Once developed, these procedures are applied to a class of linear DPS which are conventionally described by parabolic partial differential equation (PDE) models. An alternative integral equation representation for such processes provides a general and readily identifiable pseudo-modal input/output model. It is shown that this SVD-based integral equation model for DPS provides a sound basis for design of both a pseudo-modal feedback controller and a model-predictive dynamic matrix controller. Although theoretically developed for linear systems, these methods are successfully applied to nonlinear DPS by means of local linearization techniques.

Keywords—Distributed parameter systems; singular value analysis; modal control; model-predictive control; identification.

INTRODUCTION

This work is concerned with the problem of developing effective and readily implementable techniques for the identification and control of Distributed Parameter Systems—dynamic processes whose states, controls, and parameters are spatially varying. Examples of such processes include metallurgical heating, tubular reactors, packed bed reactors, absorption columns, and drying or curing operations. If the proper equations, parameters, and boundary conditions that make up the conventional PDE models of these processes are known, then there are a variety of DPS control system design procedures available. However, the specific equations and parameters for these distributed models are often either unknown, too complex to be tractable, or require sophisticated parameter and state estimation techniques. Furthermore, because of the inherent mathematical complexity of most PDE-based DPS control system design procedures, they have received only limited application.

In practice, the spatially distributed nature of such processes is generally overlooked or ignored, and conventional lumped control system design techniques are applied using approximate lumped models identified from input/output testing. Because these simple lumped models ignore the spatially varying nature of the DPS, they will often suffer from strong interactions and apparent time delays due to the underlying diffusion and convection phenomena inherent in such processes.

Therefore, there exists a need for an identification procedure that recovers a general DPS model which accounts for the spatially varying structure of the process and which is based on data obtained from simple input/output experiments. Use of integral equation models for DPS provides this desired general input/output model structure for a variety of different processes. Furthermore, use of the structure, properties, and techniques of the SVD generates a model which provides a firm basis for both modal feedback and model-predictive control strategies.

The methods presented in this paper represent an extension to those developed by Gay and Ray(1986), and are particularly relevant to the task of applying model based control strategies to DPS. The generality of the integral equation model formulation allows application of this analysis to a wide variety of spatially distributed processes in which the physical phenomena of convection and diffusion are significant. The readily implementable input/output identification procedure requires process information in a form which is similar to that needed by established predictive control techniques for lumped systems. The incorporation of additional information regarding the spatial locations of the system inputs and outputs into the model preserves the spatially distributed nature of the process. The application of SVD-based techniques presented in this work to this process information results in a model for the DPS with a naturally decoupled structure and an obvious scheme for model reduction. These inherent qualities provide a solid basis for an effective and reliable implementation of model based control to DPS.

DPS SINGULAR VALUE THEORY

In this section some basic results of singular value theory for integral equations will be presented. For the class of DPS which we shall initially investigate, we assume that the process can be represented by a linear time-invariant integral operator $\mathcal{K}$ which maps the distributed input $u(z,t)$ to the distributed output $y(z,t)$Both $u(z,t)$ and $y(z,t)$ are $\mathcal{L}^2$ (square integrable) functions defined on the region $\{z,t : 0 \le z \le 1, t \ge 0\}$. The general model for the DPS is expressed in terms of an integral equation with $\mathcal{L}^2$ kernel $k(z,\zeta,t-\tau)$:

$$y(z,t) = \int_0^t \int_0^1 k(z,\zeta,t-\tau)u(\zeta,t)\,d\zeta\,d\tau. \quad (1)$$

In operator notation, Eq. (1) can be compactly expressed as $y = \mathcal{K}u$. Because $\mathcal{K}$ is time-invariant, the Laplace transform of Eq. (1) with respect to t yields a transfer function representation of the DPS:

$$y(z,s) = \int_0^1 k(z,\zeta,s)u(\zeta,s)\,d\zeta. \quad (2)$$

Equation (2) is a Fredholm integral equation of the first kind, with a square-integrable kernel $k(z,\zeta,s)$ (in general unsymmetric and complex) which generates a linear integral operator $\mathcal{K}$. The integral operator $\mathcal{K}^*$, which is adjoint to $\mathcal{K}$, is defined by the adjoint kernel $k^*(z,\zeta,s) = \overline{k(\zeta,z,s)}$, where the overbar denotes the complex conjugate. If $\mathcal{K} = \mathcal{K}^*$, then $\mathcal{K}$ is symmetric (Hermitian), or self-adjoint. Less restrictively, if $\mathcal{K}\mathcal{K}^* = \mathcal{K}^*\mathcal{K}$, then $\mathcal{K}$ is a *normal* operator. Both Hermitian and normal operators possess a convergent bilinear expansion for the kernel in terms of a complete set of orthonormal eigenfunctions and their corresponding eigenvalues. For a Hermitian operator, these eigenvalues are always real and distinct, while a normal operator may have complex eigenvalues.

Singular Function Bases

It would be unduly restrictive to assume that an unknown DPS could be well approximated by an identification procedure based on the theory of normal operators. Instead, a more generally applicable model will be used, which extends some of the useful properties of symmetric operators, (e.g. orthonormal eigenfunctions, convergent expansions) to the general case. Such a model is based on singular value analysis of $\mathcal{K}$.

A detailed account of this theory, along with several important proofs referenced below, may be found in Smithies(1958) and Cochran(1972). This general model assumes only that the the kernel of $\mathcal{K}$ is a square-integrable function. For such an arbitrary kernel, if eigenfunctions do exist, they will not in general be orthogonal (in the sense of the usual $\mathcal{L}^2$ inner product), nor will they in general provide a convergent expansion. However, from the unsymmetric kernel of such an operator two Hermitian kernels may be constructed:

$$kk^*(z,\zeta,s) = \int_0^1 k(z,\xi,s)\overline{k(\zeta,\xi,s)}\,d\xi\ ; \quad (3)$$

$$k^*k(z,\zeta,s) = \int_0^1 \overline{k(\xi,z,s)}k(\xi,\zeta,s)\,d\xi. \quad (4)$$

By the symmetry of these two kernels, there exists for each a sequence of distinct real eigenvalues and associated orthonormal eigenfunctions (generally complex). In fact, it may be readily proven that these kernels share the same eigenvalues, $\{\sigma_i^2(s)\}$.

We denote by $\{w_i(z,s)\}$ the set of eigenfunctions for $kk^*(z,\zeta,s)$, and denote by $\{v_i(z,s)\}$ the set of eigenfunctions for $k^*k(z,\zeta,s)$. Given their mutual eigenvalues, it may be proven that these two sets of orthonormal functions are related through the kernel $k(z,\zeta,s)$ and the square roots of the eigenvalues:

$$\sigma_i(s)w_i(z,s) = \int_0^1 k(z,\zeta,s)v_i(\zeta,s)\,d\zeta. \quad (5)$$

The elements of $\{w_i(z,s)\}$ and $\{v_i(z,s)\}$ are known as the left and right singular functions of Eq. (2), corresponding to the singular values $\{\sigma_i(s)\}$. The singular functions satisfy the orthonormality relations:

$$\int_0^1 w_i(z,s)\overline{w_j(z,s)}dz = \int_0^1 v_i(z,s)\overline{v_j(z,s)}dz = \delta_{ij}.$$

By convention, the singular values are arranged in order of magnitude such that: $\sigma_1(s) \ge \sigma_2(s) \ge \ldots \ge 0$.

The left and right singular functions form orthonormal bases for the ranges of $\mathcal{K}$ and $\mathcal{K}^*$ respectively. If $\mathcal{K}$ is normal, then $kk^*(z,\zeta,s) = k^*k(z,\zeta,s)$, and the left and right singular functions of $\mathcal{K}$ are the same. In the special case of a real normal kernel with positive eigenvalues, the singular functions and singular values correspond exactly to the eigenfunctions and eigenvalues. Such kernels arise in steady-state ($s = 0$) analysis.

For the general case of a non-normal kernel, there may exist no eigenfunctions at all, or perhaps only a finite set. However, such a kernel will *always* possess a complete set of orthonormal singular functions and distinct, real singular values, which provide a mean-square convergent bilinear expansion for the kernel. It must be emphasized that the analytic determination of singular functions is a very difficult problem. In general, singular functions must be determined numerically.

General Kernel Expansion

An important consequence of this theory is that the kernel in Eq. (2) may be expressed as an expansion of singular functions and singular values:

$$k(z,\zeta,s) = \sum_{i=1}^{\infty} w_i(z,s)\sigma_i(s)\overline{v_i(\zeta,s)}. \quad (6)$$

This relation for the kernel of an unsymmetric integral operator represents the infinite-dimensional analogue of the matrix Singular Value Decomposition: $\boldsymbol{K}(s) = \boldsymbol{W}(s)\boldsymbol{\Sigma}(s)\boldsymbol{V}^*(s)$. The orthonormal singular functions correspond to the columns of the unitary matrices $\boldsymbol{W}(s)$ and $\boldsymbol{V}(s)$, and the singular values correspond to elements of the diagonal matrix $\boldsymbol{\Sigma}(s)$.

It may be shown that every function $y(z,s)$ in the range of $\mathcal{K}$ has a convergent series expansion in the orthonormal system $\{w_i(z,s)\}$, and that every function $u(z,s)$ in the range of $\mathcal{K}^*$ has a convergent series expansion in the orthonormal system $\{v_i(z,s)\}$:

$$y(z,s) = \sum_{i=1}^{\infty} w_i(z,s)\beta_i(s). \quad (7)$$

$$u(z,s) = \sum_{i=1}^{\infty} v_i(z,s)\alpha_i(s). \quad (8)$$

These singular functions may be used to construct finite-dimensional transformations which are *partially isometric* from the input space spanned by $\{v_i(z,s)\}$ to the output space spanned by $\{w_i(z,s)\}$ Such transformations are useful for reduced-order modeling and control of DPS. In fact, it can be proven (Butkovskiy, 1969; Smithies, 1958) that, for a given integer N, the *best* mean square approximation to a kernel $k(z,\zeta,s)$ by a finite-dimensional kernel of rank N is given by the N-term truncated singular value decomposition (TSVD) expansion of Eq. (6). The general existence and optimal approximation properties of the TSVD representation for the kernel makes it an excellent choice as a basis for the modeling and identification of DPS.

Decoupled SVD Control Structure

The TSVD represention of the kernel in terms of two sets of orthonormal functions also provides a theoretical basis for a dynamically decoupled control structure, with the input-output relationship shown in Fig. 1.

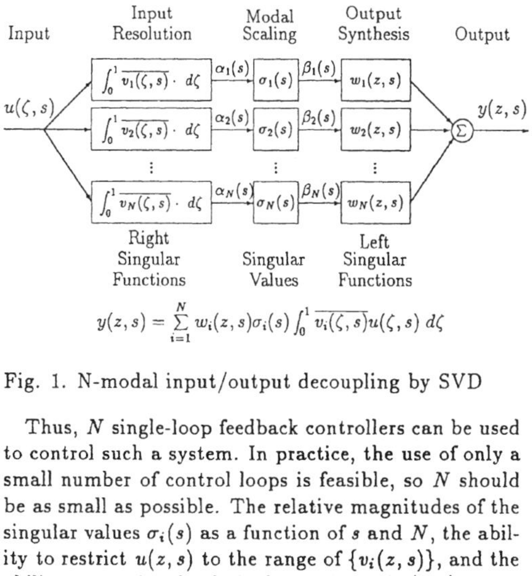

$$y(z,s) = \sum_{i=1}^{N} w_i(z,s)\sigma_i(s) \int_0^1 \overline{v_i(\zeta,s)} u(\zeta,s)\, d\zeta$$

Fig. 1. N-modal input/output decoupling by SVD

Thus, N single-loop feedback controllers can be used to control such a system. In practice, the use of only a small number of control loops is feasible, so N should be as small as possible. The relative magnitudes of the singular values $\sigma_i(s)$ as a function of s and N, the ability to restrict $u(z,s)$ to the range of $\{v_i(z,s)\}$, and the ability to restrict the desired setpoints of $y(z,s)$ to the range of $\{w_i(z,s)\}$ all have significant effects on the performance of such a finite-dimensional control system. The latter two goals can be difficult to realize when only a small number of discrete output measurements and input control functions are available. Furthermore, the dependence of the singular functions on the complex parameter s, along with the fact that they will in general be complex as well, creates problems of practical implementation.

The experience of this work has shown that such problems may be successfully overcome by carefully constructing a TSVD model which properly accounts for the spatially distributed nature of the process and which can be identified through input/output testing. Depending on the characteristics of the DPS, such a model provides the basis for design of either a pseudo-modal feedback or a model-predictive control system.

SINGULAR FUNCTION IDENTIFICATION

In order to calculate the singular functions from a limited number of spatial measurements, good numerical approximation methods must be used. The methods used in this work are similar to numerical techniques developed for solution of Fredholm integral equations of the first kind (Hansen and Christiansen, 1985; Hanson, 1971). These techniques also use TSVD expansions for approximation of the kernel.

Given the existence of an infinite-dimensional singular function expansion for the kernel of the integral equation model for the DPS, we therefore seek to construct a finite-dimensional kernel approximation with a TSVD expansion. This approximation will be constructed in a manner which facilitates identification of approximate singular functions and singular values from process input/output information.

To accomplish this, two sets of spatially independent functions are needed, one as a trial basis for the input $u(z,s)$, and one as a trial basis for the output $y(z,s)$. It is recommended that these trial basis functions be chosen so that their span can closely represent the physical control and measurement structure. The flexible, numerically stable set of functions known as B-splines have been found to work well in this capacity.

Basis Function Specification

The m trial basis functions for the input $u(z,s)$ are denoted by $\{p_1(z),\ldots,p_m(z)\}$. Similarly, the n trial basis functions for the output $y(z,s)$ are denoted by $\{q_1(z),\ldots,q_n(z)\}$. Corresponding to these basis functions are m linear functionals which determine the coefficients $a_j(s)$ of the input's projection onto the span of the input trial basis and n linear functionals which determine the coefficients $b_i(s)$ of the output's projection onto the span of the output trial basis. Thus, the distributed input and output are represented as:

$$u(z,s) \approx \sum_{j=1}^{m} p_j(z)a_j(s) \;;\; y(z,s) \approx \sum_{i=1}^{n} q_i(z)b_i(s) \quad (9)$$

In general, $n \neq m$, which allows the use of different numbers of trial basis functions for describing the spatial distribution of the inputs and outputs. For the present derivation, $n \geq m$ will be assumed.

Based on the previous theory, the kernel of the integral operator is approximated by a TSVD expansion truncated after m terms. The first m left and right singular functions are expressed in terms of the corresponding input and output trial basis functions:

$$v_r(z,s) \approx \sum_{j=1}^{m} p_l(z)v_{lr}(s) \;;\; w_r(z,s) \approx \sum_{i=1}^{n} q_i(z)w_{ir}(s). \quad (10)$$

Here $v_{jr}(s)$ and $w_{ir}(s)$ are trial basis coefficients of the singular functions for $r = 1,\ldots,m$.

Galerkin's Method Solution

A m-dimensional TSVD kernel based on the singular functions from Eq. (10) and the input and output functions from Eq. (9) are substituted into Eq. (2) to begin the solution. After rearrangement of the order of integration and summation, an approximate representation of this integral equation is obtained in terms of the trial basis functions. Solution of this equation by Galerkin's method requires the residual to be spatially orthogonal to each of the output trial basis functions, and results in the set of n equations:

$$\int_0^1 q_h(z)\sum_{i=1}^{n} q_i(z)\sum_{r=1}^{m} w_{ir}(s)\sigma_r(s)\sum_{l=1}^{m} v_{lr}(s) \int_0^1 p_l(\zeta)\sum_{j=1}^{m} p_j(\zeta)a_j(s)\, d\zeta\, dz = \int_0^1 q_h(z)\sum_{i=1}^{n} q_i(z)b_i(s). \quad (11)$$

Use of Galerkin's method to generate the n independent equations in Eq. (11), rather than simply equating corresponding coefficients of each of the n trial basis functions, is an important consideration. This assures that the singular functions determined using the SVD are spatially orthonormal, even though the trial basis functions are not. Thus, this technique allows the use of numerically well-behaved families of trial basis functions such as B-splines, which may be tailored to the specific spatial control and measurement structure of the system, in order to generate a set of orthonormal approximations to the singular functions. This technique effectively combines the flexibility and good numerical behavior of piecewise polynomial basis functions with the monotonic convergence properties of an orthonormal basis. Note that in Eq. (11) all the integrations involve known trial basis functions so that these integrals may be computed exactly. For compactness, we introduce the following matrix notation:

$$\begin{array}{ll}
\boldsymbol{p}(z) = [p_1(z),\dots,p_m(z)]^T & \boldsymbol{q}(z) = [q_1(z),\dots,q_n(z)]^T \\
\boldsymbol{a}(s) = [a_1(s),\dots,a_m(s)]^T & \boldsymbol{b}(s) = [b_1(s),\dots,b_m(s)]^T \\
u(z,s) = \boldsymbol{p}(z)^T\boldsymbol{a}(s) & y(z,s) = \boldsymbol{q}(z)^T\boldsymbol{b}(s) \\
\boldsymbol{V}(s) = (v_{lr}(s)) & \boldsymbol{W}(s) = (w_{ir}(s)) \\
v(z,s) = \left[\boldsymbol{p}(z)^T\boldsymbol{V}(s)\right]^T & w(z,s) = \left[\boldsymbol{q}(z)^T\boldsymbol{W}(s)\right]^T \\
\boldsymbol{P} = \left(\int_0^1 \boldsymbol{p}(\zeta)\boldsymbol{p}(\zeta)^T d\zeta\right) & \boldsymbol{Q} = \left(\int_0^1 \boldsymbol{q}(z)\boldsymbol{q}(z)^T dz\right) \\
\boldsymbol{P} = \boldsymbol{C}^T\boldsymbol{C} & \boldsymbol{Q} = \boldsymbol{D}^T\boldsymbol{D} \\
\boldsymbol{A}(s) = (a_{jr}) & \boldsymbol{B}(s) = (b_{ir}) \\
\boldsymbol{K}(s) = \boldsymbol{B}(s)\boldsymbol{A}(s)^{-1} & \boldsymbol{\Sigma}(s) = \mathrm{diag}\,(\sigma_r(s))
\end{array}$$

which allows Eq. (11) to be succinctly expressed as:

$$\boldsymbol{Q}\boldsymbol{W}(s)\boldsymbol{\Sigma}(s)\boldsymbol{V}^*(s)\boldsymbol{P}\boldsymbol{a}(s) = \boldsymbol{Q}\boldsymbol{b}(s). \quad (12)$$

Identification Procedure

In order to solve Eq. (12) for the matrices of singular function coefficients $\boldsymbol{W}(s)$ and $\boldsymbol{V}(s)$ and singular values $\boldsymbol{\Sigma}(s)$, experimental data from m independent input/output tests of the DPS is required. This data may be expressed as the matrices $\boldsymbol{A}(s)$ and $\boldsymbol{B}(s)$ of trial basis coefficients, corresponding to measurements of $u(z,s)$ and $y(z,s)$ respectively.

Finally, calculation of the weighting matrices $\boldsymbol{C}$ and $\boldsymbol{D}$ is required for proper scaling of the kernel matrix $\boldsymbol{K}(s)$. These matrices are determined by a Cholesky decomposition of the symmetric positive definite Gram matrices $\boldsymbol{P}$ and $\boldsymbol{Q}$ for the input and output trial basis functions. The Cholesky factors $\boldsymbol{C}$ and $\boldsymbol{D}$ are upper triangular matrices with positive diagonal elements. Use of these weighting matrices insures that the SVD algorithm generates matrices of singular function coefficients which correspond to spatially orthonormal singular functions (rather than functions whose trial basis coefficients satisfy summation orthonormality).

Substitution of the input and output basis coefficient matrices and the Cholesky factors of the Gram matrices into Eq. (12) yields, upon rearrangement, the relation:

$$\boldsymbol{D}\boldsymbol{W}(s)\boldsymbol{\Sigma}(s)\boldsymbol{V}^*(s)\boldsymbol{C}^T = \boldsymbol{D}\boldsymbol{K}(s)\boldsymbol{C}^{-1} = \widehat{\boldsymbol{K}}(s) \quad (13)$$

The right hand side of Eq. (13) consists of data from identification tests, represented in terms of basis coefficients, scaled by the Cholesky factors of $\boldsymbol{P}$ and $\boldsymbol{Q}$ to reflect the spatial distribution of the trial basis functions. After application of the matrix SVD algorithm:

$$\widehat{\boldsymbol{W}}(s)\boldsymbol{\Sigma}(s)\widehat{\boldsymbol{V}}^*(s) = \widehat{\boldsymbol{K}}(s). \quad (14)$$

Here $\boldsymbol{\Sigma}(s)$ is a diagonal matrix of singular values, and the matrices $\widehat{\boldsymbol{W}}(s)$ and $\widehat{\boldsymbol{V}}(s)$ are unitary matrices:

$$\widehat{\boldsymbol{W}}^*(s)\widehat{\boldsymbol{W}}(s) = \boldsymbol{I} \qquad \widehat{\boldsymbol{V}}^*(s)\widehat{\boldsymbol{V}}(s) = \boldsymbol{I}. \quad (15)$$

The singular function basis coefficient matrices are

$$\boldsymbol{W}(s) = \boldsymbol{D}^{-1}\widehat{\boldsymbol{W}}(s) \qquad \boldsymbol{V}(s) = \boldsymbol{C}^{-1}\widehat{\boldsymbol{V}}(s), \quad (16)$$

where only the first m columns of $\boldsymbol{W}(s)$ are required. The singular function approximations defined by these coefficients and the trial basis functions are orthonormal in the sense of the $\mathcal{L}^2$ inner product:

$$\int_0^1 w(z,s)w(z,s)^* dz = \boldsymbol{W}^*(s)\boldsymbol{Q}\boldsymbol{W}(s) = \boldsymbol{I} \quad (17)$$

$$\int_0^1 v(z,s)v(z,s)^* dz = \boldsymbol{V}^*(s)\boldsymbol{P}\boldsymbol{V}(s) = \boldsymbol{I} \quad (18)$$

Given a finite number of quality input/output measurements and two sets of well-chosen basis functions, this identification method will provide good approximations to the orthonormal singular functions of the DPS.

MODEL BASED CONTROL

The SVD-based identification procedure provides a distributed model in a form which is quite useful for control purposes. Consider Eq. (12) for vectors of input coefficients $\boldsymbol{a}(s)$ and output coefficients $\boldsymbol{b}(s)$. Multiplication of both sides by $\boldsymbol{W}^*(s)$ and use of Eq. (17) gives the decoupled open loop relation between the modal inputs $\boldsymbol{\alpha}(s)$ and the modal outputs $\boldsymbol{\beta}(s)$ through the matrix of singular values $\boldsymbol{\Sigma}(s)$:

$$\boldsymbol{\Sigma}(s)\boldsymbol{\alpha}(s) = \boldsymbol{\beta}(s) \quad (19)$$

$$\boldsymbol{\alpha}(s) = \boldsymbol{V}^*(s)\boldsymbol{P}\boldsymbol{a}(s) \quad ; \quad \boldsymbol{\beta}(s) = \boldsymbol{W}^*(s)\boldsymbol{Q}\boldsymbol{b}(s)$$

Note that application of the identification procedure at different frequencies s will generally produce different complex matrices $\boldsymbol{W}(s)$ and $\boldsymbol{V}(s)$ of singular function coefficients. One pragmatic solution is to calculate these matrices at $s = 0$, which corresponds to identification of a steady-state input/output model. In this case the singular function coefficients are *real*. Following such a steady-state identification of the singular functions, the dynamic behavior of the singular value matrix can be calculated from the scaled dynamic kernel matrix $\widehat{K}(s)$ by the equation:

$$\Sigma(s) = \widehat{W}^T(0)\widehat{K}(s)\widehat{V}(0). \qquad (20)$$

Although the steady state singular value matrix $\Sigma(0)$ is by definition a diagonal matrix, the dynamic modal response matrix formed in Eq. (20) will not be diagonal for all s. The magnitudes of the off-diagonal elements in $\Sigma(s)$ provide a measure of the dynamic interactions which will exist in such a Distributed Parameter Singular Value (DPSV) feedback control system.

DPSV Pseudo-Modal Control

The pseudo-modal structure of the identified singular function model can be used directly to design a feedback control system which accounts for the distributed nature of the inputs and outputs. This DPSV-based model provides decoupling of orthonormal combinations of system inputs and outputs, ranked in terms of the magnitude of their steady-state gains. This special structure can be used to advantage in designing reduced-order feedback control systems.

One implementation of these ideas is shown in the block-diagram of Fig. 2, where the matrix quantities correspond to those defined previously.

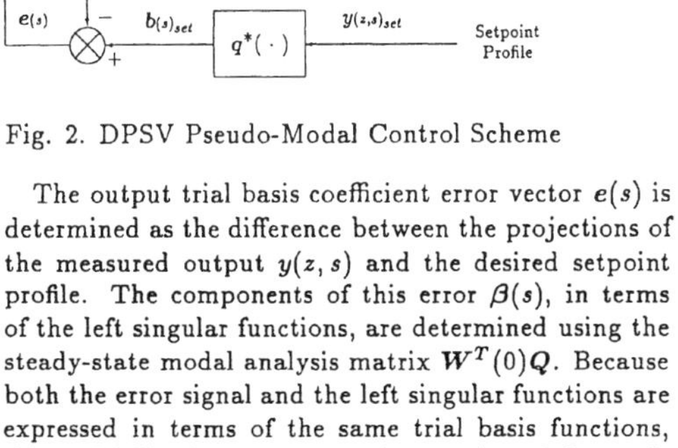

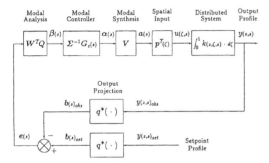

Fig. 2. DPSV Pseudo-Modal Control Scheme

The output trial basis coefficient error vector $e(s)$ is determined as the difference between the projections of the measured output $y(z,s)$ and the desired setpoint profile. The components of this error $\beta(s)$, in terms of the left singular functions, are determined using the steady-state modal analysis matrix $W^T(0)Q$. Because both the error signal and the left singular functions are expressed in terms of the same trial basis functions, this multiplication amounts to integration of the products of the approximate spatial error with each of the identified left singular functions. The vector $\alpha(s)$ of control coefficients in terms of the right singular functions, is calculated by a diagonal controller based only on the corresponding coefficient of the modal error vector $\beta(s)$, scaled by the inverse singular value $\sigma_i^{-1}(0)$. The modal synthesis matrix $V(0)$ is then used to convert these control inputs to the input trial basis coefficient vector $a(s)$. The approximation of the physical control action $u(z,s)$ is then given by the sum of each input trial basis function times its respective coefficient.

Such a feedback controller will drive the DPS to a steady state profile which is the orthogonal projection of the scaled setpoint profile onto the range of the scaled kernel matrix $\widehat{K}(0)$. Since the left singular functions provide an orthonormal basis for the range of $\widehat{K}(0)$, the residual is therefore $\mathcal{L}^2$ orthogonal to the first m left singular functions. This amounts to a steady-state solution of the least-squares problem

$$\min \|\widehat{K}(0)\widehat{a} - \widehat{b}_s\|^2 \qquad (21)$$

for the minimum-norm scaled input coefficient vector $\widehat{a}$, given the scaled setpoint coefficient vector $\widehat{b}_s$. Using the SVD, the solution of Eq. (21) is expressed as:

$$\widehat{a} = \widehat{V}(0)\widehat{\Sigma}^{-1}(0)\widehat{W}^T(0)\widehat{b}_s \qquad (22)$$

Weighting of the output coefficients by D insures that the sum of squares of the steady-state residual error signal $\widehat{e} = \widehat{b}_s - \widehat{b}$ is minimized in the sense:

$$\|\widehat{e}\|^2 = \|De\|^2 = e^TQe \approx \int_0^1 e^2(z,0)dz \qquad (23)$$

Likewise, weighting of the input coefficients by C results in a steady-state input coefficient vector with a norm minimized in the sense:

$$\|\widehat{a}\| = \|Ca\| = a^TPa \approx \left(\int_0^1 u^2(z,0)dz\right)^{1/2} \qquad (24)$$

For control purposes, such a control input solution is desirable, since it has a minimum spatial norm and thus represents the most efficient control action.

Note that these computations are performed using the matrix SVD, which is based on the normal euclidian vector norm. However, since the input and output measurements used in this computation are represented as specially weighted basis coefficients, the obtained steady-state solution actually represents a least-squares minimum-norm solution in the spatial norm, restricted to the range of the trial basis functions. If these basis functions are well chosen, this provides a good approximation to the least-squares minimum-norm solution in infinite spatial dimension.

While this approach will provide a steady-state modal decoupling controller for any linear DPS, if dynamic modal interactions or nonlinearities are significant, a more complex controller design based on the DPSV model should be considered.

DPSV Dynamic Matrix Control

The identification and feedback control techniques presented in the previous sections provide a good control scheme for systems in which there is not a large frequency dependence in the singular functions or significant nonlinearities. When these factors are significant, however, the basic ideas of the DPSV controller may be generalized to model-predictive controllers such as Dynamic Matrix Control (DMC) (Cutler, 1983). In fact, by recasting the least squares solution of the DMC controller equations in terms of the SVD, it can be shown that both techniques share a common theoretical basis.

The conventional formulation of the DMC controller equation for a n-output, m-input multivariable discrete-time system has been well-documented (Garcia and Morshedi, 1985). However, in this work we choose an alternative formulation of the DMC controller equation which makes use of the previously described basis function representation of the spatially distributed inputs and outputs, as well as the basis function weighting matrices introduced earlier. For an $n \times m$ multivariable DPS, the DMC equation is written as:

$$\widehat{e} = \widehat{K}\widehat{a} \tag{25}$$

where $\widehat{e}$ is a vector of output error predictions, and $\widehat{a}$ is a vector if calculated future control moves. Both vectors are composed of basis coefficients which have been properly scaled by means of the appropriate weighting matrices C and D. In this formulation, the $nP \times mR$ scaled dynamic matrix $\widehat{K}$ has been partitioned with respect to the sampling time, as opposed to the conventional DMC partitioning with respect to the input and output variables. P is the number of future sample times for which the output error is predicted, while R is the number of future control moves to be calculated.

$$\widehat{K} = \begin{bmatrix} \widehat{K}_{(1)} & & 0 \\ \widehat{K}_{(2)} & \ddots & \\ \vdots & \ddots & \widehat{K}_{(1)} \\ \vdots & & \vdots \\ \widehat{K}_{(P)} & \cdots & \widehat{K}_{(S)} \end{bmatrix} \tag{26}$$

Each $n \times m$ submatrix has been scaled in the manner previously described in the singular function identification algorithm:

$$\widehat{K}_{(l)} = DK_{(l)}C^{-1}. \tag{27}$$

Each of the unscaled sub-matrices $K_{(l)}$ is the matrix of discrete-time step response coefficients at sample time $l\Delta T$.

$$K_{(l)} = \begin{bmatrix} k_{11}(l\Delta T) & \cdots & k_{1m}(l\Delta T) \\ \vdots & \ddots & \vdots \\ k_{n1}(l\Delta T) & \cdots & k_{nm}(l\Delta T) \end{bmatrix} \tag{28}$$

where each element k_{ij} is the response of output measurement i to a unit step in control element j at sampling time $l\Delta T$. $K_{(S)}, \ldots, K_{(P)}$, $S = P - R + 1$ are equal to the steady-state gain matrix of the process,

The composite nP-vector $\widehat{e}$ contains predictions of scaled errors between the coefficients of the output and setpoint profile at future sampling times. The composite mR-vector $\widehat{a}$ is the calculated solution, which contains scaled coefficients of current and future control moves. These composite vectors are formed as:

$$\widehat{e} = \begin{bmatrix} \widehat{e}_{(1)} \\ \vdots \\ \widehat{e}_{(P)} \end{bmatrix} \qquad \widehat{a} = \begin{bmatrix} \widehat{a}_{(0)} \\ \vdots \\ \widehat{a}_{(R-1)} \end{bmatrix}$$

$$\widehat{e}_{(l)} = De_{(l)} \qquad \widehat{a}_{(l)} = Ca_{(l)}$$

$$e_{(l)} = \begin{bmatrix} e_1(l\Delta T) \\ \vdots \\ e_n(l\Delta T) \end{bmatrix} \qquad a_{(l)} = \begin{bmatrix} a_1(l\Delta T) \\ \vdots \\ a_m(l\Delta T) \end{bmatrix}$$

The least-squares solution of the DMC equation is conventionally given in the form of the *normal equations:*

$$\widehat{a} = (\widehat{K}^T\widehat{K})^{-1}\widehat{K}^T\widehat{e}, \tag{29}$$

which can be quite ill-conditioned. A completely reliable solution of the DMC equation for ill-conditioned or rank-deficient matrices cab can be provided by the SVD (Golub, 1983). Given the SVD of the dynamic matrix of scaled coefficients in Eq. (26) as $\widehat{K} = \widehat{W}\Sigma\widehat{V}^T$, the SVD solution of the DPSV DMC equation is:

$$\widehat{a} = \widehat{V}\Sigma^{-1}\widehat{W}^T\widehat{e} \tag{30}$$

which has the same structure as Eq. (22). An examination of this formulation of the DPSV DMC controller shows that for the case of steady-state prediction $P = 1$ and a single control move $R = 1$, a DPSV pseudo-modal controller with $G(s) = I$ results. Thus, this DMC controller formulation is a logical extension of the pseudo-modal controller.

DPSV DMC controller tuning

In the conventional DMC formulation, a control move suppression matrix is appended to the dynamic matrix in order to reduce the size of calculated control moves. This is essentially an effort to improve the conditioning of the dynamic matrix. Since the choice of such weights is a rather arbitrary and iterative process, a different and more direct approach will be taken in this work. This technique explicitly restricts the condition number of the dynamic matrix to a given value, and requires the adjustment of only a single parameter, rather than the adjustment of a penalty weight for every input. The solution in Eq. (30) may be expressed in the form of a modal expansion of mR terms:

$$\widehat{a} = \sum_{i=1}^{mR} \frac{\widehat{v}_i\widehat{w}_i^T\widehat{e}}{\sigma_i} \tag{31}$$

Truncation of this modal expansion at N terms results in a truncated set of least squares equations with a condition number $\kappa = \sigma_1/\sigma_N$. Such a method has been described as 'component selection' (Maurath, Seborg, and Mellichamp, 1985), due to the statistical designation of singular vectors as 'principal components'. However, in the spirit of the present work the term 'modal truncation' is preferred, since *all* of the orthogonal modes greater than N of the scaled dynamic matrix are removed from the least-squares solution of the DMC controller equation. The tuning process is therefore reduced to the determination of the desired order of modal truncation. As the number of modes which are truncated is increased, the condition number of the truncated dynamic matrix is reduced, and the calculated control moves become smaller.

EXAMPLES

We now illustrate the application of these techniques for examples of linear and nonlinear DPS.

Metallurgical Heating System

The first example system considered is an experimental apparatus for the heating of a long, thin aluminum slab (Mäder, 1976). The slab temperature is controlled by twenty zones of heating located above and below. There are 21 equally spaced thermocouples used to measure the temperature profile. Both ends of the slab are well insulated, and internal cooling is provided by constant temperature water flowing through channels.

This system is described by a linear parabolic PDE with constant coefficients and zero-flux boundary conditions. This system may also be described by a normal integral operator. Thus the steady-state singular functions are equal to the eigenfunctions: $\sqrt{2}\cos i\pi z,\ i = 0, \ldots, \infty$.

In order to identify the singular functions of this system, the temperature profile is represented by $n = 21$ parabolic B-spline trial basis functions, and the heating input is represented by $m = 20$ piecewise constant B-spline functions. Twenty step tests are performed in order to construct the input/output data matrix $\boldsymbol{K}(0)$. The first six left and right identified singular functions are shown in Fig. 3, along with the actual eigenfunctions. For such a linear DPS, with a large number of measurements and inputs, the singular functions are identified extremely accurately.

Closed loop control of this system by means of a reduced order DPSV feedback PI controller is shown in Fig. 4. The order of the modal controller was reduced from 20 to 6 and the setpoint profile was chosen to be in the range of the first six left singular functions, so it is able to be achieved exactly. There are no modal dynamic interactions for this system, so this type of controller is quite effective.

For discussion of singular function identification and control for a moving metal slab, described by a *non-normal* integral operator, see Gay and Ray(1986).

Tubular Chemical Reactor

The second example system considered is a nonlinear tubular chemical reactor with wall cooling (Alvarez and others, 1981). As shown in Fig. 5, we assume that the reactor has $m = 3$ independent zones of wall cooling and $n = 5$ temperature sensors. The temperature profile is represented by a parabolic B-spline trial basis and the wall cooling is represented by piecewise constant trial basis. The base temperature and conversion profiles shown result from a uniform feed and wall temperature of 500K. Figure 6 shows a set of singular functions for this reactor, which were identified based on the deviations from the base temperature profile that resulted from step changes in wall temperature.

Because this system is nonlinear, a DPSV pseudo-modal controller based on these steady-state singular functions will have dynamic modal interactions of the type shown in Fig. 7, a time-domain plot of Eq. (20). Thus, a DPSV DMC controller was identified and applied to control this system. Only temperature measurements were used in the identification and control. Figures 8 and 9 show the closed-loop response of both the temperature and conversion profiles under DMC control for a setpoint change from the base profile to a linearly increasing profile. The controller was identified with a dimensionless sample time of 0.04, a prediction horizon of $P = 50$, and $R = 10$ calculated future control moves. Tuning of this DPSV DMC controller was accomplished by modal truncation at $N = 12$, corresponding to a condition number of 100 in Fig. 10. With this number of retained modes, control actions change smoothly from base to final values without large moves. By retaining additional modes, the system responds faster, but control moves become excessively large. Conversely, as fewer modes are retained, the system responds more slowly and control moves become more restrained.

Since the specified setpoint is not completely in the range of the identified singular functions of Fig. 6, the setpoints at the sensor locations are not satisfied exactly. The steady-state resulting profile represents a least squares approximation to the setpoint, and the residual is orthogonal to the three left singular functions.

SUMMARY

The SVD-based identification and control techniques presented in this paper provide a solid foundation for model-based control of DPS. Modeling of the DPS in an integral equation representation with a kernel expressed in terms of two finite sets of spatial basis functions retains the spatially distributed nature of the process, yet allows calculations to be performed with matrix techniques. Identified singular functions provide an inherent orthonormal coordinate system for a process, even if process models and parameters are not known. The identified model can be used to design modal feedback controllers, or more complex model-predictive controllers. Reduction of the order of these models and controllers is simple, while providing a useful tuning parameter for a predictive controller.

REFERENCES

Alvarez, J., J.A. Romagnoli, and G. Stephanopoulos, (1981). Variable measurement structures for the control of a tubular reactor. *Chem. Eng. Sci.*, **36**(10), 1695-1712.

Butkovskiy, A.G., (1969). *Distributed Control Systems.* American Elsevier, New York.

Cochran, J.A., (1972). *The Analysis of Linear Integral Equations.* McGraw-Hill, New York.

Cutler, C.R., (1983). Dynamic Matrix Control: An optimal multivariable control algorithm with constraints. Ph.D. Thesis, University of Houston.

Garcia, C.E. and A.M. Morshedi, (1986). Quadratic programming solution of dynamic matrix control (QDMC). *Chem. Eng. Commun.*, **46**, 73-87.

Gay, D.H. and W.H. Ray, (1986). Identification and control of linear distributed parameter systems through the use of experimentally determined singular functions. In H.E. Rauch (Ed.), *Control of Distributed Parameter Systems 1986.* Pergamon Press, Oxford.

Golub, G.H. and C.F. van Loan, (1983). *Matrix Computations.* Johns Hopkins Press, Baltimore.

Hansen, P.C. and S. Christiansen, (1985). An SVD analysis of linear algebraic equations derived from first kind integral equations. *J. Comp. & Appl. Math.*, **12&13**, 341–357.

Hanson, R.J., (1971). A numerical method for solving Fredholm integral equations of the first kind using singular values. *SIAM J. Numer. Anal.*, **8**(3), 616.

Mäder, H.F., (1976). Modell und modale Regelung eines technisch realisierten Wärmeleitsystems. *Regelungstechnik*, **24**, 347.

Maurath, P.R., D.E. Seborg, and D.A. Mellichamp, (1985). Predictive controller design by principal components analysis. In *Proceedings of the 1985 American Control Conference.*

Smithies, F., (1958). *Integral Equations.* Cambridge University Press, Cambridge.

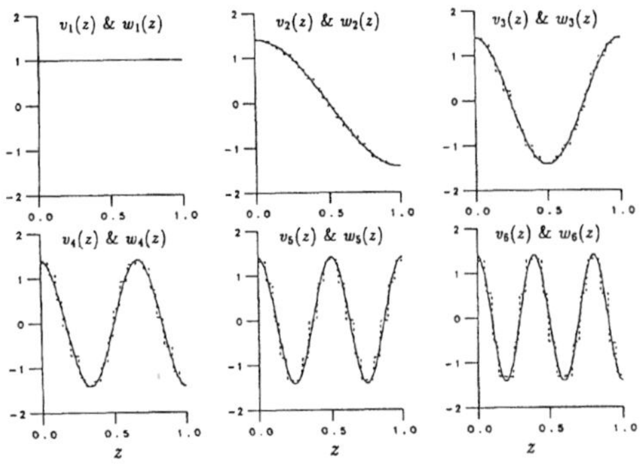

Fig. 3. Singular functions of heated slab

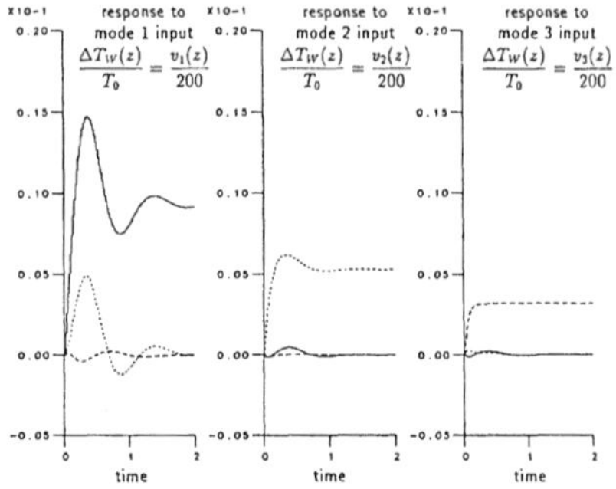

Fig. 7. Open-loop DPSV modal interactions

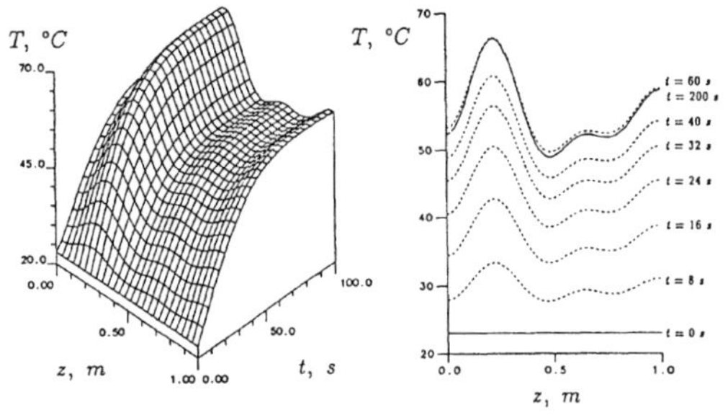

Fig. 4. Closed-loop DPSV pseudo-modal control

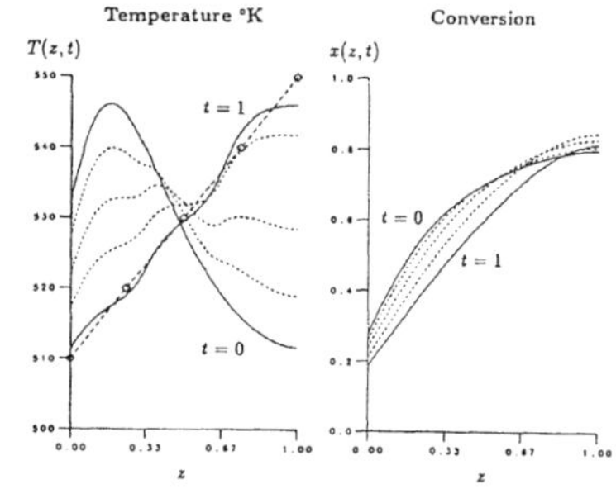

Fig. 8. 2-D response to DPSV DMC controller

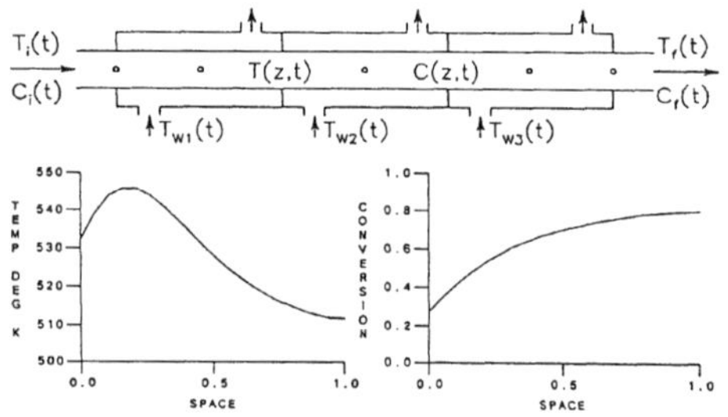

Fig. 5. Tubular chemical reactor base profiles

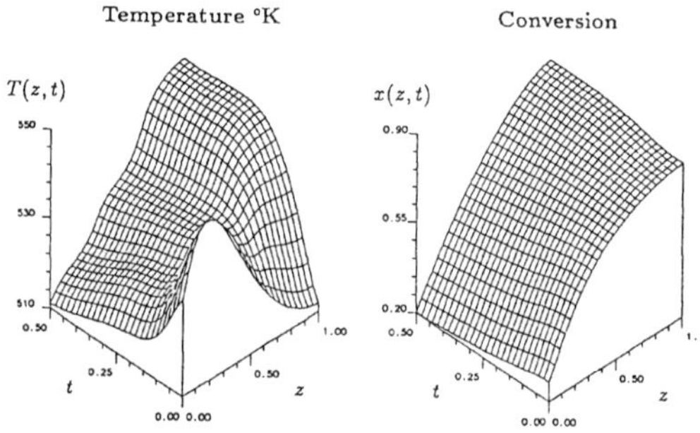

Fig. 9. 3-D response to DPSV DMC controller

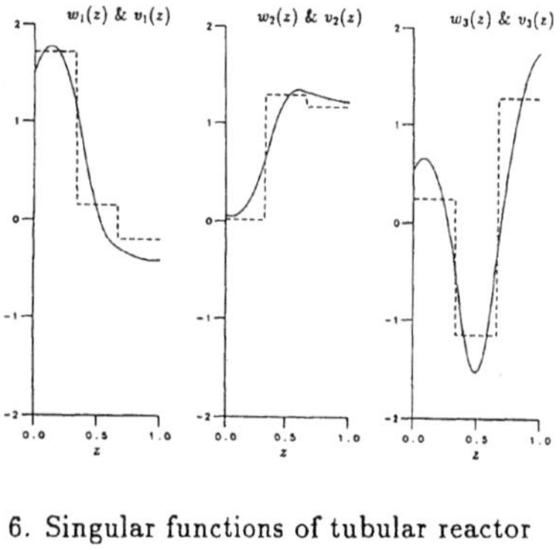

Fig. 6. Singular functions of tubular reactor

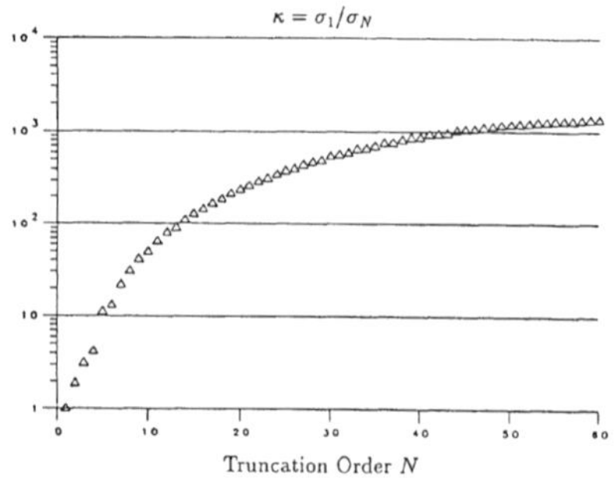

Fig. 10. Conditioning of DPSV DMC controller

MODEL-BASED CONTROL FOR SYSTEMS IN OPERATOR DESCRIPTION*

H. Itakura, W. M. Grimm** and P. M. Frank*****

***Department of Information Sciences, Chiba Institute of Technology, 2–17–1, Tsudanuma, Narashino-shi, Chiba-ken 275, Japan*
****MRT, FB9, University of Duisburg, Bismarckstr. 81, 4100 Duisburg, FRG*

Abstract. A model-based control concept is discussed for a relatively general system described by operators. A sensitivity model is also examined as well as a parameter estimation scheme via sensitivity operators.

Keywords. Process control; model-based control; operator description; sensitivity analysis; parameter estimation.

INTRODUCTION

With the increased power of readily available computer hardware with cheap microprocessors and memory chips, fairly complex control algorithms in process control systems have become feasible. They can easily be implemented digitally while in the past they had been difficult to realize with analogue devices.

One of the most interesting topics to reflect those advances is model-based process control. A special scheme in such a control configuration is that of predictive control aiming at the compensation of dead time in the process. A typical example is the well-known Smith predictor control (1957). During the years a lot of investigations have emerged addressing either the Smith type of control itself or its modifications. Special names have been given to some of them. For instance, Frank's one (1974) is the Model Principle Control, Brosilow and Tong's (1978) is the Inferential Control, Morari and his co-workers' (Economou and Morari, 1986; Garcia and Morari, 1982; Rivera and Morari, 1986) is the Internal Model Control, and so forth. In the framework of these activities, many articles have been published in journals on control theory, mathematics, process applications, and so on. Their aspects have been extended to cover design procedures and algorithms, stability issues, sensitivity analysis, parameter identification, robust stability, or design in the face of plant/model mismatch. Moreover, there is a variety of system descriptions or design tools: frequency or time domain, single- or multi-output/input, linear or nonlinear model, classical or modern type technique.

*Part of this work was done when the first author stayed at the University of Duisburg on leave from the Chiba Institute of Technology, Japan.

The scope of the work in this paper is to examine a general framework of a model-based control scheme in a conceptual operator description, where the plant input and output are related through an operator. The framework is not always limited to predictive control. The plant is assumed to consist of two cascade subsystems which are either inherent in the plant or introduced for convenience. The situation is considered where it may be desirable, but difficult, to design a controller for the overall plant directly, while it might be rather simple to control the front-end subplant, and to allow little deterioration of the performance from the back-end subplant remaining open-loop. For model-based controller design, a mathematical model has to be in hand for each of these subplants. The models are to be used for an on-line simulation of the plant dynamics. Then, the overall controller configuration has two feedback loops: An inner one and an outer one. The former is for controlling the front-end subplant and the latter one is for feeding back the plant/model mismatch of the back-end subplant. This system structure includes many model-based control schemes which have been so far investigated in the literature cited above. Alternatively, the model-based concept is applicable to a more general class of plants, as it will be examined in this paper. Included in the discussion are parameter identification considerations via sensitivity operators.

SYSTEM DESCRIPTION

The plant under consideration is assumed to have the input-output relation

$$y = P\,u \quad , \qquad (2.1)$$

where u is the input to the plant, y the output,

and P an operator which maps u to y. Furthermore, as shown in Fig. 1(a), let P be represented by the product of two operators, P_1 and P_2, i.e.,

$$P = P_2 P_1 , \qquad (2.2)$$

where P_1 and P_2 obey the associative law

$$(P_2 P_1)\, u = P_2\, (P_1 u) . \qquad (2.3)$$

We wish to control this plant by a feedback controller C, where C is also an operator which maps e to u. The error signal is

$$e := r - y \qquad (2.4)$$

and r is the set point. The situation is considered where it may be desirable, but difficult, to design C to control P directly, whereas, on the other hand, it might be comparatively easier to design C to control P_1, and to tolerate little deterioration of the performance due to P_2 remaining open-loop.

Let the mathematical models of P_1 and P_2, P_{m1} and P_{m2}, respectively, be available in reasonable accuracy. In order to design C in a model-based structure, the signal u is negatively fed back through P_{m1} and positively by $P_{m2}P_{m1}$ as shown in Fig. 1(a). It can be readily seen that if P_{m1} and P_{m2} are identical with P_1 and P_2, respectively, the problem is equivalent to that of designing the feedback controller C simply for P_1, leaving P_2 outside the loop. The resulting system structure is drawn in Fig. 1(b).

CONVENTIONAL MODEL-BASED SCHEMES

Firstly, it is observed that the above conceptual control design framework naturally includes many model-based control schemes which have been investigated so far. Some typical ones are the following:

1) When P_2 is a pure dead-time transfer function and P_1 is dead-time free and linear, the traditional Smith predictor (1957) is reached. Investigations of the stability of the overall system in the case of plant/model mismatch have been carried out by several authors. Recent interest has been focussed on the design of the controller, where robustness features are incorporated in order to cope with the mismatch (Etterich, 1988; Grimm et al., 1987, 1988; Laughlin et al., 1987; Palmor, 1986).
2) Several contributions have recently been made to the field of the so-called Internal Model Control by Morari and his co-workers (Garcia et al., 1982; Rivera et al., 1986). Although there are a lot of modified versions of their control structure, their fundamental concept is obtained by, for example, studying the case of

$$P_1 = P_{m1} = I , \qquad (3.1)$$

where I is the identity operator. Then, for linear systems the transfer-function description of the controller becomes

$$C(s) = (G_r(s)^{-1} G_m(s) - I)^{-1} , \qquad (3.2)$$

where $G_m(s)$ is a minimum-phase portion of the plant model P_{m2}. $G_r(s)$ is a transfer function of a simple form to be specified, which can be, for example, a lag of an appropriate order. The introduction of $G_r(s)$ aims at augmenting robustness in case of plant/model mismatch. Methods for designing $G_r(s)$ have been discussed in some papers (Frank, 1974; Pollard, 1986).
3) If the input-output characteristics of the plant are governed by a set of nonlinear differential equations, P_1 and P_2 are nonlinear-function operators. As a special case, where again equation (3.1) holds, Frank (1974) and Economou et al. (1986) have suggested the possibility of the design of nonlinear controllers using inverse operators in the framework of Internal Model Control.
4) P_1 and P_2 can also be partial-differential operators to describe a distributed-parameter system. For example, Vit (1979) has examined a Smith-like predictor for control of a heat-diffusion process.

ALTERNATIVE APPLICATIONS

A model-based control design configuration is fundamentally possible, whenever the plant P can be devided into two parts P_1 and P_2 in cascade with imposing additional conditions on P_1 and P_2. In the sequel, let us examine another class of plants which have a feature that is not always restricted to P_2 being dead-time.

Let the plant P be governed by the equations

$$\begin{aligned} \dot{x} &= f(x, u, t) \\ y &= g(x, t) \quad , \end{aligned} \qquad (4.1)$$

where

$$f := \begin{bmatrix} Ax_1 \\ h(c^T x_1, x_2, t) \end{bmatrix} + \begin{bmatrix} b \\ 0 \end{bmatrix} u \qquad (4.2)$$

$$x := \begin{bmatrix} x_1 \\ x_2 \end{bmatrix}$$

with x_1, b, and c being n_1-vectors, x_2 an n_2-vector, A an $(n_1 \times n_1)$-matrix, and h an n_2-vector of functions which are analytic in all arguments. The quantities y and u are scalar, and the superscript T denotes the transpose of a matrix or a vector.

This plant may be partitioned into P_1 and P_2 according to

$$P_1 : u \rightarrow z, \qquad (4.3)$$

where

$$\dot{x}_1 = Ax_1 + bu$$
$$z = c^T x_1 , \quad (4.4)$$

and

$$P_2 : z \to y , \quad (4.5)$$

where

$$\dot{x}_2 = h(z, x_2, t)$$
$$y = g(x, t) . \quad (4.6)$$

It is assumed that P_2 is stable for any bounded time-function input $z(t)$ such that $z(t) \to 0$ as $t \to \infty$. While the design of an optimum controller C for this plant,

$$P = P_2 P_1, \quad (4.7)$$

is clearly difficult, a design is very simple for the subplant P_1 using the above model-based control configuration since P_1 is linear. Thus, any design technique, either in the time domain or in the frequency domain, for a linear system can be applied to constructing C for P_1.

For example, let us try to design C in the case of periodically time-varying elements of A and b. It is assumed that periodic terms perturb the time-invariant portions. Let the model $A_m(t)$ and $b_m(t)$ of A and b, respectively, have the forms

$$A_m(t) = A_0 + \epsilon A_1(t)$$
$$b_m(t) = b_0 + \epsilon b_1(t), \quad (4.8)$$

where A_0 and $A_1(t)$ are $(n_1 \times n_1)$-matrices, b_0 and $b_1(t)$ are n_1-vectors, A_0 and b_0 are constant, and $A_1(t)$ and $b_1(t)$ are continuous and periodic in t with period L according to

$$A_1(t + L) = A_1(t)$$
$$b_1(t + L) = b_1(t) \quad (4.9)$$

for any t. ϵ is a small scalar parameter. Hence the periodic terms ϵA_1 and ϵb_1 are considered to be perturbations to the time-invariant parts A_0 and b_0, respectively.

Consider the problem to find a feedback controller $u(x_{m1},t)$ for which the quadratic performance index

$$J = \lim_{t_f \to \infty} \int_{t_0}^{t_f} (x_{m1}^T Q x_{m1} + u^2)\, dt \quad (4.10)$$

is minimized, where x_{m1} is the model state of x_1 generated by

$$P_{m1} : u \to z_m \quad (4.11)$$

with

$$\dot{x}_{m1} = A_m(t)x_{m1} + b_m(t)u$$
$$z_m = c_m^T x_{m1} , \quad (4.12)$$

and Q is an $(n_1 \times n_1)$-positive definite matrix.

Although $t_f \to \infty$, the feedback law obtained from equation (4.10) depends on time due to the time-varying coefficients $A_m(t)$ and $b_m(t)$. An exact analytical solution of the optimum control law can not be obtained generally but an approximate solution is always possible when it is expressed as a power-series with some order in ϵ. Relying on the small-parameter method (or perturbation method) (Nishikawa et al., 1976), the controller is

$$u(x_{m1}, t) = -K(t)\, x_{m1} \quad (4.13)$$

with the feedback gain

$$K(t) = K_o + \sum_{i=1}^{N} \epsilon^i K_i(t), \quad (4.14)$$

where

$$K_o := b_0^T R_0$$
$$K_i(t) := b_0^T R_i(t) + b_1(t)^T R_{i-1}(t) \quad (4.15)$$
$$(i = 1, 2, \ldots, N).$$

Here, it is known (Nishikawa et al., 1976) that R_0 is a constant positive definite matrix, which is obtained as a unique solution of the quadratic algebraic equation

$$R_0 E_o R_0 - R_0 A_0 - A_0^T R_0 = Q \quad . \quad (4.16)$$

Each $R_i(t)$ $(i = 1, 2, \ldots, N)$ is a periodic function with period L, obtained as a steady-state solution of the linear differential equation

$$\dot{R}_i = -R_i S - S^T R_i + H_i \quad (4.17)$$

with the final boundary conditions

$$\lim_{t_f \to \infty} R_i(t_f) = 0, \quad (4.18)$$

and

$$S := A_0 - E_o R_0$$
$$H_1(t) := -R_0 A_1 - A_1^T R_0 + R_0 E_1 R_0$$
$$H_i(t) := -R_{i-1}A_1 - A_1^T R_{i-1} + \sum_{j=1}^{i-1} R_j E_o R_{i-j}$$
$$+ \sum_{j=0}^{i-1} R_j E_1 R_{i-j-1} + \sum_{j=0}^{i-2} R_j b_1 b_1^T R_{i-j-2}$$
$$(i = 2, 3, \ldots, N) \quad (4.19)$$
$$E_o := b_0 b_1^T$$
$$E_1 := b_0 b_1^T + b_1 b_0^T \quad .$$

When the periodic coefficients $A_1(t)$ and $b_1(t)$ consist of a finite number of frequency components according to

$$A_1(t) = \sum_{k=1}^{m_a} [A_{ck}\cos(k\omega t) + A_{sk}\sin(k\omega t)]$$
$$b_1(t) = \sum_{k=1}^{m_b} [b_{ck}\cos(k\omega t) + b_{sk}\sin(k\omega t)], \quad (4.20)$$

the problem to obtain the desired steady-state

solutions $R_i(t)$ $(i = 2, 3, \ldots, N)$ is reduced to simply solving a set of algebraic equations with respect to constant coefficients for each frequency. These algebraic equations are derived through the harmonic balance method.

Example. Consider the first-order lag:

$$\dot{x} = -\frac{1}{\alpha}x + \frac{\beta(t)}{\alpha}u \tag{4.21}$$

with the time constant α and the gain $\beta(t)$ which is fluctuated sinusoidally with a small amplitude around the nominal value β_o. Also consider the retarded and nonlinear output

$$y = x(t - \theta) + \eta x(t - \theta)^3 . \tag{4.22}$$

In order to demonstrate the principle of the proposed design procedure, let us introduce the intermediate output simply as

$$z := x . \tag{4.23}$$

Then, x can be replaced by z in equation (4.21) which describes the subplant P_1, and

$$y = z(t - \theta) + \eta z(t - \theta)^3 \tag{4.24}$$

is the subplant P_2.

Let the sinusoidal fluctuation of $\beta(t)$ be modeled by

$$\beta_m(t) = \beta_o(1 + \epsilon \sin \omega t) . \tag{4.25}$$

Applying the perturbation technique and the harmonic balance method properly, the feedback control of first order is obtained as

$$u = -K(t)x_m(t) \tag{4.26}$$

with the gain calculated by

$$K(t) = K_o[1 + \epsilon K_1 \sin (2\omega t - \Phi)] , \tag{4.27}$$

where

$$K_o := \frac{a - 1}{\beta_o}$$

$$K_1 := \left[\frac{4 + \sigma^2}{4a^2 + \sigma^2}\right]^{1/2}$$

$$\Phi := \tan^{-1}\left\{\frac{2(a - 1)\ \sigma}{4\ a + \sigma^2}\right\} \tag{4.28}$$

$$a^2 := 1 + q\beta_o^2$$

$$\sigma := \omega\alpha_m ,$$

and α_m is a model value of α. Here, q is a positive parameter which may be adjustable to incorporate robustness for model/plant mismatch: a smaller value of q allows the system to be more robust at the sacrifice of sluggish response, and conversely, a larger value causes the system to be inevitably more sensitive to parameter uncertainty, giving quick response. A further discussion on these matters is out of the scope of this contribution.

SENSITIVITY ASPECTS

While a model-based control system can have high performance, it is generally sensitive to parameter identification errors or parameter variations. In order to estimate the sensitivity, therefore, it is sometimes useful to derive a sensitivity model (Frank, 1978).

Let P_1 and P_2 include the parameter a_1 and a_2, respectively, which are uncertain due to erroneous identification and/or are time-variant. For simplicity, let both a_1 and a_2 be scalar parameters. The parameter sensitivities of y with respect to a_1 and a_2 are formally defined by

$$\begin{aligned} y_{a_1} &= P_{a_1}u + Pu_{a_1} \\ &= P_2P_{1a_1}u + P_2P_1u_{a_1} \\ y_{a_2} &= P_{a_2}u + Pu_{a_2} \\ &= P_{2a_2}P_1u + P_2P_1u_{a_2} , \end{aligned} \tag{5.1}$$

where

$$y_{a_i} = \frac{\partial y}{\partial a_i}, \quad u_{a_i} = \frac{\partial u}{\partial a_i}, \quad P_{ia_i} = \frac{\partial P_i}{\partial a_i} \tag{5.2}$$

for $i = 1,2$. The u_{a_i}'s are fed back through the plant model $P_{m2}P_{m1}$, as well as y_{a_i}, to fulfill the equations

$$\begin{aligned} z_{ma_i} &= P_{m1}\ u_{a_i} \\ y_{ma_i} &= P_{m2}\ z_{ma_i} \\ e_{a_i} &= -d_{a_i} = -(y_{a_i} - y_{ma_i}) \\ v_{a_i} &= e_{a_i} - z_{ma_i} \\ u_{a_i} &= C\ v_{a_i} \end{aligned} \tag{5.3}$$

for $i = 1,2$, because P_{m1}, P_{m2}, and C contain no unknown parameters. Every notation with subscript a_i in the above equations is defined in a way similar to that in equation (5.2) to be self-understandable.

The resulting formal sensitivity model is shown schematically in Fig. 2. Thus, sensitivity information on any quantities in Fig. 1(a) are obtained in Fig. 2. As a special case, Garland and Marshall (1975) have derived a sensitivity model for the Smith predictor control systems, accompanied by the sensitivity points method.

The sensitivity outputs in Fig. 2 may be used, for example, to determine the value of q in equation (4.28) in order to cope with the robustness problem. When special classes of P_1 and P_2 are considered further investigations are necassary.

Parameter Identification via Sensitivity Operators

The model-based system by nature has a structure which is advantageous for correcting erroneous model parameters because the signal of the difference between the plant and model outputs is already available. The difference may express erroneous parameter identification of the model, and/or parameter time-variation in the plant, which lead to the model/plant mismatch. Now the parameter estimation scheme may be established: For P_1 and P_2, let the model parameter values a_{m1} and a_{m2} deviate from those of the plant, a_1 and a_2, by Δa_1 and Δa_2, respectively, as

$$a_1 = a_{m1} + \Delta a_1$$
$$a_2 = a_{m2} + \Delta a_2. \quad (5.4)$$

With small values of Δa_1 and Δa_2, P_1 and P_2 can be written by

$$P_i = P_{mi} + \Delta a_i P_{\Delta i} \quad (i = 1, 2), \quad (5.5)$$

where P_{ia_i} is defined in eq. (5.2), and

$$P_{\Delta i} := P_{ia_i}\big|_{a_i=a_{mi}}. \quad (5.6)$$

The plant can now be approximated by

$$P = P_{m2} P_{m1} + P_\Delta , \quad (5.7)$$

where

$$P_\Delta := \Delta a_1 P_{m2} P_{\Delta 1} + \Delta a_2 P_{\Delta 2} P_{m1}. \quad (5.8)$$

As a consequence, the first-order approximation of the difference between the plant and model outputs is

$$d = P_\Delta u \quad . \quad (5.9)$$

If a structure which measures the quantities $P_{m2}P_{\Delta 1}u$ and $P_{\Delta 2}P_{m1}u$ can be provided, Δa_1 and Δa_2 can be estimated from these values as well as from d to update a_{m1} and a_{m2} by an appropriate skillful algorithm.

A schematic diagram can be drawn as shown in Fig. 3. A special application of the above concept is the identification of system parameters when the plant is described in the frequency domain, which has been illustrated for one of the internal model principles by Itakura (1986).

CONCLUDING REMARKS

A relatively general class of systems has been considered for the introduction of a model-based concept which is not always limited to predictive control. Of course, for particular problems detailed case studies are a requisite in order to design a controller with robust stability characteristics in the face of plant/model mismatch. Although many results have been already obtained and described in the literature, further investigations are desirable with applications to real processes.

ACKNOWLEDGEMENT

The first author is grateful to the staff at the Department of Electrical Engineering, Chiba Institute of Technology, Japan, for helping him to stay in West Germany. Also he acknowledges gratefully financial support from the Japanese Private University Promotion Foundations. The second author appreciates financial support from the Stiftung Volkswagenwerk, Hannover, West Germany.

REFERENCES

Brosilow, C., and Tong, M. (1978). Inferential control of processes: Part II. The structure and dynamics of inferential control systems, AIChE J., 24, 492-500.

Economou, C.G., and Morari, M. (1986). Internal model control. 5. Extension to nonlinear systems, Ind. Eng. Chem. Process Des. Dev., 25, 403-411.

Etterich, D. (1988). Robuste Regelung von linearen Einfachsystemen mit Totzeit , Studienarbeit am Fachgebiet MRT, University of Duisburg.

Frank, P.M. (1974). Entwurf von Regelkreisen mit vorgeschriebenem Verhalten, G. Braun Verlag, Karlsruhe.

Frank, P.M. (1978). Introduction to system sensitivity theory , Academic Press, New York.

Garcia, C.E., and Morari, M. (1982). Internal model control. 1. A unifying review and some new results, Ind. Eng. Chem. Process Des. Dev., 21, 308-323.

Garland, B., and Marshall, J.E. (1975). Application of the sensitivity points method to a linear predictor control system, Int. J. Control, 21, 681-688.

Grimm, W.M., Itakura, H., and Frank, P.M. (1988). Design of robust SISO dead-time control loops, 12th IMACS World Congress on Scientific Computation, Paris (to appear).

Grimm, W.M., and Lee, P.L. (1987). Robust closed loop log modulus design of SISO Smith predictors for a special class of processes, submitted to Chem. Eng. Communications.

Ioannides, A.C., Rogers, G.J., and Latham, V. (1979). Stability limits of a Smith controller in simple systems containing a time delay, Int. J. Control, 29, 557-563.

Itakura, H. (1986). Parameter identification in process-model control system, IEEE Trans., AC-31, 1173-1175.

Itakura, H., and Nishikawa, Y (1983). Stability of a system with process model controller, Rep. of Chiba Inst. of Techn., Pt. of Sci. and Eng., 28, 21-33.

Jerome, N.F., and Ray, W.H. (1986). High-performance multivariable control strategies for systems having time delays, AIChE J., 32, 914-931.

Laughlin, D.L., Rivera, D.E., and Morari, M. (1987). Smith predictor design for robust performance, Int. J. Control, 46, 477-504.

MacDonald, N. (1985). Comments on a simplified analytical stability test for systems with delay, IEE Proc., Pt.D, 132, 237-238.

Nishikawa, Y., Sannomiya, N., and Itakura, H. (1976). Suboptimal feedback control of linear periodic systems, JOTA, 18, 271-284.

Palmor, Z.J. (1986). Robust digital dead time compensator controller for a class of stable systems, Automatica, 22, 587-591.

Parrish, J.R., and Brosilow, C.B. (1985). Inferential control applications, Automatica, 21, 527-538.

Pollard, J.F., and Brosilow, C.B. (1986). Adaptive inferential control, Proc. Amer. Control Conference, Seattle, 701-706.

Rivera, D.E., Morari, M., and Skogestad, S. (1986). Internal model control. 4. PID controller design, Ind. Eng. Chem. Process Des. Dev., 25, 252-265.

Smith, O.J.M. (1957). Closer control of loops with dead time, Chem. Eng. Prog., 53, 217-219.

Vit, K. (1979) Smith-like predictor for control of parameter-distributed processes, Int. J. Control, 30, 179-193.

Walton, K., and Marshall, J.E. (1987). Direct method for TDS stability analysis, IEE Proc., Pt.D., 134, 101-107.

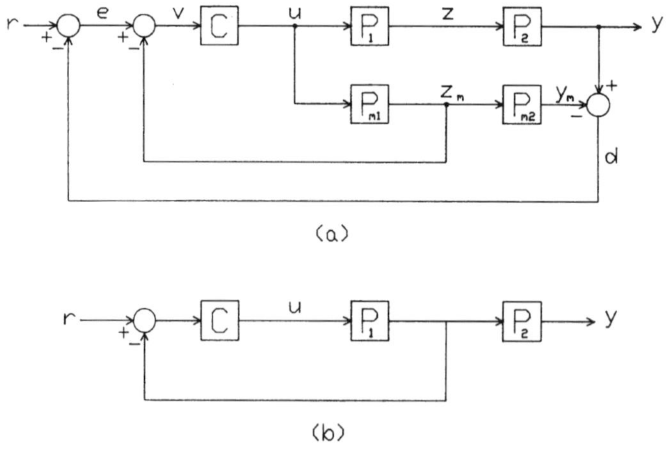

Fig. 1. (a) Model-based control system and (b) its equivalent structure when exactly $P_{m1} = P_1$ and $P_{m2} = P_2$.

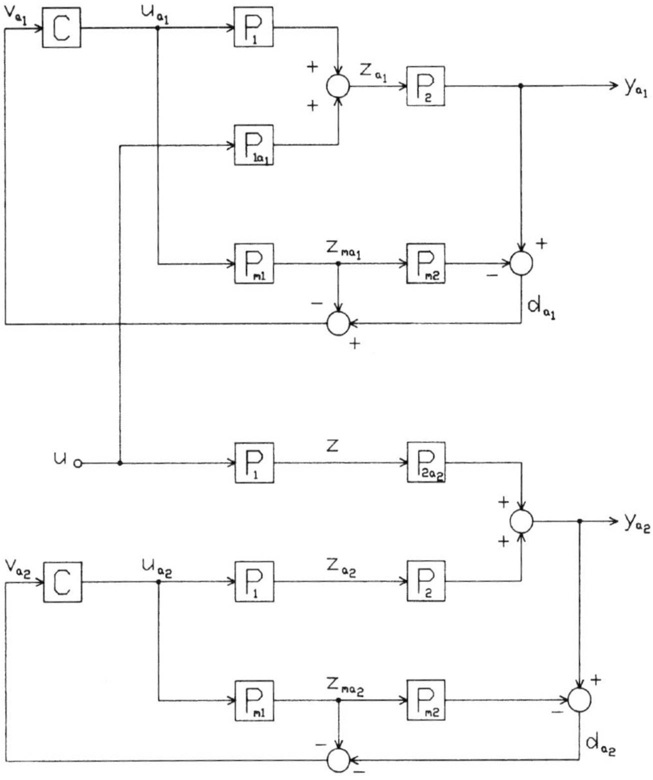

Fig. 2. Sensitivity model.

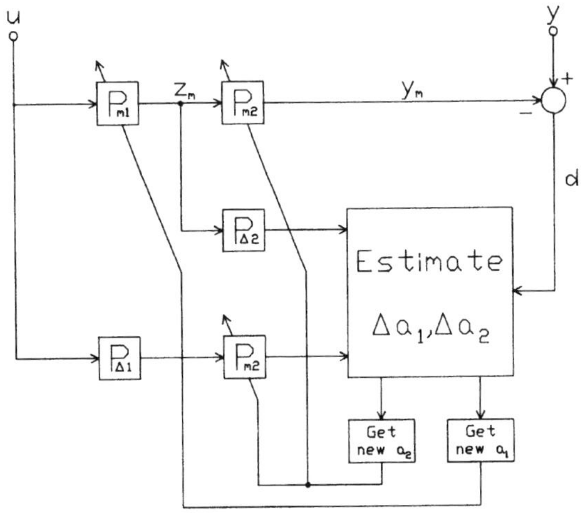

Fig. 3. Parameter estimation.

GENERIC MODEL CONTROL — THEORY AND APPLICATIONS

P. L. Lee* and G. R. Sullivan**

**Department of Chemical Engineering, University of Queensland, St Lucia, QLD 4067, Australia*

***Department of Chemical Engineering, University of Waterloo, Waterloo, Ontario N2L 3G1, Canada*

ABSTRACT

Generic Model Control (GMC) is a new control method developed by the authors that directly imbeds a process model in the control algorithm. The process model is usually a nonlinear descrption of the process derived from fundamental mass and energy balance considerations. Recent applications have employed nonlinear steady-state models and have approximated the dynamic response of the process by low-order models. The technique results in superior performance over linear based methods and has proven to be remarkable robust in the face of model and parameter uncertainties.

This paper presents a summary of GMC research to date together with simulations, pilot plant results, and industrial applications to demonstrate the approach. Since deadtime systems form an important part of chemical engineering processes a deadtime compensation version of GMC is presented. In addition, a plant/model mismatch compensation algorithm is developed that is essentially a GMC controller designed to drive the mismatch to zero.

1.0 INTRODUCTION

High performance multivariable controllers are now becoming commonplace in many industrial applications. This trend has been driven by the increasing capacity of modern process control computers together with the need for more sophisticated algorithms to contend with new integrated plant designs. Model-based controllers have drawn considerable attention in recent years with the most popular algorithms being Dynamic Matrix Control (DMC) [Cutler & Ramaker, 1980], Model Algorithmic Control (MAC) [Rouhani & Mehra, 1982] and Internal Model Control (IMC) [Morari et al, 1982 to 1986]. All of these controllers employ a linear model to represent the process and have demonstrated superior performance over traditional techniques for the control of a wide variety of chemical engineerig processes.

For the regulation of severely nonlinear chemical processes (eg. high purity distillation, nonisothermal reactor), the performance of linear model-based controllers tends to degrade. In order to address this problem, effort has been extended to "globally" linearize nonlinear systems [Brockett, 1978; Kravaris and Chung, 1987] by finding of global nonlinear transformations so the transformed system exhibits linear characteristics. On another front, when linear models are used to control highly nonlinear processes, gain scheduling is a popular technique that essentially provides new model parameters based upon a local linearization. MacDonald [1987] has successfully demonstrated this technique through the application of DMC for a high purity distillation. The research literature is abound [Seborg et al, 1986] with many applications of adaptive control where parameters are modified as process output data has been collected. In some cases, parameters have been adapted to handle slowly changing process behavious [Astrom, 1986; Huang and Stephanopoulos, 1985].

Generic Model Control (GMC) is a new control method developed by Lee and Sullivan [1988] that directly imbeds a process model in the control algorithm. The process model is usually a nonlinear description of the process derived from fundamental mass and energy balance considerations. Recent applications have employed nonlinear steady-state models and have approximated the dynamic response of the process by low-order models. This technique takes advantage of the rich library of steady-state process models available to control engineers in the process industries.

GMC can be shown to produce more classical control algorithms such as PI control by appropriate choices of the process model. The advantage that GMC offers over these techniques is the ability to directly use nonlinear descriptions of the process. In addition, unlike Dynamic Matrix control, GMC solves a single-step optimization problem while still retaining a time-horizon approach.

This paper will review the background to this new approach, present the theoretical basis, and summarize the work to date. More technical detail can be found in a series of papers by Lee, Sullivan et al [1988]. It will also review and present results obtained from industrial applications, and several laboratory and simulation studies. The theory and application will be presented that addresses common issues related to model-based controllers such as model-structure and dimensionality, dead-time compensation, robustness in light of plant/model mismatch and plant-wide applications. These results have all highlighted the utility of the method and will further amplify the basis for the technique.

2.0 CONTROL STRUCTURE

Consider a process described by the following equation:

$$\dot{x} = f(x,u,d,t) \tag{2.1}$$

$$\dot{y} = g(x) \tag{2.2}$$

where x is the state vector of dimension n, u is the vector of manipulated variables of dimension m, d is the vector of disturbance variables l, and y is the output vector of dimension p. In general, f and g are some nonlinear functions. It follows from equations 2.1 and 2.2 that:

$$\dot{y} = G_x f(x,u,d,t) \tag{2.3}$$

where $G_x = \frac{\partial g}{\partial x}$

In classical optimal control, the trajectory of y is usually compared against some nominal trajectory, $(y)^*(t)$, as a measure of system performance. As an alternative, consider the performance of the system to be such that:

$$(\dot{y})^*(t) = r^*(y) \tag{2.4}$$

where r^* represents some arbitrary function to be specified. Performance specifications using time derivatives of the output variables have been previously considered in robotic manipulators [Guo and Sardis, 1984].

Let us examine one reasonable choice of this specification function r^*. When the process is away from its desired steady state y^* we would like the rate of change of y, y, to be such that the process is returning towards steady state, ie.

$$\dot{y} = K_1(t)(y^*-y) \tag{2.5}$$

where $K_1(t)$ is some diagonal matrix.

In addition, we would like the process to have zero offset, ie.

$$\dot{y} = K_2(t)\int(y^* - y)dt \tag{2.6}$$

where $K_2(t)$ is some diagonal matrix. In the following we will consider that $K_1(t)$ and $K_2(t)$ are constant with respect to time.

Good control performance will be given by some combination of these objectives, ie.

$$(\dot{y})^* = K_1(y^* - y) + K_2\int(y^* - y)dt \tag{2.7}$$

Suitable choices of K_1 and K_2 can be made to achieve a variety of responses in y(t) as detailed later. We would like to choose u(t) such that the system follows $(\dot{y})^*(t)$ as closely as possible. This transforms into the optimal control problem, OC1,

Given $\dot{x} = f(x,u,d,t)$
$y = g(x)$
Choose u(t)
Such that $|U| < \alpha$
To minimize

$$\int_0^{t_f} \left(h(x,u,d,t)^T W h(x,u,d,t) \right) dt \tag{2.8}$$

where:

$$h(x,u,d,t) = G_X f(x,u,d,t) - K_1(y^*-y) - K_2\int_0^t (y^*-y)dt \tag{2.9}$$

and W is an appropriate positive-definite weighting matrix. If the control is feasible with respect to the constraints, and the dimensions of m and p are the same, and at least one element of u appears in each of the m equations represented by equation 2.9, and every element of u is represented in these equations, then the solution of OC1 is given by the solution of m equations in m unknowns:

$$G_x f(x,u,d,t) - K_1(y^*-y) - K_2\int_0^t (y^*-y)dt = 0 \tag{2.10}$$

Otherwise the solution reduces to the choice of u to minimize the instantaneous value of h(x,u,d,t) at every point in time. This will involve performing a single time-step on-line integration.

In general, the exact process model is rarely known, and an approximate model is introduced such that equation 2.10 becomes:

$$\hat{G}_x \hat{f}(x,u,d,t) - K_1(y^*-y) - K_2\int(y^*-y)dt \tag{2.11}$$

where $\hat{f}$ and $\hat{G}_X = \partial\hat{g}/\partial x$ represents the approximation to the true model.

The control manipulations that satisfy equation 2.11 must result in good control if the process model, $\hat{f}$ and $\hat{G}_X$, are reasonable and $(y)^*(t)$ is reasonable such that we operate within the control action constraints. Any degradation in control performance due to interactions or nonlinearities can be explained through the choice of $\hat{f}$ and $\hat{G}_X$ to be used within the control law of equation 2.11.

In circumstances where the dimensions of m and p are equal and the "difference order" of the system [Freund, 1975] is equal to 1, explicit solutions of equation 2.11 are possible. This is similar to th approach used in "nonlinear decoupling" for systems with the dfference order not equal to 1, as outlined by Freund [1973, 1982].

The control law has four desirable characteristics:

1. The control law as represented by equation 2.11 has within its structure an approximate process model.
2. Any inaccuracies introduced by this approximation will result in $\dot{y}$ not tracking $(\dot{y})^*(t)$ but this will be compensated by the integral term in the control algorithm. This integral term in the control law ensures that the controller is robust despite modelling errors.
3. It is a single step law that has time horizon characteristics as defined by equation 2.11 The solution from the last control step is a good approximation to the solution of the current control step.
4. There is no need to perform an on-line integration of the process model providing that all states are observable.

3.0 RESPONSE SPECIFICATION

Consider a multi-variable system such that:

$$\dot{y} = f(y,u,t) \quad (3.1)$$

and that the process model used is perfect, ie.

$$\hat{f} = f \quad (3.2)$$

From equation 2.11, the control law is:

$$f(y,u,t) - K_1(y^*-y) - K_2\int(y^*-y)dt = 0 \quad (3.3)$$

and

$$y = K_1(y^*-y)+K_2\int(y^*-y)dt \quad (3.4)$$

It can be seen that by different choices of K_1 and K_2 the performance specification (equation 3.4) can be altered for each variable separately. Thus, one can use these values to select any "reasonable" desired response for the system. "Reasonable" implies that the parameters are chosen in relation to the system's natural dynamic response. How well the system matches this performance index will be governed by how closely the chosen model matches the plant behaviour. This will be illustrated further in the following examples.

If Laplace transforms of equation 3.4 are taken, the resulting transfer function becomes:

$$\frac{y(s)}{y^*(s)} = \frac{2\tau\xi s+1}{\tau^2 s^2+2\tau\xi s+1} \quad (3.5)$$

where $\tau = \frac{1}{\sqrt{k_2}}$ and $\xi = \frac{k_1}{2\sqrt{k_2}}$. This system does not yield the same response as a classic second-order system [Stephanopoulos, 1984]. However, similar plots to the classic second order response showing the normalized response of the system y/y^* versus normalized time t/τ with ξ as a parameter can be produced and is shown in Figure 1. The design procedure can be specified as follows:

1. Choose ξ from Figure 1 to give desired shape of response
2. Choose τ from Figure 1 to give "appropriate" timing of response in relation to known or estimated plant speed of response
3. Calculate k_1 and k_2 for each output variable using the following equations:

$$k_1 = \frac{2\xi}{\tau} \quad (3.6)$$

$$k_2 = \frac{1}{\tau^2} \quad (3.7)$$

These values constitute the diagonal elements of the matrices K_1 and K_2.

4.0 LINEAR CONTROL IN A GMC FRAMEWORK

If the approximate process model in equation 2.11 for a SISI system is chosen such that:

$$G_x f = bu \quad (4.1)$$

then the solution of equation 2.11 becomes:

$$u = \frac{K_1}{b}(y^* - y) + \frac{K_2}{b}\int (y^*-y)dt \quad (4.2)$$

This is obviously in the form of a conventional PI control algorithm. However, the choice of the approximate model given by equation 4.1 is obviously a very crude model. The control performance as specified from the design procedure will only be achieved if the real process behaves as the process model. This is unlikely with equation 4.1 as the process model.

The following section examines one other possible choice of the process model and develops the appropriate control algorithm. In this case, we assume that disturbances are unknown, thereby leading to a feedback controller. If disturbances are modelled as part of the GMC model, then Lee and Sullivan [1988] show that a GMC analysis results in a feedforward/feedback controller with high performance characteristics.

4.1 FIRST-ORDER MODEL (FO) - SISO SYSTEM

Consider the approximate process model

$$\frac{y_{(s)}}{u_{(s)}} = \frac{K_m}{\tau_m s+1} \quad (4.3)$$

where s is the Laplace operator. This can be written in the time domain as:

$$\dot{y} = ay + bu \quad (4.4)$$

where $a = -\frac{1}{\tau_m}$ and $b = \frac{K_m}{\tau_m}$

Substituting into equation 2.11 and rearranging yields:

$$u = b^{-1}\{-ay + k_1(y^*-y) + K_2(y^*-y)dt\} \quad (4.5)$$

The velocity form of equation 4.5 is written as:

$$\Delta u = b^{-1}\{-a\Delta y + K_1\Delta(y^*-y)+K_2T_S(y^*-y)\} \quad (4.6)$$

where T_S is the sampling time. Let e be the error signal:

$$e = y^*-y \quad (4.7)$$

and noting that $\Delta e = \Delta y^*-\Delta y$, then equation 4.6 can be written in velocity form as:

$$\Delta u = b^{-1} \{(a + K_1)\Delta e + K_2T_S e-a \Delta y^*) \quad (4.8)$$

This represents a modified PI control law. The first term provides the proportional action, the second term the integral action, and the final term provides improved servo response. The final term would nearly always be zero, except when setpoint changes are introduced. "Tuning" parameters in equation 4.8 which can be determined a-priori are functions of only the model parameters a and b, and the performance parameters K_1 and K_2.

4.2 EXPERIMENTAL

For an example of a real-time application, the GMC approach has been implemented for a heated stirred tank. The process is of similar design to that described by Stephanolplous [1984].

Figure 2 shows a schematic of the process. The flow was fixed manuall and monitored using a rotameter. Inlet and outlet temperatures, and valve position are monitored by a computer. An analog signal from the computer manipulated the steam valve to control the heat flow to the coil.

The dynamics of the steam heated stirred tank can be described by the following equation:

$$\frac{dT}{dt} = \frac{F}{v}(T_{in} - T) + \frac{Q}{\rho C_p V} \tag{4.9}$$

For a first order system, the GMC control equation can be obtained using equation 4.1 with $a = -F/v$ and $b = 1/\rho CpV$. This gives the discrete form of the controlled as:

$$\Delta Q = \rho C_p V\{-\frac{F}{v}+K_1)\Delta e+K_2 T_s e-\frac{F}{v}\Delta T^*\} \tag{4.10}$$

where the change in heat, ΔQ, is translated to a change in valve position using the valve characteristic curve. Since the dead time of the system is small, a first order model was used where the time constant, F/v, was estimated to be 200 seconds from the open loop response. In order to demonstrate the tight control that can be realized, even when using an imperfect model, $\xi = 4$ was chosen. Using Figure 1, $\xi = 4$ gives a rise time of 0.55 for t/τ. The desired time to reach set point is chosen as 220 seconds. This gives a τ of 400 seconds. From equations 3.6 and 3.7, the controller constants are found to be $K_1 = 0.02\ s^{-1}$ and $K_2 = 6.25 \times 10^{-5}\ s^{-2}$.

Actual controlled runs for servo and disturbance problems using a first order model of the system are shown in Figures 4 - 6. For the system all measurements were averaged over 10 seconds and then used in the control equation and therefore a T_s of 10 seconds was used. The disturbance to the system was an injection of 250 mL of 10°C water to the stirred tank and the excellent recovery characteristics are demonstrated in Figure 3. Responses for the step changes in set point are shown in Figures 4 and 5 together with performance specification curves that would result from application of a perfect model.

4.3 LINEAR MULTIVARIABLE SYSTEMS

Consider the linear regulator problem where:

$$f(y,u,d,t) = Ay + Bu + Dd \tag{4.11}$$

and choose the approximate model:

$$\hat{f}(y,u,d,t) = \hat{A}y + \hat{B}u + \hat{D}d \tag{4.12}$$

assuming $\hat{B}$ is invertible, equation 2.11 can be solved to give the control law:

$$u = \hat{B}^{-1}K_1(y^*-Y)+\hat{B}^{-1}K_2\int(y^*-y)dt - \hat{B}^{-1}\hat{D}d - \hat{B}^{-1}\hat{A}y \tag{4.13}$$

The first two terms of 4.13 represent a multivariable feedback controller whereas the third term provides feedforward action. The final term provides state decoupling.

Substituting equation 4.13 and equation 4.11 rearranging yields the closed loop response as:

$$\dot{y} = (A-B\hat{B}^{-1}\hat{A})y + (D-B\hat{B}^{-1}D)d + B\hat{B}^{-1}(K_1(y^*-y) + K_2\int(y^*-y)dt) \tag{4.14}$$

The first two terms of equation 4.14 quantify the plant-model mismatch whereas the last two terms represent the control to compensate for the mismatch. The GMC formulation given by equation 2.11 has generalized this approach for the nonlinear case.

5.0 MODEL STRUCTURE AND DIMENSIONALITY

In all the examples considered thus far, we have assumed that the number of state variables and the number of manipulated variables are equal. Such a formulation certainly has the advantage of a simple solution structure, but it it not a prerequisite for a GMC solution. If the number of manipulated variables, m, and the number of observation variables, p, are equal then the solution for the control at each time step is given by:

$$\dot{x} = \hat{f}(x,u,d,t) \tag{5.1}$$

$$y = \hat{g}(x) \tag{5.2}$$

$$\hat{G}_x\hat{f}-K_1(y^*-y) - K_2\int(y^*-y)dt = 0 \tag{5.3}$$

This represents the solution of n + 2p equations in n + 2p unknowns (x,y,u) to yield the solution at each time step. Powerful differential-algebraic solution techniques currently available coupled with modern process control computer hardware will allow the solution to be performed on-line.

There are two situations where the complexity of the problem can be retained with a simplified solution structure. If f and g are linear functions, then an analytical integration can be performed thereby leading to a closed form solution. This problem, together with that of deadtime compensation, is addressed section 6. In the second case, if the nonlinear function, f, can be separated into a nonlinear steady-state function with assumed first-order dynamics, then it can be shown that the control calculation only involves on-line solution of the nonlinear steady-state model. The chemical engineering discipline is rich in use of simplified and complex nonlinear steady-state models and such a formulation would allow the use of any of these process models within a control framework. This approach is outlined in the next section.

5.1 GMC USING STEADY-STATE PROCESS MODELS

The process model of equation 5.1 and 5.2 assumes that an approximate dynamic model of the process can be derived, usually from fundamental mechanistic considerations. However, it is far more common to have access to models that only contain steady-state information, ie.

$$\hat{f}_{ss}(x,u,d) = 0 \tag{5.4}$$

$$y = \hat{g}(x) \tag{5.5}$$

where $\hat{f}_{ss}$ represents the approximate model of the steady-state behaviour.

These models describe the steady-state nonlinear, interactive behaviour of the process to a better accuracy than corresponding linear cause-and-effect models. However, for control purposes, we need some estimate of the plant's dynamic response. There are many different identification techniques available eg. [Harris and Billings, 1981] which could be used on-line to identify the plant dynamics. However, this section presents one possible approach to simplify this procedure. Other approaches together with more details can be found in Cott et al [1988].

5.2 DYNAMIC APPROXIMATION

It is often possible to obtain an appropriate estimate of the average time constants of the system from step response data. Although these estimates may be inaccurate at other operating conditions or at different times, the degree of approximation is often sufficient to obtain good control performance.

Thus, the response of the output variables can be approximated as:

$$\dot{y} \approx T^{-1}(y_{ss}-y) \qquad (5.6)$$

where T represents a matrix of estimated or identified time constants and y_{ss} is the ultimate value of the output variables if no further control action is taken.

Given this approximate description of the dynamic response of the system and the performance specification of equation 2.7, the ultimate response can be calculated as:

$$y_{ss} = y+T(k_1(y^*-y)+K_2\int(y^*-y)dt) \qquad (5.7)$$

The control action required to achieve this performance can be found from the nonlinear steady-state model of equations 5.4 and 5.5 with y replaced with y_{ss} as calculated by equation 5.7.

5.3 DYNAMIC EFFECTS OF DISTURBANCES

Implicit steady-state process models such as that described in equations 5.4 and 5.5 do not model the dynamic effects of the disturbance variables, d. A simple but effective method to improve the model performance is to pass the disturbance variables through dynamic "filters" to compensate for the lack of dynamic structure in the process model. These "filters" can be determined from open-loop responses and is similar to conventional feedforward control.

5.4 INDUSTRIAL APPLICATION

As an example of an industrial application, a GMC was implemented on a depropanizer distillation column at AGEC's ethylene manufacturing complex at Joffre, Alberta, Canada. The depropanizer selected is part of the ethylene byproducts recovery facilities. The byproducts distillation columns are prime candidates for product recovery and energy use optimization. Product values are subject to monthly variations, whereas energy costs vary plant loading and ambient conditions. Hence the depropanizer was chosen for GMC implementation, with strong consideration given to potential optimization implementation.

The column separates a mixed hydrocarbon feed stream into a 'C3' stream overhead (primarily propylene) and a 'C4+' stream. Product specifications are: 2 weight percent C4+ overhead; 0.3 weight percent C3 in the C4+ bottoms. The column has composition analysers on both the feed and overhead streams. Unfortunately, no reliable bottoms stream analysis is available, so a correlation between bottom tray temperature, column pressure and bottoms stream composition was developed.

The implementation of Generic Model Controllers usually involves two steps: a model parameter update and a control action calculation. The model parameter update step provides a form of feedback to the controller allowing it to account for the current operation of the column. The control action calculation step provides the control to achieve the desired produce compositions. Here we define the dimension of the measurement vector, y, to be 2, consisting of y and x, the composition of light component in the overhead and bottoms streams respectively. The manupulated variables were the reflux ratio and the reboil rate. The Generic Model Controller using the Douglas-Jafarey-McAvoy (DJM) [Jaffarey et al 1979] steady state, short cut design model for controlling the depropanizer follows the same pattern:

1. Model Parameter Update
   - measure the distillate, bottoms, reflux flowrates and the pressure and the top and bottom compositions (D,B,L,P,y,x) from the column
   - back-calculate a pseudo feed stream based on products
   - solve for the DJM model parameter:
     - N, the total number of stages

2. Control Action Calculations
   - determine the product setpoints, y* and x*
   - measure the feed flowrate and composition and the column pressure (F,z,P) from column and put through digitial filters to compensate for the lack of dynamic information in the process model
   - using the filtered data, solve the control law for the reflux ratio and reboil rate
   - implement the control actions on the column

An information flow diagram for the system is shown in Figure 6. The model parameter update strategy developed by Cott et al [1988], was adopted.

This implementation is straightforward and can be accomplished on any process control computer, or with modern instrumentation systems with enhanced function features. The tuning of the GMC for the column requires determining the following quantities:

1. the dynamic filters for F and z
2. the sampling frequency
3. the estimate time constants of the process
4. the desired trajectory of the process outputs, y and x

Over a two-week trial period, the regulatory performance of the DJM GMC proved to be at least as good as that of the previous material balance strategy. As expected, the advantage of the DJM GMC was in servo control performance. Figure 7 illustrates the performance of GMC as compared to the previous control scheme for a 0.10 mole percent change in C4+ overhead composition setpoint. The GMC accurately predicted the new steady-state operating conditions, reducing response time and effectively eliminating overshoot. In both cases, bottoms tray temperature was controlled to within 0.5 degrees Celsius and, therefore, close to bottoms composition setpoint.

6.0 MULTIVARIABLE DEADTIME SYSTEMS

Deadtime (or time delay) frequently occurs in chemical engineering process because of transportation and measurement delays. When the deadtime becomes large, conventional control systems provide quite unsatisfactory closed-loop response. Thus, in practice, compensation of deadtimes in control systems is very important.

Many alternative schemes have emerged to deal with deadtimes in systems over the last thirty years [Watanabe and Iko, 1981]. For example, the Smith Predictor [Smith, 1975] is a well-known deadtime compensation technique. It utilizes a process model to add feedback from the characteristic equation of the closed-loop system and converts the control problem with deadtime to one without deadtime. Naturally, for process model-based control, the compensation of deadtime is also one of the significant issues to be addressed. Holt and Morari [1985], Russell and Perkins [1987] discussed the effects of the deadtimes on the best response achievable by any controller for a system ie. controllability and operability characteristics (called dynamic resilience). The methods to treat deadtime in MAC and DMC are to enlarge the prediction and optimization horizons to include the effect of the inputs at present time, so that these inputs could be determined by optimization. Here we will apply the approach of Jerome and Ray [1986] to a GMC control environment.

Although the theory is presented here for linear systems, there exists a direct extension for nonlinear systems [Lee et al, 1988a]. For most n x n linear systems with multiple deadtimes, a transfer function representation adequately describes the system behaviour:

$$\begin{bmatrix} y_1(s) \\ \vdots \\ y_n(s) \end{bmatrix} = \begin{bmatrix} \frac{K_{11}e^{-d_{11}s}}{\tau_{11}S+1} & \cdots & \frac{K_{1n}e^{-d_{1n}s}}{\tau_{1n}S+1} \\ \vdots & & \vdots \\ \frac{K_{n1}e^{-d_{n1}s}}{\tau_{n1}s+1} & \cdots & \frac{K_{nn}e^{-d_{nn}s}}{\tau_{nn}s+1} \end{bmatrix} \begin{bmatrix} u_1(s) \\ \vdots \\ u_n(s) \end{bmatrix}$$

where K_{ij}, τ_{ij}, d_{ij} $(i,j = 1 \ldots n)$ are the gains, time constant and deadtimes of output variable $y_i(t)$ to input variable $u_j(t)$ respectively. The parameters can be achieved using one of a variety of identification techniques (e.g. open loop response).

Define the sub-output variables x_{ij} as

$$\dot{x}_{ij} = \frac{1}{\tau_{ij}} x_{ij} + \frac{k_{ij}}{\tau_{ij}} u_j(t-d_{ij}) \quad (i,j=1\ldots) \tag{6.2}$$

Then we can rewrite equation 6.1 in the time domain as:

$$y_i(t) = \sum_{i=1}^{n} x_{ij}(t) \tag{6.3}$$

$$\dot{y}_i(t) = \sum_{j=1}^{n} \left(-\frac{x_{ij}}{\tau_{ij}}\right) + \sum_{j=1}^{n} \frac{K_{ij}}{\tau_{ij}} u_i(t-d_{ij}) \tag{6.4}$$

$$(i = 1 \ldots n)$$

Equation 6.2 and 6.4 represent n(n+1) differential equations in n outputs and n manipulated variables. The discrete-solution of equation 6.2 is:

$$x_{ij}(t) = e^{-\frac{\Delta t}{\tau_{ij}}} x_{ij}(t-\Delta t) + (1-e^{-\frac{\Delta t}{\tau_{ij}}})K_{ij} \, u_j(t-\Delta t-d_{ij}) \tag{6.5}$$

where t is the sample time. From equation 6.4 we have:

$$\dot{y}_i(t+d_{ii}) = -\sum_{j=1}^{n} \frac{x_{ij}(t+d_{ii})}{\tau_{ij}} + \sum_{j=1}^{n} \left[\frac{K_{ij}}{\tau_{ij}} u_j(t-(d_{ij}-d_{ii}))\right] \tag{6.6}$$

Using GMC for $y_i(t+d_{ii})$, the equations to determine the control law are:

$$\dot{y}_i(t+d_{ii}) = K_{1i}(y^*(t+d_{ii})-y_i(t+d_{ii})) + K_{2i}\int_0^{t+d_{ii}} (y^*_i(t)-y_i(t))dt \tag{6.7}$$

$$(i = 1 \ldots n)$$

where K_{1i}, K_{2i} are the diagonal elements of the GMC performance specification matrices K_1 and K_2 respectively.

Suppose the following condition is satisfied:

$$d_{ii} < d_{ij} \quad (i \neq j;\ i,j = 1 \ldots n) \tag{6.8}$$

That is, the diagonal deadtime is less than the others in each row. From equations 6.6 and 6.7, we obtain:

$$u_i(t) = \frac{\tau_{ii}}{K_{ii}}\Big\{ \sum_{j=1}^{n} \frac{1}{\tau_{ij}} x_{ij}(t+d_{ii}) - \sum_{\substack{j=1 \\ j\neq i}}^{n} \frac{K_{ij}}{\tau_{ij}} u_j(t-(d_{ij}-d_{ii})) + K_{ii}(y^*_i(t+d_{ii}) - y_i(t+d_{ii})) + K_{2i}\int_0^{t+d_{ii}} (y^*_i(t)-y_i(t))dt \tag{6.9}$$

$$(i = 1 \ldots n)$$

Equation 6.9 determines the control action at each sampling time, which will make the system output follow the desired trajectory.

One may question the validity of condition 6.8. However, for a wide variety of chemical engineering systems, the equations can be ordered such that 6.8 is true. Physically, this can be interpreted that the minimum deadtime associated with each manipulated variable corresponds to an unique output variable. Further discussions on rearrangement can be found in the papers of Holt and Morari [1985], Russell and Perkins [1987], and Jerome and Ray [1986].

Note that this multivariable problem now has independent solutions for each manipulated variable that rely on past values of the other manipulated variables. Equation 6.9 also requires the prediction of $x_{ij}(t+d_{ii})$ and $y_i(t+d_{ii})$ using the present measurement of $y_i(t)$ and past data. Equation 6.3 and 6.5 can be used to accomplish this as:

$$x_{ij}(t+d_{ii}) = e^{-\frac{\Delta t}{\tau_{ij}}} x_{ij}(t-\Delta t+d_{ii}) + (1-e^{-\frac{\Delta t}{\tau_{ij}}}) K_{ij}u_j(t-\Delta t-(d_{ij}-d_{ii})) \qquad (6.10)$$

$$y_i(t+d_{ii}) = y_i(t) + \sum_{j=1}^{n} (x_{ij}(t-d_{ii}) - x_{ij}(t)) \qquad (6.11)$$

The whole procedure can be summarized as Algorithm 6.1.

Algorithm 6.1:

At each sampling time instant

Step 1 Update $x_{ij}(t+d_{ii})$ by equation 6.10
Step 2 Update $y_i(t+d_{ii})$ by equation 6.11
Step 3 Calculate control action $u_i(t)$ ($i = 1 \ldots n$) by equation 6.9
Step 4 Implementing $u_i(t)$, shift to next sampling

6.0 MULTIVARIABLE DEADTIME APPLICATION

The above algorithm has been tested through simulation for a binary distillation column that separates a mixture of methanol and water [Wood and Berry, 1973]. This system was also studied by Garcia and Morari [1985a], and Arulalan and Deshpande [1987]. The process transfer functions are:

$$\begin{bmatrix} Y_1(s) \\ Y_2(s) \end{bmatrix} = \begin{bmatrix} \frac{12.8e^{-s}}{16.7s+1} & \frac{-18.9e^{-3s}}{21.0s+1} \\ \frac{6.6e^{-7s}}{10.9s+1} & \frac{-19.4e^{-3s}}{14.4s+1} \end{bmatrix} \begin{bmatrix} U_1(s) \\ U_2(s) \end{bmatrix} + \begin{bmatrix} \frac{3.8e^{-8.1s}}{14.9+1} \\ \frac{4.9e^{-3.4s}}{13.2s+1} \end{bmatrix} d(S) \qquad (6.12)$$

Clearly condition 6.8 is satisfied.

As shown in Figure 1, if we choose $\xi = 3.0$ the relative rise time is 0.63. We would like the closed-loop responses rise time close to the average time constant, say 15.75, which gives $\tau = 25$. This implies $K_1 = .24$, $K_2 = 0.0016$ for both output variables (overhead composition and bottom composition). The responses of the outputs corresponding to a step change in the in Figure 8, which follows the specified trajectory. Figure 9 shows the output responses and input moves for a step change in the setpoint of the bottom composition. Figure 8 and 9 also present the results obtained by Simplified Model Predictive Control [Arulalan and Deshpande, 1987]. The advantage of the new method is that performance is guaranteed without any on-line tuning or complicated design procedure.

Of course, any desired trajectory could be selected as shown in Figure 1. For example, if we require fast responses, $\xi = 10$, $\tau = 10$ could be chosen. In this case, responses are similar to those obtained by IMC with a filter parameter $\alpha = 0$ [Garcia and Morari, 1985a].

7.0 SUMMARY

We have shown that GMC provides a suitable framework for development of model-based controllers. Within the GMC structure, the control engineer can concentrate on provision of a realistic process model for the control algorithm, with guaranteed perspecified performance and no on-line tuning. Direct extension to compensate for process deadtime has been presented. Future work will address process/model mismatch the inclusion of constraints into the GMC formulation.

8.0 ACKNOWLEDGEMENTS

The authors would like to thank B. Cott, B. Howie, M. Brown, B. Doerr and M. Whaley at Waterloo together with W. Zhou at Queensland. In addition, we would like to acknowledge the industrial access provided by Alberta Gas Ethylene Co. Ltd. and the assistance of R. Durham. Some of this work was supported through grants from the Natural Sciences and Engineering Research Council of Canada. Finally, the authors appreciate the ongoing support of IBM Australia and IBM Canada through provision of process control hardware and software facilities.

9.0 REFERENCES

Arulalan, G.R., and P.B. Deshpande. (1987) Simplified Model Predictive Control, I. & E.C. Res., 26, 347.

Astrom, K.J.. (1986). Adaptation, Auto-Tuning and Smart Controls, Chemical Process Control - CPC III, Proceedings of the Third International Conference on Chemical Process Control, Asilomar, California, Jan. 1986, 427.

Brockett, R.W. (1978). Feedback Invariants for Nonlinear System, Proc. 7th IFAC World Cong., Helsinki.

Cott, B.J., R.G. Durham, P.L. Lee, and G.R. Sullivan. (1988a). Process Model-Based Engineering Part 1 - Model Selection and Parameter Update, (Submitted).

Cott, B.J., R.G. Durham, P.L. Lee, and G.R. Sullivan. (1988b). Process Model-Based Engineering Part 2 - Process Control, (Submitted).

Cott, B.J., R.G. Durham, P.L. Lee, and G.R. Sullivan. (1988c). Process Model-Based Engineering Part 3 - Process Optimization (Submitted).

Cutler, C.R., and B.L. Ramaker. (1980). A Computer Control Algorithm, Joing Automatic Control Conference Preprints, Paper WP5-B, San Francisco.

Economou, C.G., M. Morari, and B.O. Palsson. (1986). Internal Model Control. 5. Extension to Nonlinear Systems, I & E.C. Proc. Des. Dev., 25, 403.

Economou, C.G., and M. Morari. (1986). Internal Model Control. 6. Multiloop Design, I. & E.C. Proc. Des. Dev., 25, 411.

Freund, E. (1973). Decoupling and Pole Assignment in Nonlinear System, Electronic Letters, 9, 373.

Freund, E. (1975). The Structure of Nonlinear Systems, Int. J. Control, 21, 443.

Freund, E. (1982). Fast Nonlinear Control Concepts for Nonlinear Computer Controlled Manipulators, Int. J. of Robotics Res., 1, 65.

Garcia, C.E., and M. Morari. (1982). Internal Model Control. 1. A Unifying Review and Some New Results, I. & E.C. Proc. Des. Dev., 21, 308.

Garcia, C.E. and M. Morari. (1985a). Internal Model Control. 2. Design Procedure for Multivariable Systems, I & E.C. Proc. Des. Dev., 24, 472.

Garcia, C.E., and M. Morari. (1985b). Internal Model Control. 3. Multivariable Control Law Computation and Tuning Guideline, I & E.C. Proc. Des. Dev., 24, 484.

Guo, G.L. and G.N. Sardis. (1985). Optimal/PID Formulation for Control of Robotic Manipulators, IEEE Int. Conf. on Robotics and Automation, S. Louis.

Harris, C.J., and S.A. Billings. (1981). Self-Tuning and Adaptive Control: Theory and Application, IEE Control Eng. Series 15, London, Peter Peregrinus, Ltd.

Holt, B.R., and M. Morari. (1985). Design of Resilient Processing Plants. V. The Effect of Deadtime on Dynamic Resilience, Chem. Eng. Sci., 40(7), 1229.

Huang, H., and G. Stephanopoulos. (1985). Adaptive Design for Model Based Controllers, Proceedings of the 1985 American Control Conference, 3, 1520.

Jaffarey, A., J.M Douglas, and T.J. McAvoy. (1979). Short-cut techniques for distillation design and control. 1. Column design, I. & E.C. Proc. Des. Dev., 18(2), 197-202.

Jerome, N.F., Wh.H. Ray. (168). High Performance Multivariable Controllers for System Having Time Delays, AIChE Journal, 1986, 32(6), p. 914.

Kravaris, C., and C. Chung. (1987). Nonlinear State Feedback Synthesis by Global Input/Output Linearization, AIChE Journal, 33(4), 592.

Lee, P.L., R.B. Newell, and G.R. Sullivan. (1988). Generic Model Control - A Case Study (Submitted).

Lee, P.L., and G.R. Sullivan. 1988. Generic Model Control (GMC), Computer and Chemical Engineering (In press).

Lee, P.L., G.R. Sullivan, and W. Zhou. (1988a). A New Multivariable Deadtime Control Algorithm, (Submitted).

Lee, P.L., G.R. Sullivan and W. Zhou. (1988b). Process/Model Mismatch Compensation for Model-Based Controller (Submitted).

McDonald, K. (1987). Performance Comparison of Methods for On-Line Updating of Process Model for High Purity Distillation Control, AIChE Spring National Meeting, Houston.

Rivera, D.E., M. Morari, and S. Skogestad. 1986. Internal Mode Control. 4. PID Controller Design, I. & E.C. Proc. Des. Dev., 25, 252.

Rouhani, R., and R.K. Mehra. (1982). ModelAlgorithmic Control (MAC); Basic Theoretical Properties, Automatica, 18, 401.

Russell, L.W. and J.D. Perkins. (1987). Towards a Method for Diagnosis of Controllability and Operability Problems in Chemical Plants, Chem. Eng. Res. Des., 65, 453.

Seborg, D.E., T.F. Edgar, and S.L. Shah. (1986). Adaptive Control Strategies for Process Control: A Survey, AIChE Journal, 32(6), 881.

Smith, O.J.M.. (1957). Closer Control of Loops With Dead Time, Chem. Eng. Prog., 53, 217.

Stephanopoulos, G. (1984). Chemical Process Control. An Introduction Theory and Practice, Prentice-Hall Inc., Englewood Clifss, New Jersey.

Watanabe, K., and M. Ito. (1981). A Process-Model Control for Linear Systems with Delay, IEEE Trans. Autom. Control, AC-26, 1261.

Wood, R.K., and M.W. Berry. (1973). Terminal Composition Control of a Binary Distillation Column, Chem. Eng. Sci., 28, 1707.

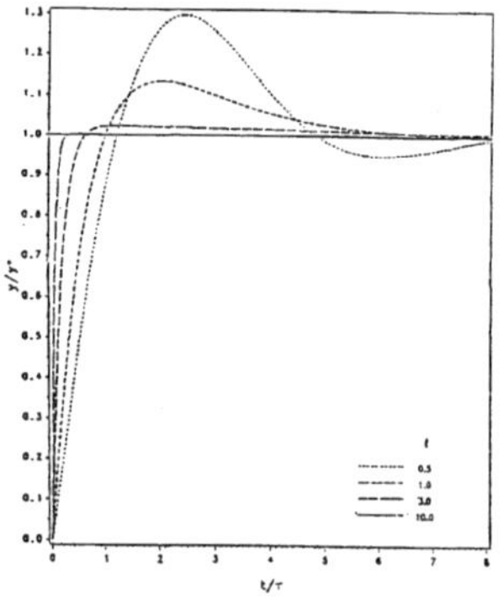

Fig. 1. GMC response specification with parameters ξ and t/T.

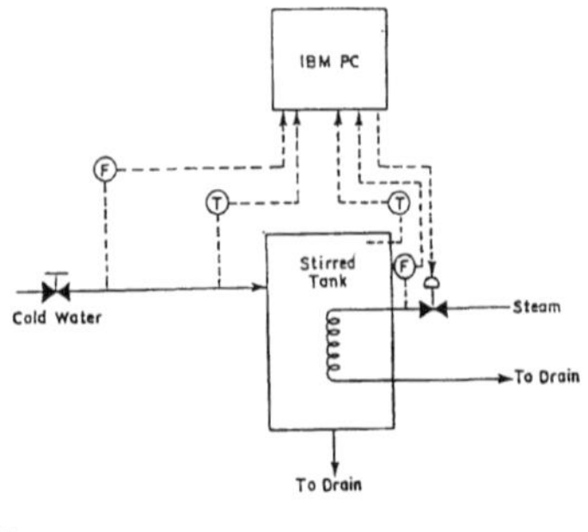

Fig. 2. Stirred tank heater process schematic.

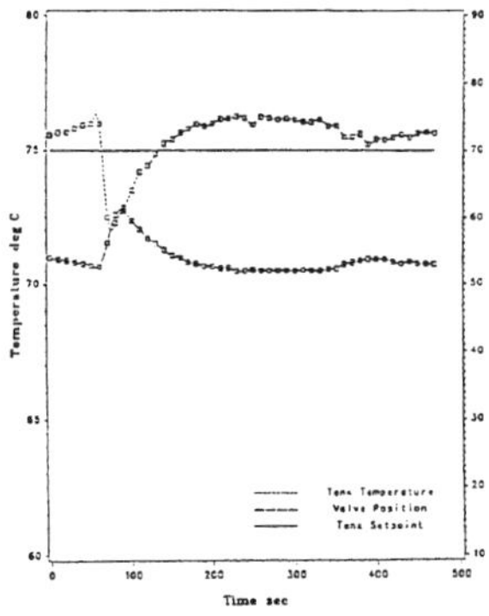

Fig. 3. Stirred tank heater controller response to a disturbance. GMC constants: $\xi = 4$, $r = 400$ s.

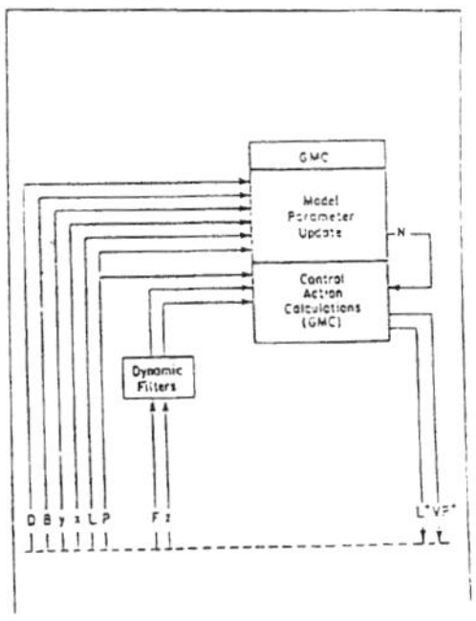

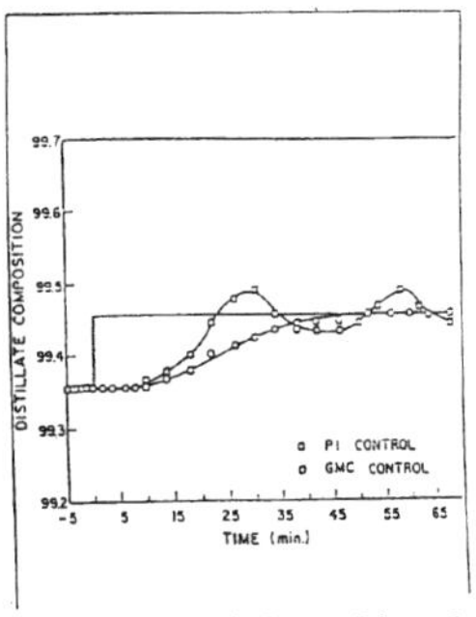

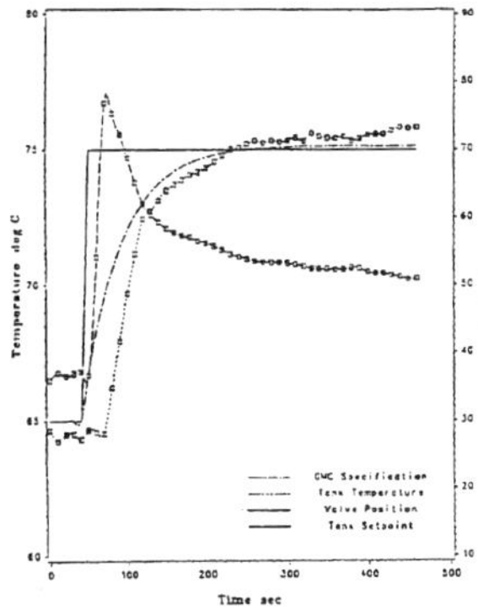

Fig. 4. Stirred tank heater controller response to a setpoint increase. GMC constants: $\xi = 4$, $r = 400$ s.

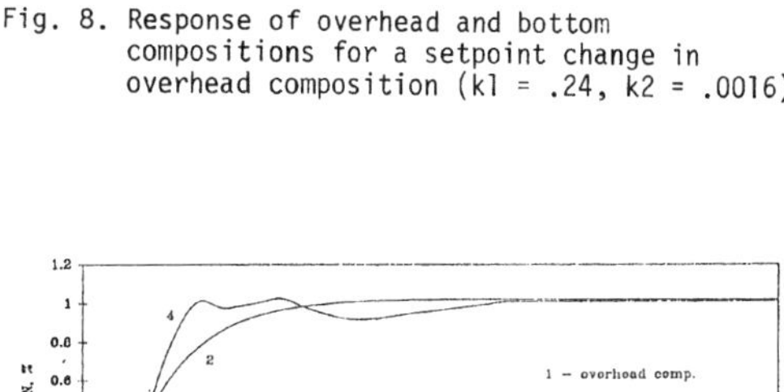

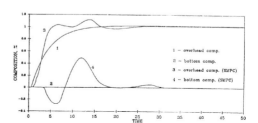

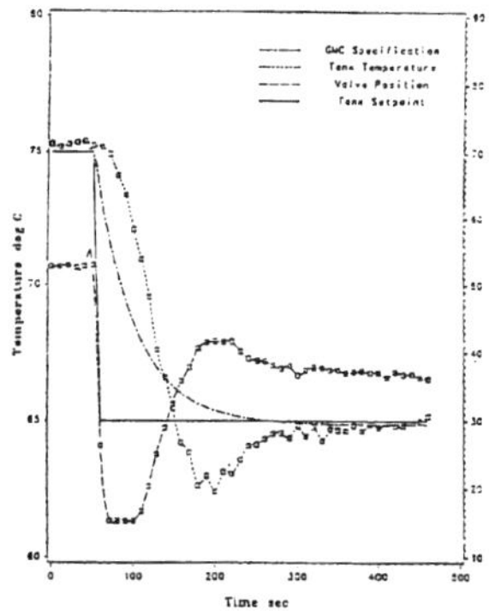

Fig. 5. Stirred tank heater controller response to a setpoint decrease. GMC constants: $\xi = 4$, $T = 400$ s.

Fig. 6. Information flow diagram for GMC controller

Fig. 7. Industrial application. Overhead composition change.

Fig. 8. Response of overhead and bottom compositions for a setpoint change in overhead composition (k1 = .24, k2 = .0016).

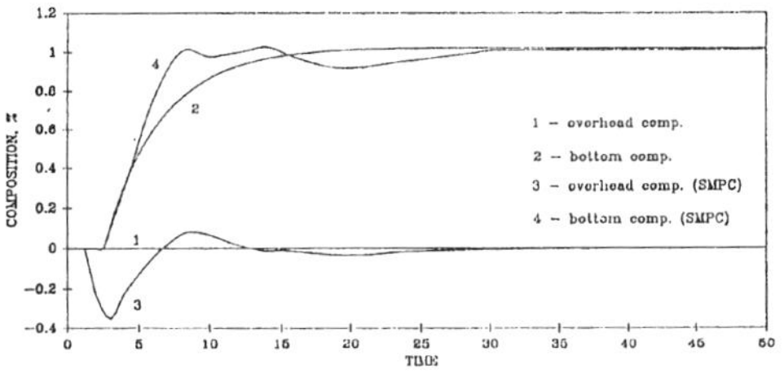

Fig. 9. Response of overhead and bottom compositions for a setpoint change in bottom composition (k1 = .24, k2 = .0016).

INTEGRATED MODEL BASED CONTROL OF MULTIVARIABLE NONLINEAR SYSTEMS

Babu Joseph*, Shi-Shang Jang and Hiro Mukai*****

**Department of Chemical Engineering, Washington University, St. Louis, Missouri, USA*

***Department of Chemical Engineering, National Tsing Hua University, Hsinchu, Taiwan,*

****Department of Systems Science and Mathematics, Washington University, St. Louis, Missouri, USA*

A new algorithm, called Integrated Model Based Control, is proposed for the control of complex unit operations. The algorithm utilizes a process model based on first principles as opposed to mathematically convenient forms. All available process measurements are utilized in an on-line identification step of the algorithm. The algorithm allows us to approach the control problem in an integrated fashion taking into account overall control objectives, process operating constraints and the inherent interaction among the process variables. Most importantly, the algorithm allows the integration of process control with the economics of process operation.

A feasibility study was conducted using simulated implementation of the algorithm on a distillation column. The results are very encouraging and suggest further exploration of the concepts.

INTRODUCTION

Regulatory control of complex unit operations such as distillation and fluid catalytic cracking reactors is often reduced in practice to the feedback regulation of selected measured output variables. This is justified because it brings simplicity to an otherwise chaotic problem. Each output is paired with an appropriate manipulated input variable selected to minimize the interaction among the various control loops. In the past, before the advent of digital computer control systems, this was the only feasible way to approach the control problem. However, the decision to deal with the control problem via decomposition to single loop control strategies has led to some inherent problems. These include:

(i) The problem of interaction, mentioned above,
(ii) the difficulty of dealing with process constraints in a systematic way (often ad hoc strategies are employed),
(iii) the need to reduce the overall control objectives to the regulation of selected process measurements resulting in compromises in control quality, and
(iv) the inability to utilize knowledge about the process in the form of process models, on-line sensor data, etc.

The availability of on-line process computers spurred the development of many new algorithms for computer control (Cutler and Ramaker, 1978; Prett and Gillette, 1978). These algorithms utilize on-line process models and reduce the control problem to a problem in minimizing a measure of the error in regulation of the controlled output. More recently, this approach has been extended to include process constraints (Garcia, 1984).

The new algorithms rely upon linear input/output models identified on-line. They also rely upon the measurements of the output being controlled with no way of incorporating other process knowledge available in three forms: first, in the form of process models based on an understanding of the dominant physical and chemical events taking place inside the process unit, and second, in the form of historic data from secondary sensors, and third, in the form of past experience in operating the plant (experiential knowledge).

Assuming the present trend with advances in computer hardware, software and sensor technology continues, the challenges facing the control engineer are: (i) how to integrate the tremendous wealth of information coming from the plant with the knowledge he has about the process and (ii) how to utilize this integrated knowledge in the development of an intelligent control system for the plant?

The objective of this paper is to suggest an approach to deal with some of the issues posed above. Needless to say, this is only a beginning and considerably more work to be done before the concept of 'intelligent systems' can be fully realized. The paper is divided into three parts. The first part presents the theory behind the proposed approach. The theory is developed further in the context of controlling a distillation column next. The last part discusses the results of this application study.

THEORY

The proposed approach is based on an important premise: a sufficiently descriptive model of the process relating the manipulated inputs to the process measurements is available. Process models of varying complexity have been developed in the past mainly in the context of design. However, these models are rarely used in control primarily because of the difficulty of matching observed plant behavior. Instead of these, most control engineers have relied upon simple linear input/output relationships. In fact, this pattern is followed even in the new model based computer control algorithms cited above. Unfortunately, these models tend to be non-parsimonius and their linearity often limits the range of validity. The need to relate each input with each output also contributes to many parameters in the model. Adaptive techniques have been proposed to extend the validity of the process model but these have not found widespread applications in the process industry.

We are proposing to retain the nonlinear structure of the model at the risk of increased computational load. The criticism that these models are not reliable is answered by incorporating model adaptation.

Briefly, the problem can be stated as follows:

Given an approximate model of the process and a set of present and past process measurements, determine the changes to be made in the plant manipulated inputs that will meet the control objectives of the process.

The term *control objective* needs to be defined further. In the simplest sense, the term control objective might be taken to be keeping selected output variables at some preset values. For a complex unit operation it may be more appropriate to think of minimizing a composite objective function reflecting the costs associated with the operation of the process. In addition, operational constraints might also be stated. For example, in distillation the control objective might be to remove a certain fraction of feed impurity, or to produce a stream of specified purity (a constraint) while minimizing the heat input (the objective function). A precise quantitative definition of the control objective may not be readily obtainable. But simply choosing to maintain the output(s) constant may not always be the best choice and sometimes may even be infeasible.

Mathematical Statement of the Problem

Given a process model of the form

$$\overset{\circ}{x} = f(x, m, p, t)$$

where

x = state variables
m = manipulated variable
p = unmeasured disturbances and unknown parameters

and operational constraints of the form

$$h(x, m, p, t) \leq 0$$

and, a set of measurements, z

$$z = g(x, m, p, t) + \text{noise}$$

Determine the control policy m(t) such that the control objective

$$\Phi = \int_t^{t + T_h} \phi\,(x, m, p, t)\,dt$$

is minimized.

The variables z includes all available measurements on the process (continuous, intermittent, as well as sampled). The variables p include uncertain parameters on the process, unmeasured input disturbances which are expected, as well as those parameters in the process which can be expected to change with time (e.g., heat transfer coefficient, catalyst activity, tray efficiency, etc.).

We make a further assumption that the structure of the model is adequate to cover the expected region of operation. This implies that no events take place that change the operating characteristics of the process which invalidated the assumptions used in the model. In such cases, one must build mechanisms to detect such transitions and to change the model used.

The choice of process model is an important issue. In general, we can expect to find a wide spectrum of process models of varying levels of complexity and detail. The cost associated with these consists of two parts. The first is the development cost which can be substantial even for well characterized systems. The second is the cost of computing which can become significant for large dimensional systems. Both of these costs must be weighted against the potential benefits (payback) from improved control. One desireable characteristic of any control algorithm is the ability to accept process models of varying complexity (a characteristic lacking in current approaches to the control problem.

Integrated Model Based Control Algorithm (IMBC)

The structure of the Integrated Model Based Control (IMBC) is shown in Fig. 1. There are two components to this algorithm: model identification and the control computation block. Although represented as a block diagram, the computations within each block are complex.

The identification is carried out in real-time for the following reasons:

(i) All process models are approximations to the real world and hence, it is important to make some adjustments to the model continuously in order to have a reasonably accurate predictive model for use in the control block.
(ii) Some process parameters change with time and the on-line identification can track these variations.
(iii) There may be unknown or unmeasured disturbances to the process and the identifier can be set up to track these if we know ahead of time how they might affect the observations.

Let us say that

$$\bar{z}(1),\ \bar{z}(2)\ \ldots\ \bar{z}(N_i)$$

are a set of past observations, z(1) being the most recent. The identification problem can be stated as

$$\min_{x_o,\, p} \sum_{j=1}^{N_i} \cdot \left|\left| \bar{z}(j) - z(j) \right|\right|^2$$

subject to the constraints

$$\begin{aligned} \overset{\circ}{x} &= f(x, m, p, t) \\ x(t(N_i)) &= x_o \\ z_j &= g(x, m, p, t(j)) \\ t(j) &= \text{time of } j^{th} \text{ observation} \\ t(N_j) - t(1) &= \text{sampling Horizon, H} \end{aligned}$$

This is an equality constrained optimization problem which is readily solved using standard nonlinear programming packages with some minor modification to account for the differential equations constraint.

A popular approach to solving this type of problem is to use the Extended Kalman Filter (Jazwinski, 1970). This approach requires some stringent specifications on the statistics of the noise in the input perturbation and the measurements. A comparison study of the two approaches have been done and is presented in Jang et al. (1986). Our results indicate that the above direct approach is superior to the extended Kalman Filter in terms of speed of convergence and more importantly, robustness. The Kalman Filter offers the advantage of computational efficiency due to its recursive nature.

We made some assumptions in the above formulation which are stated next. It is assumed that the variable p remains constant over the time frame under consideration. If p is rapidly fluctuating, its effect will be filtered out by the process. If p is slowly varying then the assumption is reasonable. If p is subject to random fluctuations such as step increases, then it is appropriate to start the identification using data after a change in p has been detected.

Note that the inclusion of x_0, the initial state of the process as an identified variable. Using large dimensional models will require the identification of a large number of initial states which in turn would require large numbers of observations.

Another assumption is that the process is identifiable. There are two prerequisites for this. The first is the well known criteria of observability. This criteria ensures that there are sufficient numbers of measurements on the proces to follow the changes in the states and parameters. The second prerequisite is that the input forcing functions to the process are sufficiently rich to excite all the observable modes of the system. If the data is not sufficiently rich with variations, then identification will not work. This is another way of saying that, the data should reflect some sensitivity to the changes in p and x_0. In the past, authors have recommended adding random noise to the input to ensure that the observations are sufficiently rich. In the example problems studied, we have found that the perturbations introduced by the control signal itself, are sufficient to ensure identifiability for the examples studied. Another precaution we like to recommend, is to turn off the identification step after the process has reached a calm state.

The Control Block

The function of the control block is to compute the manipulated input which optimize some control objective while meeting the constraints. Mathematically, we solve

$$\min_{m(t)} \Phi = \int_{t}^{\bar{t} + T_h} \phi(m, y, p, t)\, dt$$

subject to

$$\overset{\circ}{x} = f(x, m, p, t)$$

$$y = w(x, m, p, t)$$

$$h(x, m, p, t) \leq 0$$

In practice, instead of minimizing over m(t), we discretize the problem and minimize over m_1, m_2, ... m_{No}

where

$$m(t) = m_i, \bar{t} + (i-1)(\Delta T_o) \leq t \leq t + i(\Delta T_o)$$

One may view ΔT_o as the discrete control interval, (which need not be the same as the discrete observation interval).

Note the similarity of this problem to the identification problem. Both can be solved using nonlinear programming packages. However, the control problem requires the satisfaction of the constraint h at all times. This is not easily accommodated. One pragmatic approach is to implement

$$h_i = h(x_i, m_i, p, t_i) \leq 0$$

for i = 1, 2, .. N_o, where $t_i = \bar{t} + i\,\Delta T_o$.

This ensures point satisfaction of the constraint. If constraint violation is detected, then one may add more points on the curve. There may be other ways of strictly enforcing the constraint, at least for special situations. We are presently exploring this further.

Finally, we would like to point out that in the case of linear process models and quadratic objective functions, the above algorithm reduces to the QDMC algorithm proposed by Garcia (1984). The effect of the parameters such as the control interval, sampling interval, control horizon, etc., have been discussed in detail (for linear systems) by Garcia (1982). The lack of a sufficiently general mathematical theory for the analysis of nonlinear systems makes it difficult to make the same analysis for the algorithm proposed above. Nevertheless, one can expect the same trends to be valid for the nonlinear case and useful guidelines for practical applications can be obtained from the linear analysis.

APPLICATION TO DISTILLATION COLUMN CONTROL

The feasibility of the approach suggested above was tested by application to the control of a binary distillation column. The simulated study was conducted as follows. First, a detailed model of the column was constructed using stage-by-stage mass and energy balance and tray hydraulics equations. The model was used to simulate plant behavior and will be referred to as the plant model.

IMBC was implemented using reduced order models constructed using the collocation approach of Cho and Joseph (1981). This will be referred to as the process model. By employing different models in the plant simulation and the optimization stage, we were able to evaluate the proposed approach in a situation very similar to actual implementation. The difference between the models reflects the inaccuracies of the process models. Figure 2 shows a schematic of the test strategy.

The Plant Model

The stage-by-stage mass and energy balance model employed here has been proved to be a reasonably good approximation of distillation column dynamics. The main assumptions used are (i) each tray is perfectly mixed, (ii) there is a definite relationship between the vapor and liquid phases on each tray, (iii) the vapor and liquid leaving each tray is in thermal equilibrium, and (iv) the vapor holdup is negligible. The resulting equations are given below.

Component Balance:

$$\frac{d}{dt}(M_j, x_{ij}) = L_{j-1}x_{i,j-1} + V_j y_{ij} - L_j x_{ij} - V_{j-1} y_{i,j-1}$$

Total Balance:

$$\frac{dM_j}{dt} = L_{j-1} + V_j - L_j - V_{j-1}$$

Energy Balance:

$$\frac{d}{dt}(M_j h_j) = L_{j-1}h_{j-1} + V_j H_j - L_j H_j - V_{j-1}H_{j-1}$$

Flow Hydraulics Equation:

$$M_j = m(L_j)$$

Vapor-Liquid Relationship:

$$y_{i,j-1} = y(x_{ij}, T_j)$$

Bubble-Point Relations:

$$\sum_{i=1}^{c} y(x_{ij}, T_j) = 1$$

The simulation program using this model was developed at Monsanto. The program uses a semi-implicit integration scheme with automatic step size correction. The model is based on the work of Brosilow and Ballard (1981).

The Process Model

A number of considerations go into the selection of the process model used in the identification and control stages. Developing or selecting a suitable model can be the most time consuming step in this approach. We have not yet developed quantitative measures to judge the quality and suitability of a process model. Control performance measures, stated economically will need to be developed for this purpose. This area is currently under investigation.

For the distillation example, the use of a model similar to the one stated above was ruled out because of its high dimensionality. This type of a model will introduce a large number of variables in the optimization problem enough to tax (and hence, possibly lead to the failure of nonlinear programming package). Coupled with this was the consideration of speed which is critical in on-line applications.

The process model we selected is based on the model reduction strategy proposed by Cho and Joseph. this model reduction approach reduces the number of equations required to model the column by an order of magnitude. Anoter advantage of this strategy is the availability of a spectrum of models of varying accuracy. Details of the model development are given in the series of papers by Cho and Joseph (1981a, b) and Srivastava and Joseph (1986; 1987).

For comparison purposes, we summarize the equations used in the process model below.

Component Balance:

$$M_j \frac{dx_{ij}}{dt} = \sum_{k=1}^{n+2} A_{jk}[(L_k x_{ik} - V_k y_k) - x_{ij}(L_k - V_k)]$$

Total Balance:

$$\frac{dM}{dt} = \sum_{k=1}^{n+2} A_{jk}(L_k - V_k)$$

Energy Balance:

$$M_j \frac{dh_j}{dt} = \sum_{k=1}^{n+2} A_{jk}[(l_k h_k - V_k H_k) - h_j(L_k - V_k)]$$

Flow Hydraulic Equation:

$$M_j = m(L_j)$$

The Bubble-Point Relationship:

$$\sum_{i=1}^{c} y_{ij} = 1.0$$

The Equilibrium Relationship:

$$y_{ij} = y\left(\sum_{k=1}^{n+2} \ell_k(z_j + \Delta z)x_{ik}, T(z_j + \Delta z)\right)$$

The details of the derivation of the above model are not relevant. The important point to note is that the original set of equations in the plant model are written for each tray. The equations in the reduced model are written at each collocation point. By using a small number of collocation points, significant reduction in order is achievable.

RESULT OF THE EXAMPLE STUDY

Integrated Model Based Control was applied to a binary distillation column for the separation of benzene and toluene. A schematic of the column is shown in Fig. 3. The nominal steady state operating conditions of the column are shown in Table 1. The time constant of the top and bottom compositions to feed disturbances are of the order of 1 hour.

Time Domain Decomposition of Control Objectives

The distillation column has associated with it a large number of control objectives. This includes maintaining total material balance (which translates into control of levels in the accumulator and the reboiler and the pressure in the column) and maintaining quality (which translates into controlling the component material balance). These control variables have widely differing time constants associated with them. For this reason, these control objectives are decomposed into two sets. The faster responding level and pressure control loops were done using conventional feedback loops. IMBC was applied to the longer time constant control objectives only.

For the plant simulation, pressure in the column was assumed constant (assuming perfect control). PI controllers were used for controlling levels in the reboiler and accumulator drum.

Results of Identification Studies

The following variables were chosen as the identified variables (selected arbitrarily):

1. Feed composition
2. Feed flow rate
3. Initial values of state variables in the process model (namely, the liquid composition at each collocation point. Holdups were not included).

The following variables were used as measurements (selected arbitrarily):

1. The temperature on Tray 1, and/or
2. The composition of the bottom stream.

Note that since this is a binary column, the temperature is a good indication of the composition. Nevertheless, measuring both together has some advantages since the response times of two measurements are different. Also, due to the noise inherent in all measurements, having two measurements can be expected to yield better control loop performance. A number of runs were carried out to determine the effect of the following variables:

1. The size (order of the process model used)
2. The sampling time for the observations
3. The sampling horizon H, the time period over which the data is collected and used in the identification step
4. The level of noise in the measurements.

To summarize the results, we define a quantity called the Maximum Error in Identification (MEI):

MEI = Maximum difference between the identified value and true values of a variable expressed as a percentage of the true value.

Table 2 summarizes the effect of sampling horizon and the order of the process model (measured in terms of the number of collocation points used in the reduced order model). The accuracy (and size of the model) increase with the number of collocation points used. The increase in the number of collocation points leads to improved accuracy of the identified variables as expected. However, there appears to be a limit on the accuracy achievable with this type of process model.

Increasing the horizon should increase the number of observations available for the identification. One would expect this to improve the accuracy of the identification step. This is indeed true for the 2-point collocation model. But for the 3-point model, the MEI actually increased slightly. The sampling time was kept constant and hence a longer horizon includes a larger number of points over which the least-square fit is made. This would imply that the MEI alone is really not a complete indicator of the goodness of the model. A wider variety of tests is required to verify the model accuracy.

Table 3 shows the result of adding noise to the measurements on the error in identification. As expected, the addition of noise increases the MEI except for one case (row 2 in Table 3). The exception is not significant and is probably reflecting the chance cancellation of one error by another. More importantly, the errors do not increase significantly. In fact, the addition of a measurement reduces the error to the same level as those reported in Table 2 (without noise).

In these tests, no attempt was made to reduce the noise by time series filtering of the measurements. The noisy data was fed directly to the identification algorithm. The noise is filtered out the least-square fit of the data.

Figure 4 shows the remarkable accuracy with which the system states are approximated. Figure 4a shows the approximation to the initial state used and Fig. 4b shows the estimated composition profile after 2 hours of sampling and identification.

Figure 5a shows the fit between estimated bottoms composition and the actual composition as a function of time. Fig. 5b shows the fit for the temperature on the 1st tray. The largest errors are encountered in the initial period of operation (right after the plant was perturbed).

Results of the Application of IMBC

To implement IMBC, one must define the control objectives for the operation of the column. Various options are possible e.g., maintain top and bottom composition constant or minimize energy or maximize the removal of a certain impurity within the constraints of operation, etc.

For the purposes of this feasibility study, the exact choice of control objective is immaterial. To illustrate the flexibility of the approach, we chose the following objective

$$\max \Phi = \int_{\bar{t}}^{\bar{t} + T_h} (P_T F_T + P_B F_B - C_Q Q_H)\, dt$$

subject to

$$Q_{min} \leq Q_H \leq Q_{max}$$

$$x_B,\ x_T \geq 0.90$$

In words, we seek to maximize the value of products minus the cost of energy consumption in the reboiler subject to purity constraints. Further, the value of products were related linearly to the purity of the product.

$$P_T = \$1.0 + (1.0 - x_t)\ 10$$

$$P_B = \$1.0 + (1.0 - x_B)\ 10$$

Tests were conducted as follows. Plant was started up at steady state conditions and a disturbance through monitoring the difference between the measurements and the predicted values. When a disturbance is detected, data collection is started and after sufficient data has been collected, the identification algorithm is executed (2 hours in this case). In the period 0-2 hours, IMBC does not have enough data to detect the disturbance and is working under the assumption as though no change has taken place.

Figure 6 shows the result of applying IMBD when the column is subject to a 10% increase in the feed flow rate. For comparison purposes, the result of not taking any control action is also shown. A horizon time of 2 hours was used. During the first 2 hours, the IMBC is controlling the process using the outdated model. At the end of 2 hours, the disturbance is identified and the controller brings the process to the new optimum operating condition. Note that the new optimum calls for a much purer bottom product while the top product is reduced in purity.

Figure 7 shows the result of applying IMBC with simultaneous upsets in feed flow and feed composition. In this case, a longer time horizon (H = 3 hours) is needed to identify the disturbances. Again the optimizer calls for a purer top product while increasing the impurity in the bottoms. The uncontrolled response of the bottoms shows the severity of the disturbance.

Finally, Fig. 8 illustrates the ability of IMBC to deal with recurring disturbances. Here two disturbances in feed flow are imposed one at t = 0 (10% increase) and another at t = 2.0 hrs. (another 10% increase). A horizon time of 1 hour was used. The bottom product purity is severely affected but kept under the constraint specified in the control objective.

These, and other similar results reported in Jang (1986), clearly indicate the capability of the algorithm to perform under adverse conditions. We did not undertake to characterize the performance in terms of the parameters in the algorithm. Such studies are reported in a related paper (Joseph and Jang, 1987).

CONCLUSIONS

Control using linear models is already a practical reality. This paper presented an approach to integrated process control using nonlinear models. The advantages of this approach are:

(i) the ability to integrate all types of plant measurements through the process model identification step
(ii) the ability to deal with the control problem in an integrated way as opposed to piecemeal control of selected output variables
(iii) ability to incorporate process constraints
(iv) the natural way in which interaction among the process variables are accommodated.

The feasibility of this framework was clearly demonstrated by the application study. Despite the differences between the plant model and the process model, the algorithm performed well in identifying the process and in meeting the control objectives.

Many questions remain unanswered. These include the selection of process models, the selection of measurements and their location, the stability characteristics of the algorithm, and the selection of parameters in the algorithm.

The driving force behind the use of advanced control techniques is one of economics. The advantages of on-line optimizing control is now recognized and practiced to some extent by industry. The approach presented here should be of use in the integration of plant control with economics of operation.

NOTATIONS

- x - state variable
- m - manipulated variable
- p - unknown disturbances and unknown parameters
- h - constraints
- z - observation
- Φ - objective function
- ϕ - operating cost
- t - time
- y - output
- N_o - no. of stages looking ahead
- M_j - holdup on stage j
- x_{ij} - composition of component i on stage j
- L_j - liquid flow rate of stage j
- v_j - vapor flow rate of stage j
- h_j - enthalpy of liquid phase on stage j
- H_j - enthalpy of vapor phase on stage j
- y_{ij} - composition of component i on stage j
- T_j - temperature of stage j
- A_{jk} - numerical value of $d\ell_k/dt$ evaluated at collocation point z_j
- ℓ_k - Lagrange polynomial for collocation point z_k
- z_j - distance on collocation number j
- H_o - search horizon

REFERENCES

Brosilow,C.B. and D.Ballard(1978)."Dynamic Simulation of Multicomponent Distillation Columns",AIChE Meeting,42a,Miami.

Cho,Y.S. and B.Joseph(1983a)."Reduced-Order Steady State and Dynamic Models for Separation Processes",Part I:Development of the Model Reduction Procedure",AIChE Journal,29,2,261.

Cho,Y.S. and B.Joseph(1983b)."Reduced-Order Models for Separation Processes,Part II:Application to Multicomponent Distillation",AIChE Journal, 29,2,271.

Cutler,C.R. and B.Ramaker(1979)."Dynamic Matrix Control Algorithm",Paper Presented at 86th National Meeting of AIChE,Houston.

Garcia,C.E.(1984)."Quadratic/Dynamic Matrix Control of Nonlinear Processes: An Application to Batch Reation Processes",Paper Presented at AIChE Meeting, San Francisco.

Garcia,C.E.(1982)."Internal Model Control",Ph.D. Thesis, University of Wisconsin, Madison.

Jang,S.S.,B.Joseph and H. Mukai(1987)."On-line Optimization of Constrained Multivariable Processes",AIChE Journal,33,26.

Jang,S.S.(1986)."On-line Optimization of Chemical Processes,Ph.D. Thesis, Washington University, St. Louis.

Jang,S.S.B. Joseph and H. Mukai(1986)."A Comparison of Two Approaches to On-line Parameter and State Estimation of Nonlinear Systems", I&EC,PDD.,25,809.

Jazwinski,A.H.(1970).Stochastic Processes and Filtering Theory,Academic Press,New York.

Prett,D.M. and R.D. Gillette(1979)."Optimization and Constrained Multivariable Conrol of a Catalytic Cracking Unit",Paper Presented at 86th National Meeting of AIChE,Houston.

Srivastava,R.K. and B.Joseph(1987a)."Reduced-Order Models for Separation Processes,Part III: Treatment of Columns with Multiple Feeds and Sidestreams",Computers and Chemical Engineering Journal.To appear.

Srivastava,R.K. and B.Joseph(1986)."Reduced-Order Models for Separation Processes,Part IV: Selection of Collocation Points",Computers and Chemical Engineering Journal,9,601.

TABLE 1 Operating Parameters for the Benzene-Toluene Column

Column Specifications	
Number of stages	= 18
Column Diameter	= 4.5 ft.
Reflux Drum Diameter	= 4 ft.
Sump Diameter	= 4.5 ft.
Nominal Reflux Level	= 3 ft.
Nominal Sump Level	= 4 ft.
Tray Area (at 4.5 ft. Diameter)	= 15.90 ft^2
Fraction Downcomer	= 0.15
Height of Liquid over Weir (Trays 2-10)	= 0.98 in.
(Trays 11-17)	= 0.65 in.
Height of clear liquid (Trays 2-10)	= 1.68 in.
(Trays 11-17)	= 1.78 in.
Feed Specifications	
Number of Feeds to Column	= 1
Feed Location, Tray No.	= 10
Feed Rate	= 200.0 lbmole/hr
Enthalpy of Feed	= 110.085 K.Btu/lbmole
Feed compositon, mole fraction of comp. 1	= 0.65
mole fraction of comp. 2	= 0.35
Components	
Number of Components	= 2
Component 1	= Benzene
Component 2	= Toluene
Light Key Component	= Component 1
Heavy Key Component	= Component 2
Controllers	
1. Accumulator Level Controller	
Lower limit of level sensor	= 0 ft.
Upper limit of level sensor	= 6 ft.
Manipulated variable	= reflux flow rate
Lower limit of manipulated variable	= 0.0 lbmole/hr
Upper limit of manipulated variable	= 600.0 lbmole/hr
Controller, proportional Band	= 20.0
Reset, hour	= 0.5
Accumulated level set point	= 3.0 ft.
2. Sump Level Controller	
Lower limit of level sensor	= 0 ft.
Upper limit of level sensor	= 8.0 ft.
Manipulated variable	= bottoms flow rate
Lower limit of manipulated variable	= 0 lbmoles/hr
Upper limit of manipulated variable	= 150.0 lbmoles/hr
Controller, proportional band	= 150.0
Reset, hours	= 1.0
Sump level set point	= 4.0 ft.

Number of Collocation Points Used in the Model	Identification Horizon, Hours	Maximum Error in Identification Percent
1	1.0	20.0
2	1.0	12.6
3	1.0	3.0
1	1.5	14.0
2	1.5	3.0
3	1.5	3.2

TABLE 2 The Maximum Error in Identification

Measurement	Standard Deviation of Noise	Collocation Points Used	Max Error in Identification
Bottom Composition	2%	2	9.4
"	2%	3	1.8
"	5%	2	13.0
"	5%	3	9.7
Bottom Composition and Temp on 1st Tray	5% on comp. 5°K on temp.	3	2.5

TABLE 3 Effect Of Noise On The Measurement On The Maximum Error In Identification (H_O = 1.5 hr.)

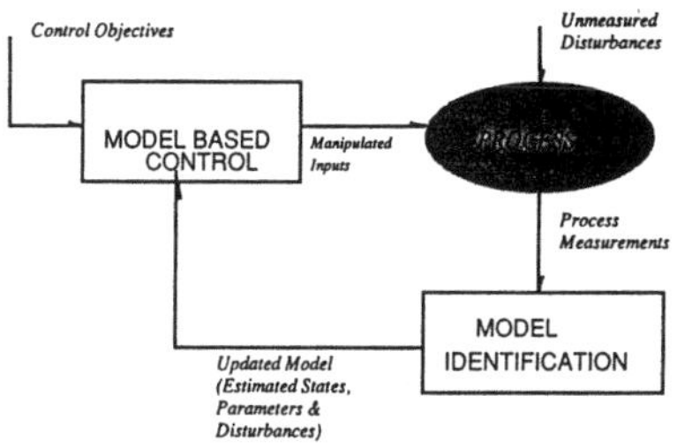

Fig. 1. Block structure of integrated model based control (IMBC).

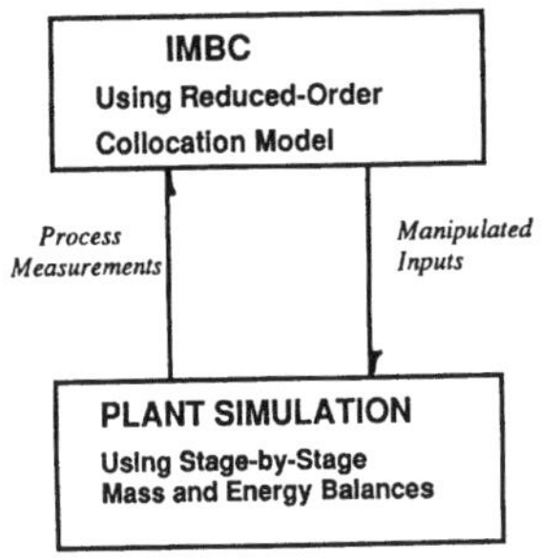

Fig. 2. Strategy used in the application study.

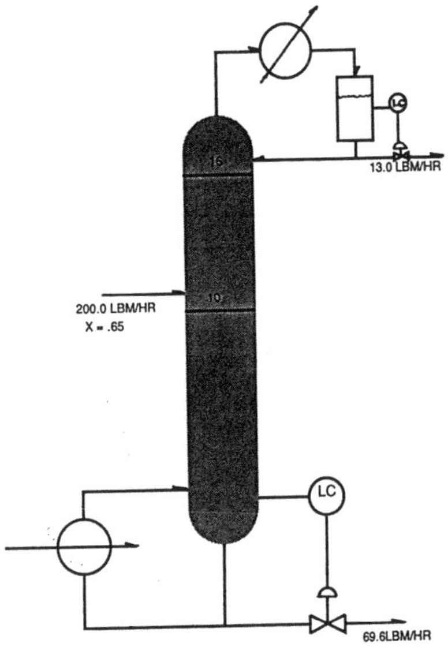

Fig. 3. Schematic of benzene/toluene distillation column used in the control study.

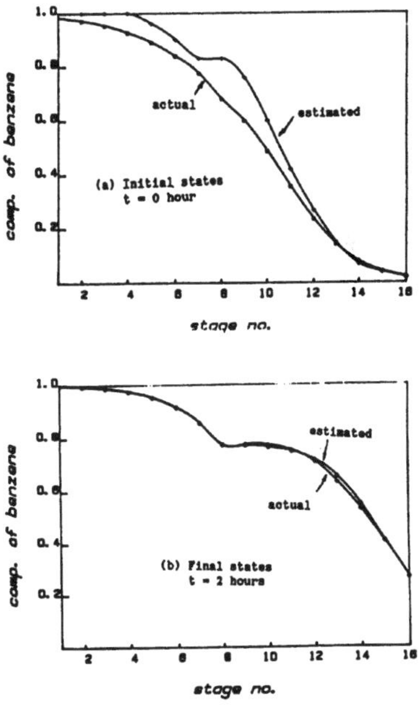

Fig. 4. Convergence of the estimated composition profile to the actual values using 2 hours of data on bottoms compositions alone.

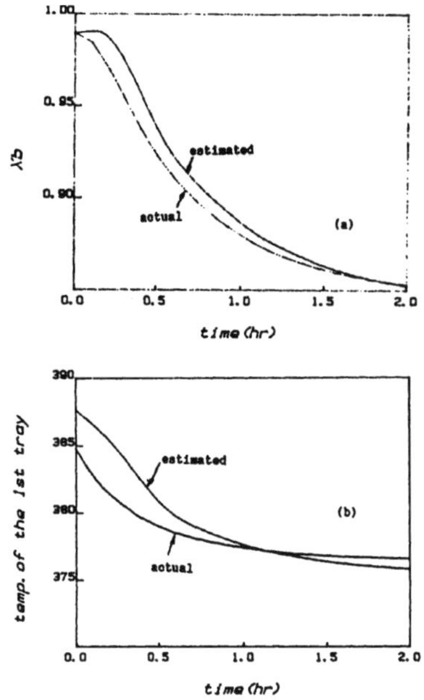

Fig. 5. Comparison of estimated and actual composition and temperatures as a function of time. Data over a 2 hour horizon was used.

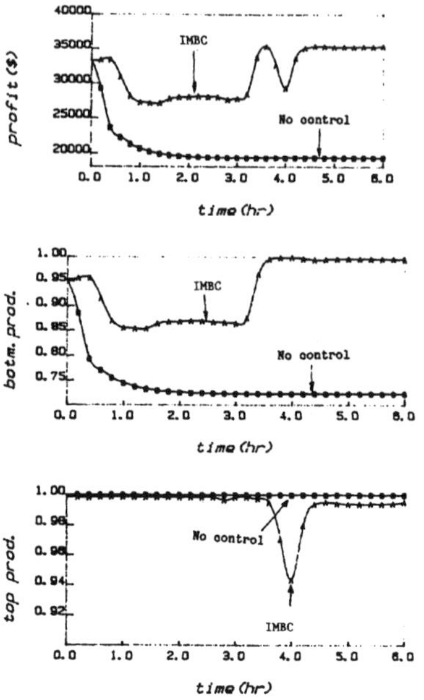

Fig. 7. Performance of IMBC when column is subject to simultaneous disturbances in the feed flow and feed composition.

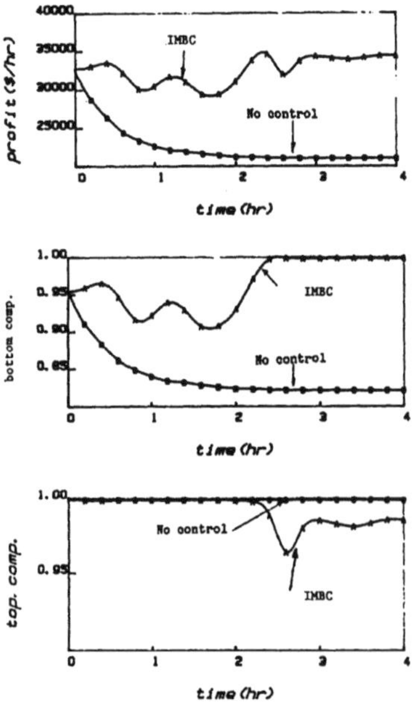

Fig. 6. Performance of IMBC when column is subject to an increase in the feed flow rate.

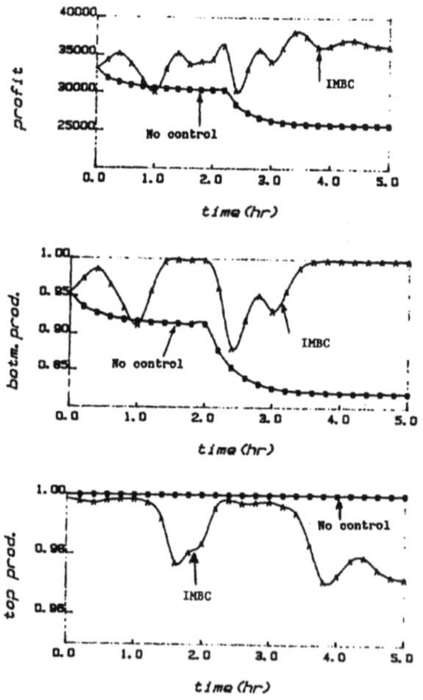

Fig. 8. Performance of IMBC when subjected to sequential disturbances in feed flow rate. Feed changed from 200 lbm/hr to 220 lbm/hr at t=0, and to 240 lmb/hr at t=2.0 hrs.

MODEL-PREDICTIVE CONTROL AND SENSITIVITY ANALYSIS FOR CONSTRAINED NONLINEAR PROCESSES

J. W. Eaton, J. B. Rawlings and T. F. Edgar

Department of Chemical Engineering, The University of Texas at Austin, Austin, Texas 78712, USA

Abstract. This paper describes methods for feedback control and sensitivity analysis of systems modeled by nonlinear differential-algebraic equations. An optimal control policy for the nominal plant is determined using a simultaneous optimization and simulation approach. The sensitivity of the optimal solution with respect to the model parameters is then computed. The feedback from the measurements is used to update the important model parameters. Confidence intervals are also placed on the optimal control profiles by considering second order variations in the Lagrangian. Computational results are shown for two process models of interest in chemical engineering.

Keywords: Feedback control; Sensitivity analysis; Nonlinear programming; Optimal control.

Introduction

The open loop control of a chemical process may be stated naturally as a nonlinear optimization problem subject to differential-algebraic constraints. Recently, several researchers have demonstrated that this class of optimization problems may be solved efficiently by the application of an infeasible path optimization strategy in connection with weighted residual techniques [1,11,3]. Using these methods, the model is solved simultaneously with the optimization. The significant improvement in efficiency over other methods is achieved by not requiring the constraint equations to be satisfied at each iteration in the infeasible path optimization. In this study, orthogonal collocation on finite elements and successive quadratic programming are used in order to find the optimal control profile which satisfies the model constraints for the nominal plant.

Any practical control algorithm must include feedback to account for imperfect modelling (*i.e.* unmodelled disturbances, plant/model mismatch, etc.). Feedback is incorporated in this algorithm through parameter estimation as shown in Figure 1. The sensitivity of the optimal solution with respect to changes in model parameters is also computed in order to determine which parameters have the greatest effect on the optimization and must therefore be known most accurately.

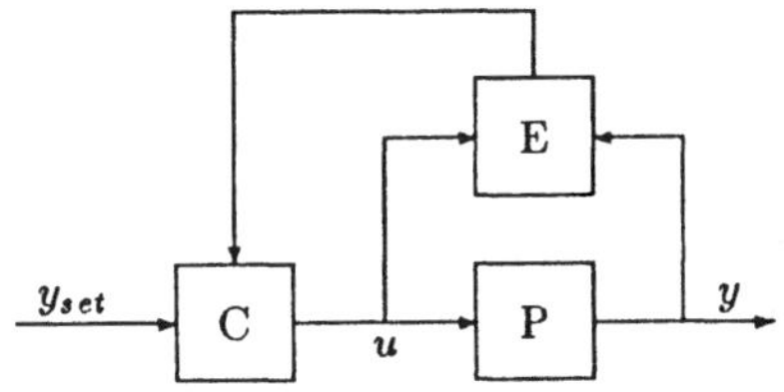

Figure 1: feedback structure

Jang and coworkers [7] propose the standard optimal control formulation coupled with least squares parameter estimation. No guidelines are given for how to select the important, identifiable parameters, and the computational efficiency of their method is not discussed.

In this work, sensitivity estimates are obtained from the first order necessary conditions with only a small increase in computational effort. Confidence intervals are also computed from the Hessian of the Lagrangian. This allows approximate bounds to be determined which indicate the relative degree to which the control moves affect the objective function.

Solution Methods

Consider the following nonlinear program:

$$\begin{aligned}
\min \quad & \Phi(x(t), u(t), p, t) \\
\text{s.t.} \quad \frac{dx}{dt} - f(x(t), u(t), p, t) &= 0 \\
h(x(t), u(t), p, t) &= 0 \qquad (1) \\
g(x(t), u(t), p, t) &\geq 0 \\
x(t_0) &= x_0
\end{aligned}$$

where $x(t)$ and $u(t)$ are vector valued state and manipulated variable functions respectively, and p is a vector of model parameters. Note that the objective function Φ is not required to take any special form, though the particular choice of the objective function will obviously have an effect on the problem's solution and the effort required to obtain it. In process control problems it is most often stated as a function of the deviation of state variables from their setpoints over some future time horizon in continuous processes, or as a function of state variables at a fixed final time in batch processes.

There are many methods for computing solutions to (1). The classic methods, such as control vector iteration and conversion to a two point boundary value problem, are derived using the necessary conditions of the maximum principle [10,2]. Another straightforward method is to simply use an ODE solver in combination with an optimization algorithm and solve the problem sequentially. Using this approach, the differential equations are numerically integrated over the required interval and the objective function is evaluated. Gradient information, if required by the optimization algorithm, may be obtained using finite differences or by the integration of additional sensitivity equations.

These methods are generally inefficient because they require that the model equations be solved at each iteration within the optimization. A more efficient solution of (1) may be obtained by combining weighted residual techniques with an infeasible path optimization such as successive quadratic programming. It is important to note that the greater efficiency of this approach is a result of combining the weighted residual technique with the infeasible path optimization. Simply using weighted residuals to convert the differential equation constraints to algebraic constraints is not sufficient because a feasible path optimization will require that the algebraic constraints be satisfied at each iteration – the model is solved accurately at each iteration.

For this work, orthogonal collocation on finite elements has been used in connection with successive quadratic programming to solve (1). Using this technique, the original problem may be restated as

$$\begin{aligned}
\min \quad & \Phi(x, u, p, t) \\
\text{s.t.} \quad Ax - f(x, u, p, t) &= 0 \\
h(x, u, p, t) &= 0 \qquad (2) \\
g(x, u, p, t) &\geq 0 \\
x(t_0) &= x_0
\end{aligned}$$

where A is a matrix of collocation weights [14,8,13].

This approach offers several advantages over traditional methods. First, state, manipulated variable, and rate of change constraints may be incorporated naturally in the problem statement simply by imposing bounds on the variables at the collocation points. In order to handle the same class of constraints using other methods, one must either add penalty terms to the objective function [2] or include additional state variables [12]. Second, one may choose to implement either continuous or discrete manipulated variables by simply changing the order of the collocation. Finally, the solution of the optimization problem provides useful sensitivity information at little additional cost. This point is discussed further below.

Parametric Sensitivity

Given the nonlinear program (2), it is desired to find the sensitivity of the optimal solution with respect to changes in parameter values p. These sensitivities will prove useful in determining which of the model parameters have the greatest effect on the solution of the dynamic optimization and must be known most accurately. Sensitivity results for problems of this class may be found in Fiacco [4] and are summarized below.

At the optimum, the following necessary conditions hold

$$\begin{aligned} \nabla_x L(x,p) &= 0 & \\ g_i(x,p) &= 0 & i = 1,r \\ h_i(x,p) &= 0 & j = 1,m \\ v_i g_i(x,p) &= 0 & i = r+1,n \end{aligned}$$

where the first r inequality constraints are in the active set and L is the augmented Lagrangian used by Gill et. al. [6].

$$L(x,v,w,p,\rho_i) \equiv \Phi(x,p) - \sum_{i=1}^{n}(v_i - \frac{1}{2}\rho_i g_i)g_i - \sum_{j=1}^{m}(w_j - \frac{1}{2}\rho_j h_j)h_j$$

where the ρ_i are positive constants. By differentiating the necessary conditions and noting that at the optimum the following conditions hold:

$$\begin{aligned} h_j &= 0, & w_j &\neq 0 & j &= 1,m \\ g_i &= 0, & v_i &\neq 0 & i &= 1,r \\ g_i &> 0, & v_i &= 0 & i &= r+1,n \end{aligned}$$

the set of sensitivity equations may be written as

$$\begin{bmatrix} \nabla_{xx}L & -\nabla_x \bar{g}^T & \nabla_x h^T \\ \nabla_x \bar{g} & 0 & 0 \\ \nabla_x h & 0 & 0 \end{bmatrix} \begin{bmatrix} \nabla_p x \\ \nabla_p \bar{v} \\ \nabla_p w \end{bmatrix} = - \begin{bmatrix} \nabla_{xp} L \\ \nabla_p \bar{g} \\ \nabla_p h \end{bmatrix}$$

where the superbar denotes the set of binding inequality constraints.

The solution of this set of equations at the optimum yields the sensitivity of the solution of the approximate nonlinear program to changes in parameter values.

Second Order Sensitivity

Consider the Taylor series expansion of the Lagrangian

$$\delta L = \nabla_x L \delta x + \frac{1}{2}\delta x^T \nabla_{xx} L \delta x.$$

For independent variations in x, the change in L at the optimum becomes

$$\delta L = \frac{1}{2}\frac{\partial^2 L}{\partial x_i^2}\delta x_i^2$$

because $\nabla_x L = 0$ at the optimum. Therefore, the change in a given decision variable for a given variation in L is

$$\delta x_i = \left[2\delta L\left(\frac{\partial^2 L}{\partial x_i^2}\right)^{-1}\right]^{\frac{1}{2}}.$$

Thus one may determine the amount the manipulated variable profile may change while keeping the objective function within some specified bound.

Computational results

The first example applies the controller to a continuous process with and without plant/model mismatch. Manipulated variable constraints and parametric sensitivity are also considered. Control of a batch reactor is shown in the second example. Approximate confidence bounds which are placed on the manipulated variable show variations in the sensitivity of the objective function with time.

Example 1

For the following model of a second order reaction taking place in a CSTR

$$\frac{dC_A}{dt} = \frac{F}{V}(C_{Ain} - C_A) - kC_A^2$$

it is desired to find the control profile which will minimize

$$\Phi = \int_0^t (C_{Aset} - C_A)^2 dt$$

for a step change in the setpoint C_{Aset}, where the inlet flow rate F is the manipulated variable. In each of the following cases, the plant parameters were

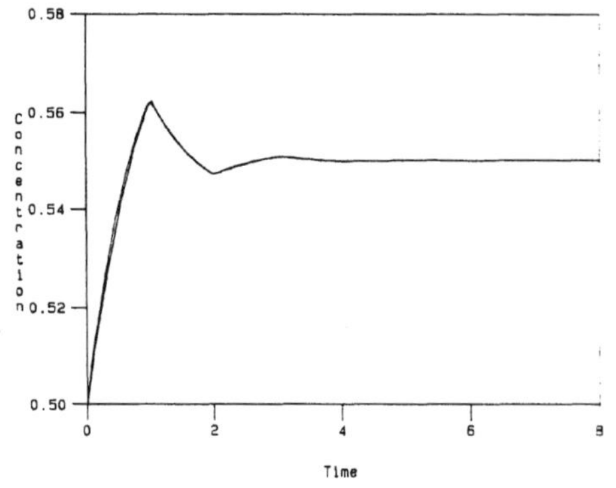

Figure 2: Example 1 case 1. System response

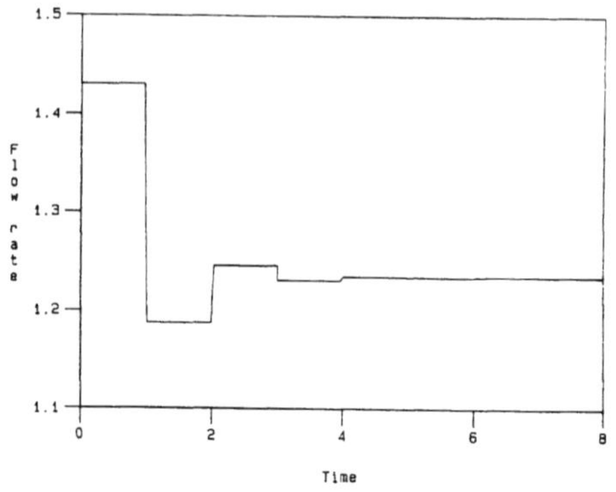

Figure 3: Example 1 case 1. Manipulated variable

$$
\begin{aligned}
V &= 10.0 \\
C_{Ain} &= 3.0 \\
k &= 1
\end{aligned}
$$

Case 1. Figures 2 and 3 show the system response and optimal manipulated variable profile for a step change in the set point for 0.5 to 0.55. The optimum profile was computed based on five equal piecewise constant control moves ($\Delta t = 1.0$). The profile was then implemented for three steps by integrating the model by using a standard ODE solver [9]. This procedure was then repeated several times with each prediction step beginning at the end of the previous implementation step. The oscillatory behavior of the solution may be removed by shortening the implementation time. Figures 4 and 5 show the response of the system and the manipulated variable profile for an implementation time of 0.5.

Case 2. For this case, the model parameters were $V = 10.5$ and $k = 1.10$ while the plant parameters remained the same as for case 1. A simple parameter estimation scheme in the form of an inferred disturbance (see, for example, [5])

$$d = y_p - y_m$$

was used to adjust the manipulated variable to account for the model error and force the plant to the desired setpoint.

Figure 6 shows the response of the system for the same setpoint change as for the first

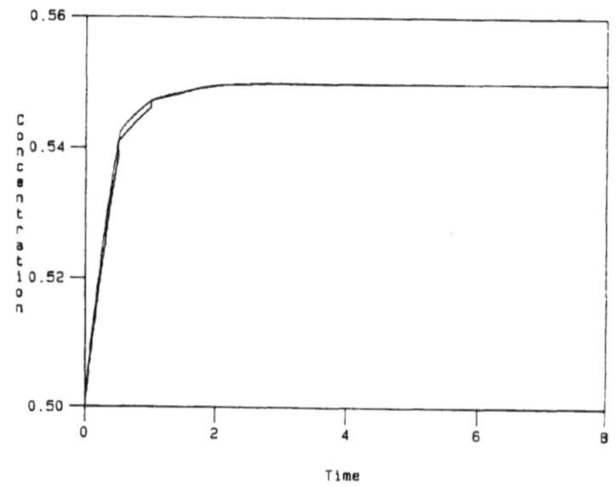

Figure 4: Example 1 case 1. System response

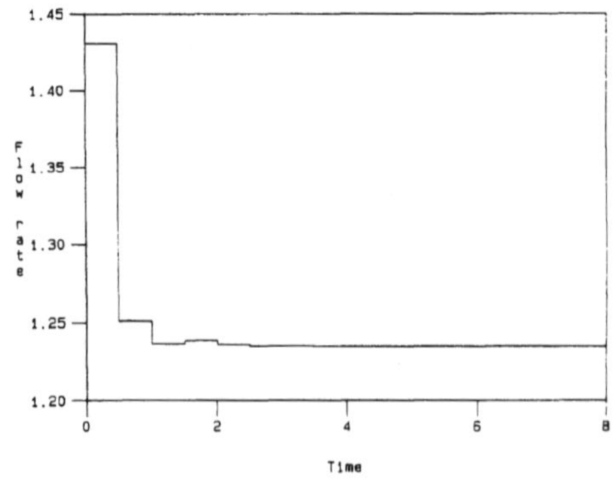

Figure 5: Example 1 case 1. Manipulated variable

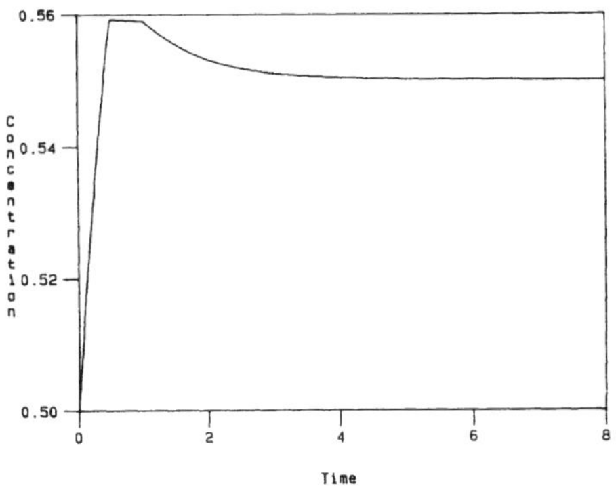

Figure 6: Example 1 case 2. System response

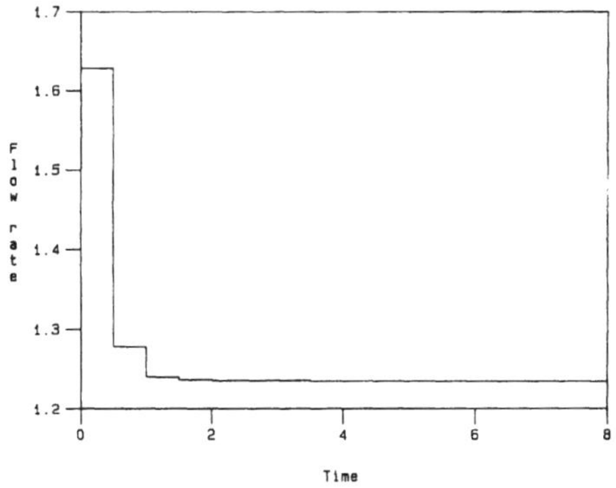

Figure 7: Example 1 case 2. Manipulated variable

0.5. The optimal manipulated variable profile for this case is shown in figure 7

Case 3. Figures 8 through 10 show the system response, manipulated variable profile, and sensitivity profiles when a constraint on the manipulated variable is encountered. Note that the sensitivity of the optimal profile with respect to changes in model parameters is zero over the period in which the constraints are active, and that it is more sensitive to the reaction rate constant than to the tank volume, as expected.

Example 2

In this example, taken from Ray [10], (problem 3.1), the control profile is computed in order to obtain the maximum concentration of product B at the end of a batch reaction for the following irreversible reactions

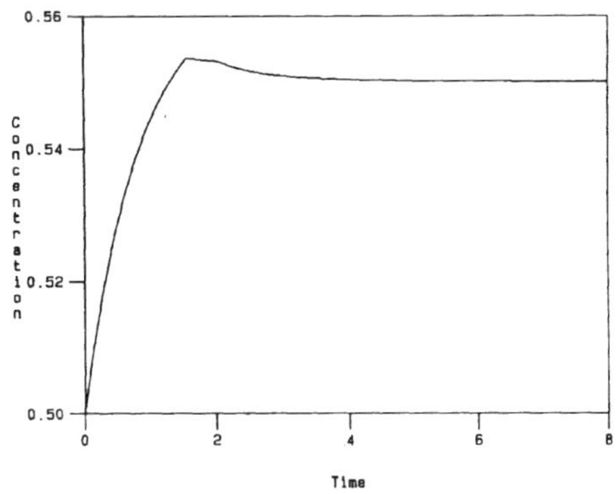

Figure 8: Example 1 case 3. System response with manipulated variable constraint

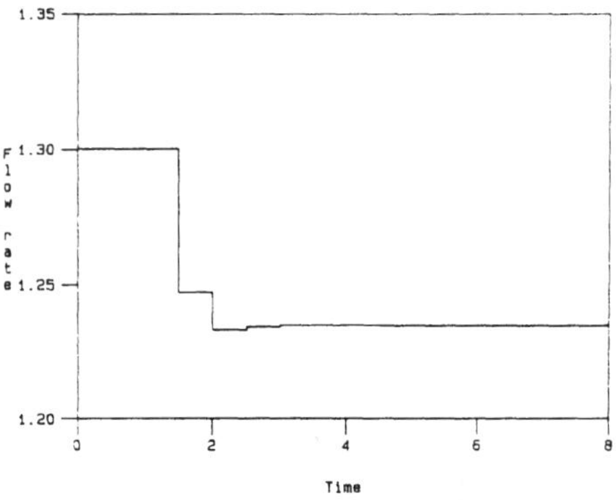

Figure 9: Example 1 case 3. Manipulated variable

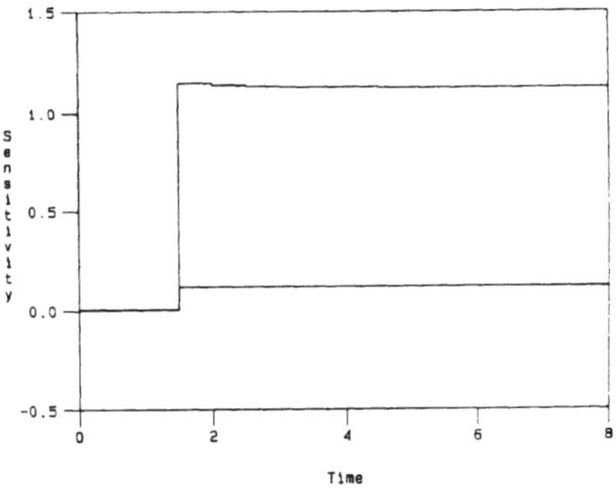

Figure 10: Example 1 case 3. Sensitivity of the optimal manipulated variable profile with respect to model parameters

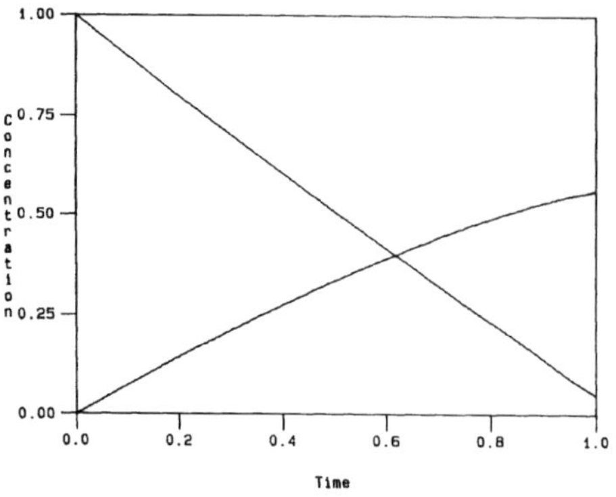

Figure 11: Example 2. State variables

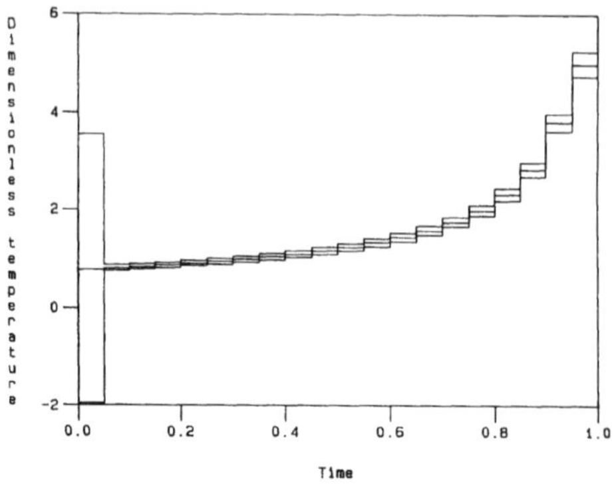

Figure 12: Example 2. Manipulated variable with confidence bounds

$$\mathrm{A} \longrightarrow \mathrm{B}, \qquad \mathrm{A} \searrow \mathrm{C}$$

The final dimensionless form of the model as given by Ray is

$$\begin{aligned} \frac{dx_1}{dt} &= -(u + \alpha u^p)x_1 & x_1(0) &= 1 \\ \frac{dx_2}{dt} &= ux_1 & x_2(0) &= 0 \end{aligned}$$

$$x_1 = \frac{\mathrm{A}}{\mathrm{A_0}} \quad x_2 = \frac{\mathrm{B}}{\mathrm{A_0}}$$

$$0.0 \leq u \leq 5.0$$

The values of α and $p = 2$ are 0.5 and 2 respectively.

Figure 12 shows the optimum manipulated variable profile and ten percent confidence bounds computed using 20 piecewise constant control moves. With the exception of the first control move, the bounds indicate that the objective function becomes increasingly insensitive to the manipulated variable as the reaction nears completion.

Conclusion

An efficient method of solving nonlinear optimal control problems has been extended to provide sensitivity of the optimal solution with respect to model parameters. In addition, a method for determining confidence intervals for the optimal control profiles has been demonstrated.

References

[1] L. T. Biegler. Solution of dynamic optimization problems by successive quadratic programming and orthogonal collocation. *Computers and Chemical Engineering*, 8(3/4):243–248, 1984.

[2] A. E. Bryson and Y. Ho. *Applied Optimal Control.* Hemisphere Publishing, New York, 1975.

[3] J. E. Cuthrell and L. T. Biegler. On the optimization of differential-algebraic systems. *AIChE Journal*, 33(8):1257–1270, 1987.

[4] A. V. Fiacco. *Introduction to Sensitivity Analysis in Nonlinear Programming.* Academic Press, New York, 1983.

[5] C. E. Garcia and D. M. Prett. Advances in industrial model-predictive control. In M. Morari and T. J. McAvoy, editors, *Chemical Process Control - CPCIII*, pages 245–293, Third International Conference on Chemical Process Control, Elsevier, Amsterdam, 1986.

[6] P. E. Gill, W. Murray, M. A. Saunders, and M. H. Wright. *User's Guide for SOL/NPSOL: A Fortran Package for Nonlinear Programming, Technical report SOL 83-12.* Technical Report, Systems Optimization Laboratory, Department of Operations Research, Stanford University, 1983.

[7] Shi-Shang Jang, B. Joseph, and H. Mukai. On-line optimization of con-

strained multivariable chemical processes. *AIChE Journal*, 33(1):26–35, 1987.

[8] M. L. Michelsen and J. Villadsen. A convenient computational procedure for collocation constants. *The Chemical Engineering Journal*, 4:64–68, 1972.

[9] L. R. Petzold. A description of dassl: a differential-algebraic system solver. In R. S. Stepleman, editor, *Scientific Computing*, page 65, North-Holland, Amsterdam, 1983.

[10] W. H. Ray. *Advanced Process Control.* McGraw-Hill, New York, 1981.

[11] J. G. Renfro, A. M. Morshedi, and O. A. Asbjornsen. Simultaneous optimization and solution of systems described by differential/algebraic equations. *Computers and Chemical Engineering*, 11(5):503–517, 1987.

[12] R. W. H. Sargent and G. R. Sullivan. The development of an efficient optimal control package. In J. Stoer, editor, *Lecture Notes in Control and Information Sciences 7*, pages 158–168, 8th IFIP Conference on Optimization Techniques, Springer–Verlag, 1978.

[13] J. Villadsen and M. L. Michelsen. *Solution of Differential Equation Models by Polynomial Approximation.* Prentice-Hall, Englewood Cliffs New Jersey, 1978.

[14] J. V. Villadsen and W. E. Stewart. Solution of boundary-value problems by orthogonal collocation. *Chemical Engineering Science*, 22:1483–1501, 1967.

CONCENTRATION PROFILE ESTIMATION AND CONTROL IN BINARY DISTILLATION

W. Marquardt

Institut für Systemdynamik und Regelungstechnik, Universität Stuttgart, D-7000 Stuttgart 80, Pfaffenwaldring 9, FRG

Abstract. A new physically motivated approach to model based control and estimation of binary distillation is suggested. A nonlinear reduced model which relies on nonlinear wave propagation phenomena in separation processes is used as a starting point for multivariable control system synthesis applying state-space methods. Apart from controlling product compositions using only few temperature measurements along the column, profile estimation is accomplished with moderate additional effort. The resulting linear control system is evaluated in nonlinear simulations of a high purity distillation column. Satisfactory control performance is exhibited.

Keywords. Process control, nonlinear systems, distributed parameter systems, state-space methods, system analysis, observers, optimal control, distillation columns.

Introduction

Modelling of separation columns on the basis of conservation laws and thermodynamics of mixtures leads to very complex dynamic models exhibiting severe nonlinearities. Depending on the fundamental assumptions about the process structure the result is either a nonlinear partial differential or a nonlinear difference-differential equation model. The first is more suitable for packed columns, whereas the latter matches the reality of staged columns better. Neither of the two models can be used directly as a starting point for control system design based on system theory. This is due to the lack of general system theoretic concepts for system analysis and control system synthesis for nonlinear distributed parameter systems as well as for nonlinear large-scale lumped parameter systems. These shortcomings in control theory are in contrast with a growing demand for tighter control of separation processes to assess reproducable product quality.

In order to overcome the theoretical difficulties and to achieve high quality control of separation processes, various model-based approaches are known from recent publications.

One class of methods stands out for the use of relatively simple models, which give the dependence of the outputs of the process on its inputs. In most cases these models are linear and discrete or continuous in time. The parameters of such models are fit to a real process or to a detailed nonlinear simulation model. Models derived in this way are rather simple but often reproduce only a crude approximation of the real nonlinear process. They are used for multivariable control system design by well-known frequency domain techniques (see for example Tyreus, 1979; Ogunnaike and co-workers, 1983; Levien, Morari, 1987) or for the design and implementation of predictive control systems (see for example Martin-Sanchez, Shah, 1984; McDonald, McAvoy, 1987; Prett, Garcia, 1987). A model of plant uncertainties has been incorporated recently in the synthesis technique to achieve better robustness of the control system (see for example Morari, 1987a).

An alternative approach to control system design is characterized by the development of reduced models, which should represent at least the main dynamic features of the system behavior. In a second step the less complex reduced model is used for control system synthesis using system theoretic concepts. This way, a-priori knowledge about the physics and the internal structure of the process are being incorporated in control system design. The first approach to be mentioned here starts with linearization of the rigorous nonlinear process model around a prescribed steady-state. The order of the resulting large-scale linear state-space model is reduced by one of the well-known methods (Föllinger, 1984). Then, this low order linear model forms the basis for control system synthesis using classical state-space methods (Lehmann, 1987). In case of strong nonlinearities, which are exhibited by most high purity separation processes, this promising technique may produce poor results. This is due to linearization errors. Furthermore, the few state variables of the reduced model reveal only limited information about the global dynamics of the physical process.

The disadvantage may be overcome by a physical approach to model reduction. This is in conceptual contrast to most common model reduction techniques, which is often based on mathematical formalisms. The state variables of such a reduced model are chosen to represent the essential dynamic behavior of the distillation process. Retzbach (1986) uses the location of sharp temperature fronts as state variables of a reduced model to control an extractive distillation plant. This approach is generalized here: apart from the location of the concentration or temperature profile a few additional state variables are introduced to characterize the spatial shape of the profile. A nonlinear reduced model based on these few state variables yields an appropriate representation of the global system dynamics. Thus, complete process knowledge may be included systematically in a model based control system design by state-space methods. Basically, the low order reduced model could be used together with nonlinear control system design methods. Because of the practical problems associated with these techniques so far the development of a linear state-space controller for the linearized reduced model is more feasible.

After having stated the control problem under consideration, the physically motivated model reduction technique is summarized briefly. The resulting reduced model is compared to a rigorous model with respect to steady-state and dynamic behavior. Two different binary distillation columns yielding high and low purity products respectively are used as examples. After linearization at a given operating point the reduced model forms the basis for designing state feedback control and concentration profile estimation. Different controllers are evaluated and compared in nonlinear simulations.

The Control Problem

The distillation column under consideration (see Fig.1) shows five controlled variables. These are the overhead pressure, the liquid levels in the reflux drum and the reboiler as well as the product concentrations. Possible manipulated variables are the distillate and bottoms product withdrawal rate, the reflux rate, and the reboiler and condensor duties. We are only interested in controlling product compositions. The main disturbances are introduced by feed changes. Ideal behavior of overhead pressure and level control is assumed, which is justified because of the widely differing time constants of concentration and flow dynamics. The manipulated variables chosen for product composition

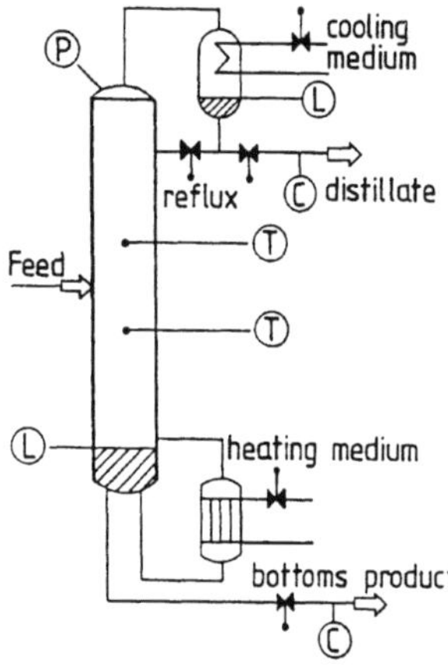

Fig. 1. The control problem.

control are the reflux rate and the reboiler duty. Two temperatures, in the stripping and rectifying section respectively, are used as measured variables instead of time-delayed and expensive concentration measurements, which are usually employed for dual composition control. In addition to the control of product compositions, an estimate of the complete spatial concentration profile should be provided for the supervision of the global plant behavior. The temperature measurements are used for the solution of this estimation problem.

Model Reduction

The dynamic behavior of distillation columns is characterized by the propagation of concentration and temperature profiles in the column sections. The location of intense mass transfer moves in the direction of the reboiler or condensor as a consequence of feed disturbances and of control action. During these transients the typical profile shape is preserved. An observer moving with a coordinate system linked to the profile would hardly recognize any variations (see Fig. 2). This typical dynamic behavior is shortly summarized to nonlinear wave propagation. The wave propagation properties are investigated theoretically under idealizing assumptions in Marquardt (1988). These properties are used for developing a reduced model for a section of the distillation column. The reduced model of a complete column is set up by combining the reduced models of the rectifying and stripping section with the models of the boundary systems, which are given by the feed tray, the reboiler and the condensor.

The starting point for model reduction of a column section is a non-equilibrium distributed parameter model, which ignores hydrodynamics and the energy balance (Marquardt, 1986,1988). It consists of two nonlinear hyperbolic partial differential equations for the liquid and the vapor concentration of the less volatile component. The spatially continuous structure of this rigorous model is appropriate to describe packed columns. It is also suitable for approximate modeling of staged columns. Sufficient accuracy may be obtained by adjusting the model parameters, for example the mass transfer coefficient, to the real plant.

Due to the propagation of the spatial profiles the model equations of the column section are transformed in a moving coordinate system linked to the wave front. The position of the front is one of the state variables of the reduced model. Since the shape of the profile is preserved during transients, a suitable spatial trial function is chosen in the moving coordinate system. It approximates the spatial concentration profile. This function depends not only on the spatial coordinate of the moving coordinate system but also on a few time varying shape parameters. They allow time dependent scaling of the profile shape during transients. These shape parameters are further state variables of the reduced model. State equations for determining the shape parameters are deduced by a method of weighted residuals (Finlayson, 1972). The model equation for calculating the wave front location is given by a global material balance of the column section.

This reduction technique results in the linear implicit differential-algebraic system of equations

$$\mathbf{K}(\mathbf{s})\dot{\mathbf{s}} = \mathbf{r}(\mathbf{s}) \quad (1)$$

with suitable initial conditions. The singular matrix $\mathbf{K}$ and the vector function $\mathbf{r}$ are complex functions of the state variables $\mathbf{s}$, typically consisting of eight dynamic state variables and few algebraic variables. The eight dynamic state variables are the compositions in the distillate and bottoms product x_D and x_B, the locations z_S and z_R of the wave fronts and the two shape parameters in every column section. These shape parameters describe the maximal gradient and some asymmetry measure of the profile. The number of state variables in the reduced model and its internal structure do not depend on the operating or construction data of the column, but are determined mainly by the phase equilibrium of the mixture. This nonlinear low order model reflects the characteristic structural properties of the separation process in its state variables. Hence, this reduction technique leads to some kind of "minimal realization" of the nonlinear, distributed parameter system description. A more detailed discussion of the model reduction topic may be found elsewhere (Marquardt 1986, 1988).

The rigorous and the reduced model are compared in Fig. 2 by means of two different distillation columns. Both columns are separating the same feed and are working under comparable operating conditions. Column H produces high purity products of about 99.9 mole percent whereas column L separates the feed into two product streams of about 95 mole percent. The com-

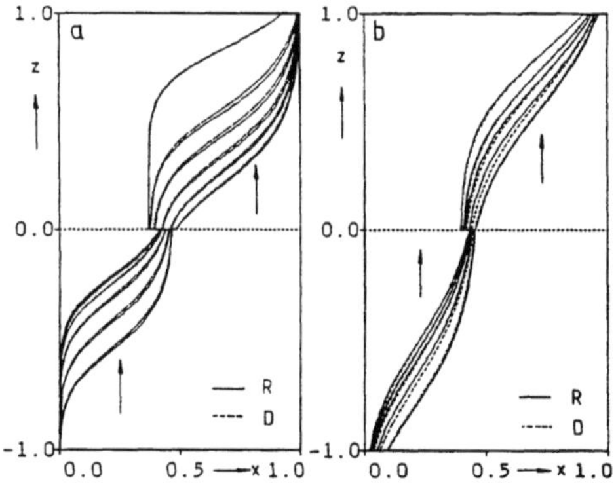

Fig. 2. Dynamic behavior after step disturbance of vapor rate: 1.5571 → 1.62; D: rigorous model, finite differences; R: reduced model; a: column H, plot times 0,2,4,6,8,20; b: column L, plot times 0,2,4,6,20.

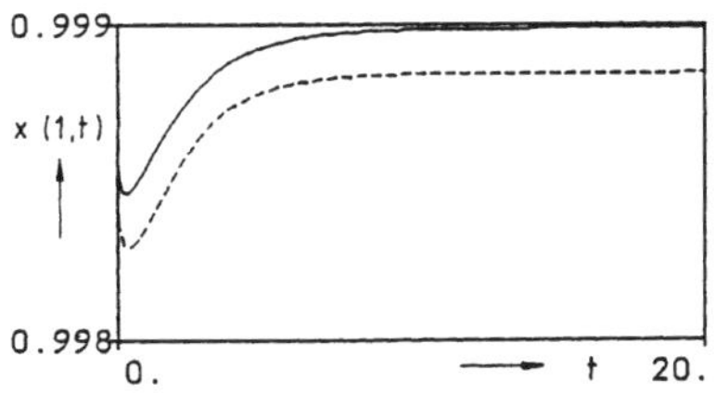

Fig. 3. Simultaneous step disturbance in both manipulated variables; column H; reflux rate: 1.0 → 1.1; vapor rate: 1.6445 → 1.7445.

plete data of the columns are given in Marquardt (1988). The approximation of both steady states, at the beginning and at the end of the simulation is very good. Differences between both profiles remain small in the whole region of all meaningful operating conditions. Hence, the reduced model reveals exact steady state gains even in the case of large disturbances. This property cannot be exhibited by any linear reduced (or rigorous) model. During transients the degree of correspondence between simulation results obtained by the reduced and the rigorous model is comparable to the steady state accuracy in most cases. Fig. 3 shows the dynamic behavior of the reduced and the rigorous model of column H after a simultaneous step change of the same magnitude in both manipulated variables. Such disturbances occur especially during closed loop control. Morari (1987b) points out, that this type of disturbance produces faster dynamic responses than a variation of only one manipulated variable. If the steady-state error of about 0.015 mole percent is neglected, both models show nearly the same behavior. Simulations of column L exhibit similar results.

It should be stressed here that all simulations are carried out without any tuning of either the reduction technique or the reduced model parameters. Hence, the reduced model may be set up in a rather straightforward manner without trial and error procedures.

System Analysis

A system analysis is performed by studying the reduced models (1) of the two distillation columns H and L after numerical linearization at the steady state. The resulting linear time invariant system in the vicinity of a prescribed operating point is given

$$\Delta\dot{\mathbf{x}} = \mathbf{A}\Delta\mathbf{x} + \mathbf{B}\Delta\mathbf{u} + \mathbf{X}\Delta\mathbf{d}; \qquad \Delta\mathbf{x}(0) = \mathbf{0} \tag{2}$$

$$\Delta\mathbf{y} = \mathbf{C}\Delta\mathbf{x} + \mathbf{D}\Delta\mathbf{u} + \mathbf{Y}\Delta\mathbf{d}, \tag{3}$$

$$\Delta\mathbf{z} = \mathbf{U}\Delta\mathbf{x} + \mathbf{V}\Delta\mathbf{u} + \mathbf{W}\Delta\mathbf{d}. \tag{4}$$

The vector $\Delta\mathbf{x}$ consists of the deviations from the steady state value of the eight distillation column dynamic state variables given above. The control vector $\Delta\mathbf{u}$ is made up by the two manipulated variables Δr and Δv, whereas the disturbance vector $\Delta\mathbf{d}$ includes the deviations Δf and Δx_F from the steady state value of the feed rate and concentration. The dimensionless variables Δr, Δv and Δf are defined as

$$\Delta r = \frac{R - R_{ss}}{R_{ss}}, \qquad \Delta v = \frac{V - V_{ss}}{R_{ss}}, \qquad \Delta f = \frac{F - F_{ss}}{R_{ss}}.$$

The variables F and R are the reflux and feed rate respectively, whereas V stands for the vapor rate leaving the reboiler. The measurement vector $\Delta\mathbf{y}$ consists of scaled temperatures at suitable locations along the column. For convenience the mildly nonlinear concentration dependence of the temperatures is approximated by a linear function. The vector $\Delta\mathbf{z}$ is made up of six algebraic variables, which result from the model reduction method. They are insignificant from the point of view of control system analysis and design. The full matrices **A**, **B**, **X**, **C**, **D**, **Y**, **U**, **V**, and **W** have appropriate dimensions. They can be supplied by the author on request.

Comparison of nonlinear and linearized reduced model

A comparison of the nonlinear reduced model with the linearized one is given in Fig. 4 and 5 for the rectifying section of columns H and L. The time history of one of the shape parameters p (here the maximum spatial gradient of the concentration profile), the location of the front, and the distillate concentration are shown in the diagrams a, b and c respectively after a large step change of the reflux rate of $\pm 0.1 r_{ss}$. The abscissa t is a dimensionless time, related to the mean residence time of the liquid phase in the rectifying section. The shape parameter p and the location z_R are exhibiting a mildly nonlinear behavior for both columns, whereas the nonlinearity in the product concentration x_D is more severe. The nonlinearities revealed by the reduced model of column H are stronger than those shown by column L. The same qualitative difference with respect to the degree of nonlinearity is also shown by rigorous distillation models. However, the nonlinearities of the reduced model are not as severe as those of the detailed model. This can be shown by comparison of a nonlinear and its related linearized rigorous model (see for example Lehmann, 1987 or Marquardt, 1985). The nonlinearity of the reduced model is diminished due to the use of a set of transformed state variables which are highly adapted to the physical process structure. As it will become apparent in the next section, the locations z_R and z_D as well as the concentration profile gradient are the most important state variables with respect to control system design. Therefore, the linearized reduced model is expected to serve as an adequate basis for control system design by linear methods.

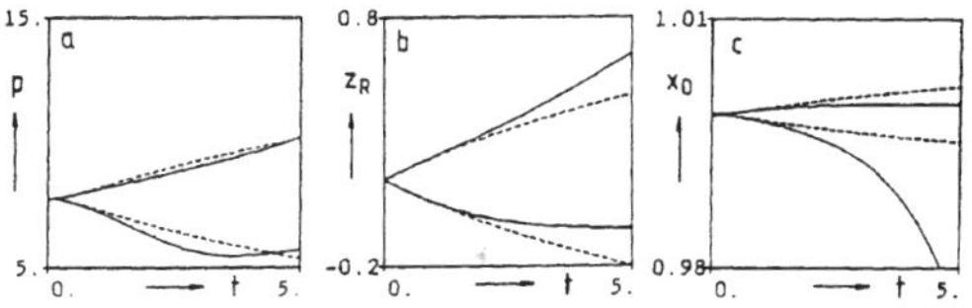

Fig. 4. Comparison of nonlinear and linearized reduced model for a step disturbance in reflux rate of column H: 1.0 ± 0.1.

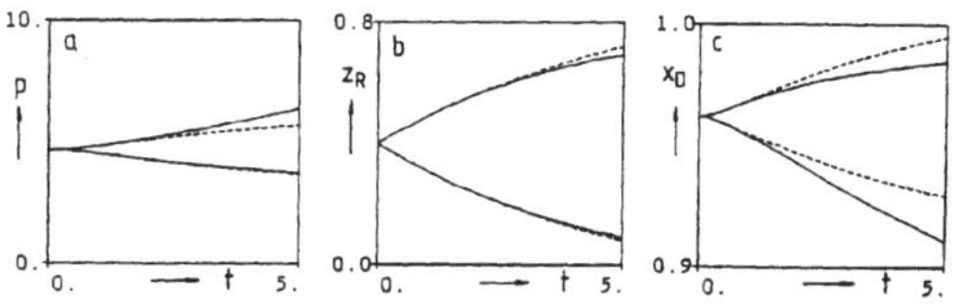

Fig. 5. Comparison of nonlinear and linearized reduced model for a step disturbance in reflux rate of column L: 1.0 ± 0.1.

Modal Analysis of the Linear Model

The results obtained by a modal analysis of the linearized reduced models of column L and H are given in Table 1. All eigenvalues are real and indicate the qualitatively different dynamic behavior of the two columns, which is exemplified in the simulation of Fig. 2. After disturbances or control actions column L reaches its new steady state in a shorter time than column H does. In the case of column H, one eigenvalue close to zero is apparent. Such an eigenvalue is missing in the spectrum of column L. This property is not very surprising in view of the high sensitivity of the concentration front in the stripping section of column H. The fronts in the rectifying section of column H and in both sections of column L are much more damped. The ratio of the largest to the smallest eigenvalue differs by a factor of about ten in the two examples. The stiffness ratio increases with increasing product quality. If the spectrum of the reduced model is compared to that of a rigorous model (Nothaft, 1984), the same qualitative properties are found. Hence, this fact emphasizes once more the correct representation of the dynamic properties of distillation columns in the reduced model.

TABLE 1. Eigenvalues, measures of controllability, disturbability and observability of the linearized reduced models. Dominant eigenvalues are marked by †.
E ... eigenvalue ,
O ... measure of observability (x 0.001) ,
C ... measure of controllability (x 0.001) ,
D ... measure of disturbability (x 0.001) .

	Column H				Column N			
	E	O	C	D	E	O	C	D
1	$-0.02^{\dagger}$	14.4	3774.	3444.	$-0.18^{\dagger}$	14.3	342.	0.1
2	$-0.29^{\dagger}$	14.2	682.	176.	$-0.56^{\dagger}$	0.92	284.	7.0
3	-1.00	0.007	1056.	8.2	-0.85	0.02	2970.	165.
4	-2.18	0.0000	0.17	0.68	$-1.23^{\dagger}$	0.002	2037.	1699.
5	-3.18	0.0008	2791.	281.	$-2.30^{\dagger}$	0.0002	1934.	4281.
6	$-3.29^{\dagger}$	13.9	449.	5689.	-3.49	27.2	13.7	5.8
7	-4.80	0.0008	7128.	32.3	-5.84	0.0006	25270.	6333.
8	-8.73	0.04	0.08	10.4	-9.01	5.7	10.7	224.

Measures of Observability, Controllability and Disturbability

The investigation of observability and controllability of the system state is of fundamental importance for control system design of multivariable systems using state-space methods. If Kalman's algebraic criteria are applied, complete controllability, disturbability and observability are found for both linear models. The last property does not depend on the number and location of temperature measurements if the relation between temperature and concentration is unique. For control system design, and especially for the choice of best measurement locations, more detailed information about these structural system properties would be useful. For this purpose several structural measures, most of which are modal based, have been developed. A critical review has been given recently by Litz (1983). Despite possible misleading results in the case of eigenvalues lying close together, the modal measures of Müller and Lückel are used to quantify the structural properties. However, according to Juen (1982) and Litz (1983), these measures are employed in a modified form. The exponential weighting with the real part of the eigenvalue of the mode under consideration is neglected. These modified measures are given in Table 1 for both columns. For this study, the two temperature measurements are located in the middle of each column section.

The measures of observability are relatively small in both models. Three to four modes are substantially better observable than all others (H: 1,2 and 6; N: 1,6 and 8). The measures of controllability are always larger. Further, they are larger than the measures of disturbability in almost all cases. Two less controllable and disturbable modes become apparent. Due to the modal nature of the computed measures, no conclusion concerning the properties of the system states can be drawn without further investigation. For an evaluation of the structural properties of the system states, additional information about the contribution of a mode k to a state variable i is necessary. If the measurement matrix **C** is chosen as the identity matrix, this relation is given by the structural dominance measures D_{ik} of Litz (1979, eq. 23) or that of Bonvin and Mellichamp (1982). After proper scaling of the columns of the matrix formed by the measures D_{ik}, the value of the matrix element i, k indicates the influence of mode k on state variable i.

In the model of column H the eigenvalue 1 clearly dominates all state variables. Further, the eigenvalues 2 and 6 are of certain significance. The contribution of all other eigenvalues is negligible. The eigenvalues 1,2 and 6 show high structural measures in Table 1. Therefore, a sufficient degree of observability and controllability is anticipated. This type of analysis shows that the linear system may be reduced to an order of three by one of the well known reduction methods without any problems. The structural dominance measures of column N are not as clearly segregated as those of column H. The highest dominance measures are obtained for eigenvalues 4 and 5. Eigenvalue 4 primarily influences the states of the rectifying section whereas the fifth eigenvalue is mainly responsible for the stripping section dynamics. Less significant but not negligible are modes 1 and 2. A linear model reduction to an order of four is also possible in this case. In summary, column L is as controllable as column H but less observable than column H. This property might be explained with the less distinctive wave propagation phenomena in column L. The observability of column L can be improved by means of more feasible measurement configurations.

Measurement Location

The measures of observability are very suitable for the choice of best measurement locations in the distributed parameter system. Table 2 shows the measures for column H with seven different measurement configurations. As expected, the observability is enhanced if the number of measurements is increased. However, unexpectedly, the measures are not diminished significantly if only one measurement is taken in the stripping section of the column instead of two measurements in both sections. If the only measurement is located in the rectifying section the measures drop remarkably. This qualitative difference results from a significantly lower sensitivity of the profile in the rectifying section compared to that of the stripping section if feed rate disturbances are encountered. Measurements in the vicinity of the feed tray or the reboiler and condensor result in low measures of observability. The same is true for product concentration measurements. Similar investigations can be carried out to select the most appropriate set of manipulated variables for dual composition control.

TABLE 2. Measures of observability (x 0.001) for column H and differing measurement configurations. The numbers in the top row of the table stand for the measurement locations. The dominant eigenvalues are marked by †.

	E	0.5 -0.5	0.2 0.6 -0.2 -0.6	1.0	0.02 -0.02	0.5	-0.5	x_D x_B
1†	-0.02	14.4	15.8	0.18	0.3	0.1	14.3	0.003
2†	-0.29	14.2	22.5	2.3	1.1	2.3	11.2	0.005
3	-1.00	0.007	0.06	4.2	0.02	0.003	0.004	0.0001
4	-2.18	0.0000	0.001	4.3	0.0002	0.0001	0.0000	0.0000
5	-3.18	0.0008	0.003	0.0007	0.0002	0.0001	0.0007	0.0001
6†	-3.29	13.9	16.9	20.0	0.02	13.9	0.0001	16.8
7	-4.80	0.0008	0.0008	0.0009	0.0000	0.0000	0.0008	0.0000
8	-8.73	0.04	0.07	0.001	0.005	0.0001	0.04	1.6

Interaction Measures

The influence of the manipulated variables on the state variables is quantified by comparison of the sensitivities of the state variables with respect to the system inputs and the steady-state direct gain interaction according to Johnston and Barton (1985). These measures show the direct influence of the manipulated variables on the system states. For the columns under consideration, both manipulated variables influence the principal states (i.e. the location of the concentration profile) of the rectifying and the stripping section in nearly the same way. Hence, a dual composition control with two independent control loops is not feasible.

Furthermore it is shown that the product compositions are excited by disturbances almost exlusively in an indirect manner. The most important signal paths from the inputs to the product compositions are those via the front locations. Fast stabilization of the product concentrations is accomplished if the front location in both column sections can be controlled tightly.

System Analysis Summary

This kind of linear system analysis employing well accepted methods results in additional information about the reduced model. The structural properties discussed so far approve most of the heuristic experience which has been made in a large number of detailed dynamic simulations of open and closed loop distillation columns. Hence, the criteria of this section applied to the linearized reduced process model allow a quantitative evaluation of the process dynamics and the discrimination of alternative control system structures.

Control System Synthesis

This section is dedicated to the synthesis and evaluation of a control system for column H. Though almost every control system design technique could be used together with the reduced column model, quasi-classical state-space methods are preferred here. They allow the solution of the combined concentration profile estimation and product composition control problem. The achieved results justify a linear approach using the results of the last section. The design of the feedback matrices for the estimator and the regulator is accomplished by means of the CACSD package KEDDC (Schmid, 1985). All nonlinear simulations are done with the process flowsheet simulator DIVA (Marquardt and co-workers, 1987). The same control system structure has also been implemented successfully on column L.

State Estimation and Proportional Feedback

All state variables must be estimated for state feedback implementation if only temperatures are measured along the column. A full order identity observer is used for that purpose. It is extended by a linear equation to determine the algebraic variables $\Delta \mathbf{z}$:

$$\Delta\dot{\hat{\mathbf{x}}} = \mathbf{A}\Delta\hat{\mathbf{x}} + \mathbf{B}\Delta\mathbf{u} + \mathbf{E}_x(\Delta\mathbf{y} - \Delta\hat{\mathbf{y}}); \quad \Delta\hat{\mathbf{x}}(0) = \Delta\hat{\mathbf{x}}_0 \quad (5)$$

$$\Delta\hat{\mathbf{y}} = \mathbf{C}\Delta\hat{\mathbf{x}} + \mathbf{D}\Delta\mathbf{u}, \quad (6)$$

$$\Delta\hat{\mathbf{z}} = \mathbf{U}\Delta\hat{\mathbf{x}} + \mathbf{V}\Delta\mathbf{u}. \quad (7)$$

Two temperature measurements are used. They are located in the middle of both column sections. The disturbance variables $\Delta\mathbf{d}$ are unknown and are therefore not included in eqs.(3)-(7). The feedback matrix $\mathbf{E}_x$ is determined by minimization of a quadratic functional. This approach leads to a stationary Kalman-Filter (Hippe, Wurmthaler, 1985). The number of trial simulations can be drastically reduced if the results of the last section are employed when choosing the (diagonal) weighting matrices.

In the first step, simple proportional feedback

$$\Delta\mathbf{u}(t) = -\mathbf{K}_x\,\Delta\hat{\mathbf{x}}(t) \quad (8)$$

using the estimated system state is employed. The feedback matrix $\mathbf{K}_x$ is determined once more by minimizing a quadratic functional. The weighting matrices were chosen to be diagonal, with high values for the location of the profile and the product concentration in both column sections. Fig.6 shows the time histories of the real and the estimated product concentrations after a step change in feed concentration. The apparent steady-state offsets are unavoidable in this case, even for perfectly estimated state variables. The estimation errors are mainly caused by the unknown disturbance and also by modeling errors.

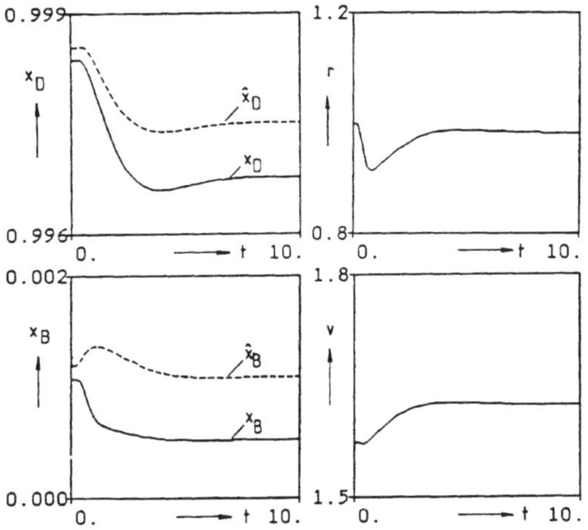

Fig. 6. Controlled column H after step disturbance in z_F: 0.45 → 0.5. Proportional feedback.

Proportional-integral Feedback

The control behavior can be improved by asymptotic compensation of the influence of the disturbance on the controlled variables. However, in contrast with conventional distillation column control systems, the task is to compensate the deviations in the non-measurable product compositions instead of those in the temperature measurements. Although this problem cannot be solved exactly, it is possible to eliminate the deviations between the set-point and the estimated product concentration. In the case of proper state estimation, this deviation decreases to very small values. In order to solve the asymptotic compensation problem, two different approaches are investigated here (Smith, Davison, 1972).

A first, simple approach relies on an integral feedback of part of the estimated system state (PI-control according to Davison) according to

$$\Delta\mathbf{u}(t) = -\mathbf{K}_x\,\Delta\hat{\mathbf{x}}(t) - \mathbf{K}_I\mathbf{R}\int_0^t \Delta\hat{\mathbf{x}}(\tau)d\tau - \mathbf{K}_I\mathbf{R}\mathbf{C}. \quad (9)$$

The matrix $\mathbf{R}$, containing only the elements 0 and 1, determines the number and type of states which are fed back integrally. The integration constants $\mathbf{C}$ are usually taken as zero. Determination of the feedback matrices in eq.(9) is accomplished in the same way as in the case of simple P-control by properly extending the system model. The set-point deviations vanish in those variables, which are chosen via matrix $\mathbf{R}$, if system (2)-(4) is controllable and if the matrix

$$\begin{pmatrix} \mathbf{A} & \mathbf{B} \\ \mathbf{R} & \mathbf{0} \end{pmatrix}$$

is of full rank. These conditions are fulfilled in the case of integral control of the product compositions. Fig.7 shows the simulation results after the feed concentration disturbance of the above. The set-point deviation vanishes totally in the estimated concentrations. However, due to the estimation errors in the system states, rather large deviations remain in the real concentrations.

A second, more sophisticated method for asymptotic compensation of system disturbances is due to Johnson. He suggested the estimation of the disturbance variables with an extended observer. These estimated values can be used for feedforward control, which is designed to compensate the disturbances in a few selected variables. The system equations (2) are extended using a two-dimensional model for the disturbances $\Delta\mathbf{d}$:

$$\Delta\dot{\mathbf{d}} = \mathbf{0}, \qquad \Delta\mathbf{d}(0) = \mathbf{0}. \quad (10)$$

These variables are estimated together with the system states

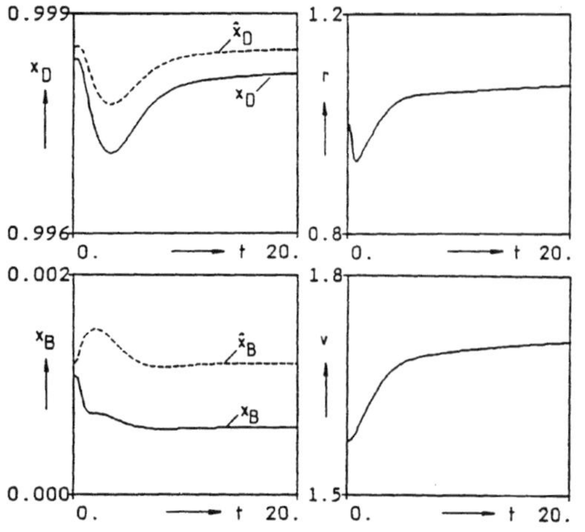

Fig. 7. Controlled column H after step disturbance in x_F: 0.45 → 0.5. Proportional-integral feedback of product compositions.

by an extended estimator. The state and disturbance may be observed if system (2)-(4) is observable and if the matrix

$$\begin{pmatrix} \mathbf{A} & \mathbf{X} \\ \mathbf{C} & \mathbf{Y} \end{pmatrix}$$

is of full rank. In this case, the extended control law is given as

$$\Delta \mathbf{u}(t) = -\mathbf{K}_x \hat{\mathbf{x}}(t) - \mathbf{K}_d \hat{\mathbf{d}}(t). \tag{11}$$

$\mathbf{K}_x$ is determined by pole placement or by optimization using a quadratic functional. The feedforward gains $\mathbf{K}_d$ are obtained as

$$\mathbf{K}_d = (\mathbf{R}(\mathbf{A} - \mathbf{B}\mathbf{K}_x)^{-1}\mathbf{B})^{-1}\mathbf{R}(\mathbf{A} - \mathbf{B}\mathbf{K}_x)^{-1}\mathbf{X}, \tag{12}$$

if asymptotic compensation in the states $\mathbf{R}\Delta\hat{\mathbf{x}}$ must be accomplished. The matrix $\mathbf{K}_d$ exists only if the system (2)-(4) is controllable and if the matrix

$$\mathbf{R}(\mathbf{A} - \mathbf{B}\mathbf{K}_x)^{-1}\mathbf{B}$$

is quadratic and regular. This condition limits the number of control variables with vanishing set-point deviation to the number of manipulated variables. For possible asymptotic compensation in more control variables the interested reader is refered to Smith and Davison (1972). The conditions for existence of matrix $\mathbf{K}_d$ are fulfilled if the deviations in the product concentration are intended to be compensated asymptotically. A block diagram of the resulting control system is shown in Fig. 8.

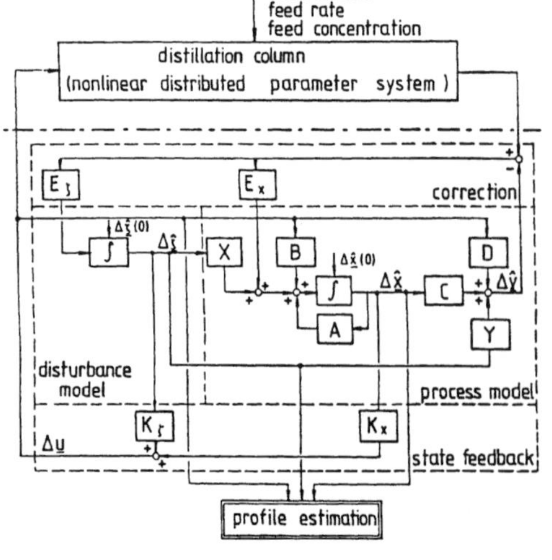

Fig. 8. Block diagram of the control system including disturbance estimation, feedforward control and estimation of concentration profile.

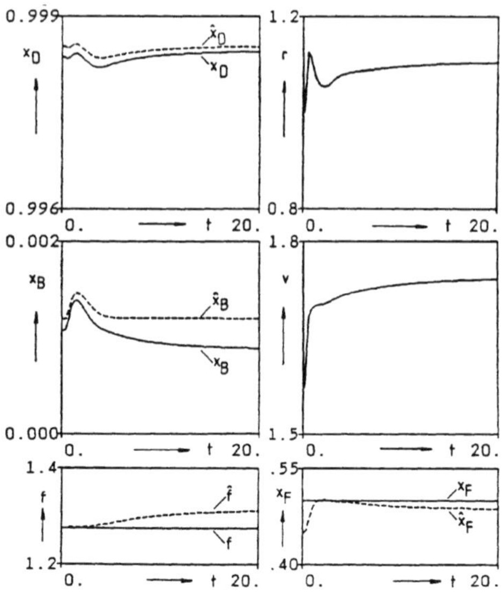

Fig. 9. Controlled column H after step disturbance in x_F: 0.45 → 0.5. Proportional-integral feedback. Disturbance estimation and feedforward control.

Fig. 9 depicts the time histories of the real and estimated product concentrations and disturbances as well as those of the manipulated variables. Despite an imperfect disturbance estimation the control system gives tight control of the product concentrations without any concentration measurements. Disturbance estimation can be improved by employing the nonlinear reduced model for state estimation. However, only a minimal effect on the control behavior can be observed in this case. The advantage of an observer, which is based on the nonlinear model, must be seen in the enhanced robustness properties of the control system. In the case of very large disturbances - i.e. 40 % step changes in the feed variables - instability might occur if the linear observer is not designed properly.

Concentration Profile Estimation

Besides the dual composition control of the distillation column, estimation of the whole concentration profile may be established with only negligible additional effort. A nonlinear equation, which determines the concentration profile as a function of the reduced model state variables, is obtained from the model reduction procedure outlined above. The concentration profile may be computed easily using the estimated states $\Delta\hat{\mathbf{x}}$ and $\Delta\hat{\mathbf{z}}$ (see also the block diagram in Fig. 8). Sufficient accuracy in the concentration profile requires the estimation of the disturbances, as discussed in the last section.

The concentration profiles for a step disturbance in the case of an uncontrolled plant are depicted in Fig. 10. It is clearly seen that the achievable quality of the estimator is limited by the use of the linearized reduced model. The estimated profile does not move as far as the real one does. It stops propagating if the limitations of the linear model are hit. Despite the significant estimation errors, the results are very useful for the supervision of the process in the case of gross system faults, i.e. the shut down of the control system. For that purpose, the estimated profiles could be monitored on a graphic display of common distributed control systems. The estimation quality can be improved significantly if the nonlinear reduced model is used for state estimation.

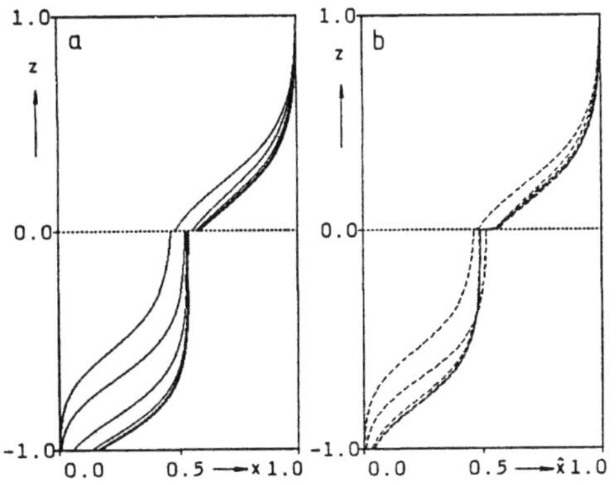

Fig. 10. Concentration profiles of uncontrolled column H after step disturbance of x_F: 0.45 → 0.5; a: plant; b: linear estimator.

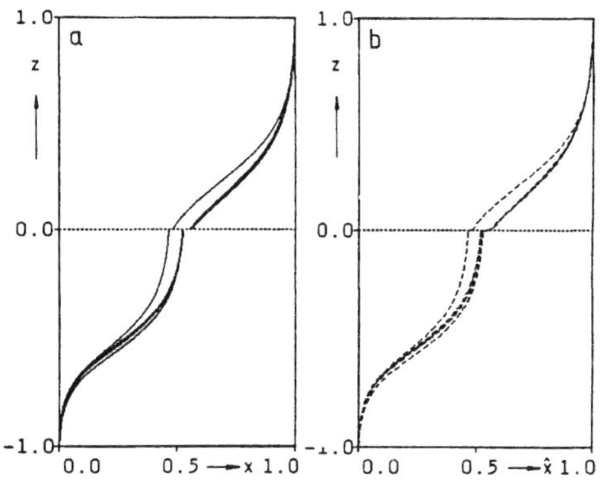

Fig. 11. Concentration profiles of PI-controlled column H after step disturbance of x_F: 0.45 → 0.5; a: plant; b: linear estimator.

Fig. 11 shows the real and the estimated profiles of the controlled column for the step responses given in Fig. 9. The quality of estimation is satisfactory in this case. The concentration profiles of the controlled plant reveal the advantage of a loose proportional control of the profile shape. Hence, steady-state errors are tolerated intentionally. Usual control systems - especially for columns yielding pure products - are designed to force the steady-state errors in a measured temperature in the sensitive region of the column profile to zero (for example Luyben, 1969; Fuentes, Luyben 1983; Longwell, 1982; Retzbach, 1986). Simulations and theoretical system analysis (Marquardt, 1988) reveal that the shape of the concentration profile is altered significantly if the internal flow rates are changed, for example, as a consequence of control action. If the profile is fixed at one spatial point in the column, the product concentrations must change. This, however, is what control should prevent. This effect becomes more pronounced with decreasing product qualities and with increasing magnitude of the disturbances. In the case of integral feedback of more than one (temperature) measurement in a column section, unstable control behavior may result. This has been shown in simulation studies (Rüchardt, 1986) and in experiments on a laboratory plant (Lang, Lehmann, Rüchardt, 1986). The instability is caused by PI-control, which tries to force the system into a state that is physically unfeasible.

Conclusions

A mathematical model, which is well adapted to the structural and physical properties of the separation process, proves an excellent starting point for model based control system design. Satisfactory control performance can be achieved by use of few and cheap measurements. At the same time, there is moderate effort for implementation of the linear controller and the profile estimator. In the future, the rigorous process model may be used instead of a nonlinear reduced model if sufficient computing power is available on-line on the process and if design methods for model based measurement and control systems are further developed for large-scale nonlinear processes. The concept of loose proportional control of the spatial profiles with the help of the whole system state solves some of the fundamental problems still encountered in the control of separation processes.

Even in the case of control systems, which are based on fairly accurate process models, concentration measurements are unavoidable if high demands are made on control performance. However, on-line simulation of a process model of adequate accuracy allows a drastic reduction of the number of concentration measurements for an update of the model. Hence, expensive analyzers, for example on-line gas chromatographs, can be employed for sequential concentration measurements of bottoms and distillate composition of one or more distillation columns at the same time. If a continuous-discrete Kalman-Filter is used for state estimation the rather large time-delay between every new measurement is of minor influence on the estimation accuracy.

References

Bonvin, D., Mellichamp, D.A. (1982). A unified derivation and critical review of modal approaches to model reduction. *Int. J. Cont.* 35, 829-848.

Finlayson, B.A. (1972). *The method of weighted residuals and variational principles.* Academic Press, New York.

Föllinger, O. (1982). Reduktion der Systemordnung. *Regelungstechnik rt* 30, 367-377.

Hippe, P.; Wurmthaler, Ch. (1985). *Zustandsregelung.* Springer-Verlag, Berlin.

Johnston, R.D.; Barton, G.W. (1985). Structural interaction analysis. *Int. J. Control* 41, 1005-1013.

Juen, G. (1982). Anmerkungen zu den Strukturmaßen für die Steuer-, Stör- und Beobachtbarkeit linearer zeitinvarianter Systeme. *Regelungstechnik rt* 30, 64-66.

Lang, L.; Lehmann, S.; Rüchardt, Ch. (1986). Personal communication.

Lehmann, S. (1987). Distillation control by output feedback designed via order reduction. *10th IFAC World Congress on Automatic Control,* Munich, FRG, July 27-31, 1987. Preprints, Vol.2, 297-302.

Levien, K.L.; Morari, M. (1987). Internal model control of coupled distillation columns. *AIChE J.* 33, 83-98.

Litz, L. (1979). Ordnungsreduktion linearer Zustandsraummodelle durch Beibehaltung der dominanten Eigenbewegungen. *Regelungstechnik rt* 27, 80-86.

Litz, L. (1983). Modale Maße für Steuerbarkeit, Beobachtbarkeit, Regelbarkeit und Dominanz - Zusammenhänge, Schwachstellen, neue Wege. *Regelungstechnik rt* 31, 148-158.

Luyben, W.L. (1969). Feedback control of distillation columns by double differential temperature control. *Ind. Eng. Chem. Fund.* 8, 739-744.

Marquardt, W. (1985). Model reduction techniques for separation columns. *Int. Conf. "On Industrial Modelling and Control",* June 6-9, 1985, Hangzhou, China. Proceedings, Vol. 1, 134-141. Zhejiang University Press.

Marquardt, W. (1986). Nonlinear model reduction for binary distillation. *IFAC Symposium DYCORD '86 - 'Dynamics and Control of Chemical Reactors and Distillation Columns.* December

1986, Bournemouth, UK.

Marquardt, W.; Holl, P.; Butz, D.; Gilles, E.D. (1987). DIVA - A flow-sheet oriented dynamic process simulator. *Chem. Eng. Technol.* 10, 164-173.

Marquardt, W. (1988). *Nichtlineare Wellenausbreitung - ein Weg zu reduzierten dynamischen Modellen in Stofftrennprozessen.* Ph.D. thesis, Universität Stuttgart.

Martin-Sanchez, J.M.; Shah, S.L. (1984). Multivariable adaptive predictive control of a binary distillation column. *Automatica* 20, 607-620.

McDonald, K.A., McAvoy, T.J. (1987). Application of dynamic matrix control to moderate and high-purity distillation towers. *Ind. Eng. Chem. Res.* 26, 1011-1018.

Morari, M. (1987a). Robust process control. *Chem. Eng. Res. Des.* 65, 462-479.

Morari, M. (1987b). Personal communication.

Nothaft, A. (1984). *Modale Analyse und Ordnungsreduktion komplexer Modelle von Destillationskolonnen.* M.Sc. thesis, Institut für Systemdynamik und Regelungstechnik, Universität Stuttgart, unpublished.

Ogunnaike, B.A.; Lemaire, J.P.; Morari, M.; Ray, W.H. (1983). Advanced multivariable control of a pilot-plant distillation column. *AIChE J.* 28, 632-640.

Prett, D.M.; Garcia, C. (1987). Design of robust process controllers. *10th IFAC World Congress on Automatic Control,* Munich, FRG, July 27-31, 1987. Preprints, Vol.2, 291-296.

Retzbach, B. (1986). *Mathematische Modelle von Destillationskolonnen zur Synthese von Regelkonzepten.* Fortschr.-Ber. VDI Reihe 8 Nr.126, VDI-Verlag, Düsseldorf.

Rüchardt, Ch. (1986). Entwurf von Mehrgrößenregelungen mittels optimaler Ausgangsrückführung am Beispiel einer Destillationskolonne. *3. DFG-Kolloquium "Messen, Steuern, Regeln von Systemen mit komplexer Struktur".* TU Munich, March 1986.

Schmid, Chr. (1985). KEDDC - a computer-aided analysis and design package for control systems. In: Jamshidi, M.; Herget, C.J.: *Computer-aided control systems engineering.* North-Holland, Amsterdam. 159-180.

Smith, H.W., Davison, E.J. (1972). Design of industrial regulators. Integral feedback and feedforward control. *Proc. IEE* 119, 1210-1216.

Tyreus, B.D. (1979). Multivariable control system design for an industrial distillation column. *Ind. Eng. Chem. Process Des. Dev.* 18, 177-182.

AN APPROACH TO MULTIVARIABLE PROCESS IDENTIFICATION

S. B. Jørgensen

Institute for kemiteknik, Technical University of Denmark, DK2800, Lyngby, Denmark

ABSTRACT: Chemical prosess models applicable for multivariable control are not generally available due to the often unique, nonlinear and instationary dyanmic process behaviour. A process knowledge based qualitative modelling method is proposed, which leads to an input-output model structure, wherein the parameters may be estimated using either on- or off-line data. These data must be obtained using suitably designed experiments, in order to provide reliable estimates of the process coupling parameters.

The applicability of the model structuring method is demonstrated in an identification example of a simulated two stage evaporator. The selection of a reasonable model structure amongst a set of candidate models, is demonstrated using experimental data from a fixed bed reactor with recycle of unconverted reactant. Subsequently the obtained model structure was used succesfully for adaptive control of the reactor.

Keywords: Process modelling, Fixed bed reactor, knowledge based, state space, discrete time, Multivariable control, Adaptive Control.

INTRODUCTION

One of the crucial limitations in application of multivariable model based control in the chemical process industry is the validity of the applied model in describing the dynamics of the overall process and of the internal couplings among various inputs and outputs. This limitation is initimately coupled to the general lack of reliable dynamic models for chemical processes and to the instationary behavior of many processes.

In order to identify a structured process model it is convenient to define a model as consisting of a model structure with sets of parameter values covering the possible operating regions. For application of multivariable model based control it is essential that the model structure may describe the possible process dynamics in the desired operating regions and over the frequency range of interest. For constant parameter control design it is essential to apply knowledge about the structural model uncertainties in the control design in order to obtain reasonably robust yet alert control designs. If the model uncertainty is unstructured then quite conservative control designs may result. In the case of adaptive control it is most important to minimize unmodelled dynamics within, and to eliminate the effect of unmodelled dynamics outside, the frequency range of interest. Thus there is a need for identification methods which produce results with a limited manpower expenditure.

In this paper a method for development of sparse linear input-output model structures is proposed. The model development method is based upon usage of qualitative and quantitative process knowledge, to structure (continuous or) discrete time models. The method may be directly applied to low order lumped processes. The method may also be applied to high order lumped and distributed processes by using knowledge of the characteristic process dynamics to reduce the model order. One of the advantages of using process knowledge in the model order reduction is that the resulting model structure covers the internal couplings within an operating region. Once the model structure is determined model parameters may be estimated from experimental (or simulated) data using appropriate on- or offline estimation methods. These methods provide information about the parameter uncertainties which combined with the knowledge about model structure uncertainties may be effectively utilized in the contol design.

The model structure determination approach taken is thus based upon process knowledge, in this respect this paper differs from the works of Gevers and Wertz (1985) and Willems (1986a-c). The starting point of their modelling is the experimental data. Both types of approaches have their advantages, but it seems most relevant a priori to use available process knowledge to fix the modelstructure, whenever possible with a reasonably limited effort. The work presented in this paper represents a further development of the method presented in Jørgensen et al. (1985). The structure of the paper is as follows: The basis for the identification method and the method is presented in section 2. The following section contains an brief presentation of the example processes with a development of candidate model structures. The identification results are subsequently presented and discussed. Finally the conclusions are drawn.

2. A MULTIVARIABLE IDENTIFICATION METHOD

To perform multivariable control it is essential to use models which describe the fundamental multivariable process behavior. Such type of behaviour may be illustrated by the process gain and phase relationship which is simple and unique in the univariate case but becomes more complex in the multivariable generally non-square case, where multiple gains are possible for each output and e.g. the maximum and the minimum gains can be utilized to bracket the possible process behaviour. In addition the directionality becomes essential, i.e. which combination of inputs yield the maximum and minimum gains respectively. To obtain multivariable process behaviour by combining single step responses or transfer functions renders it difficult to account for process interactions and especially for the directional aspects, Andersen et al. 1988.

One type of traditional process identification method is based upon development of conservation law based, possibly partial, differential equations. A linearised, possibly reduced order model may be used for control design purposes. This method provides a modelstructure for the particular operating point. whether a reduced order modelstructure is feasible over the desired operating region requires special investigataion. If the parameters in the conservtion law based model are determined by independent experiments: then there is the possibility that particular combinations of inputs may exite the process behavior such that the original parameters only provide inaccurate descriptions of the internal couplings of the process. The conservation law based identification procedure may, ofcourse, be utilized on data which contain the directionally relevant information. This however means that several inputs must be perturbed simultaneously, thus the (often nonlinear) parameter estimation problem in the original model becomes cumbersome. An additional disadvantage is that the model development is very demanding in terms of man power.

The main conclusion from these observations is that when individual submodels for various phenomena are combined into larger process models then parameter values describing these phenomena and especially their interactions must be reinvestigated. Thus determination of parameter values in multivariable models requires a multivariable experimental method.

In accordance with these observations it is required that a modelbased identification method provides:

I: A modelstructure, which can describe the process dynamics within the desired operating region(-s).

II: An experimental design, to be used for model parameter estimation, which provides directionally relevant information.

III: Procedures to discriminate among candidate models and to evaluate the model validity.

The proposed multivariable identification method consists of the following main steps:

I.1: Formulate a qualitative diagram depicting the characteristic dynamics of the process quantities (or their variables) and their interactions.
I.2: Select desirable outputs and inputs.
I.3: Formulate a structural state space representation.
I.4: Convert the representation to discrete time state space.
I.5: Transform the representation to input-output form with measurable output vector. Include unknown disturbance description.

II.1: Design experimental procedure.

III.1: Estimate model parameters either off- or on-line.
III.2: Evaluate model validity. If the model validity is unsatisfatory then return to I.

These steps will be briefly outlined:

A <u>DYnamic Quantity Interaction Diagram</u> (DYQUID) is basically just a formalization of process knowledge normally applied in mathematical modelling. Therefore it is quite straight forward to construct DYQUID's for model builders. To construct a DYQUID requires identification of the streams and phases of a process. For each of these one must identify the conserved process quantities which are relevant for the process in question (see Table 1). The hold-ups for the conserved quantities are often directly identifiable and constitute the first items to be drawn using the symbols in Table 2. Note that lumped (i.e. well mixed) hold-ups are indicated using a circle, whereas a distributed quantity hold-up is shown with an oblong symbol, where the longest direction indicates the prevailing direction of the quantity distribution. The radius of the lumped and the smallest dimension of the distributed quantities can advantageously be utilized to indicate the relative size of the inverse residence times for the hold-ups. This imfomation is most useful in the model reduction phase.

A quantity may be transported in a phase, transfered between phases, converted into another quantity chemically or physically. The rates of these elementary processes may be influenced by (other) quantities or their intensive properties, or by external conditions. These terms are defined in table 1. The above effects make up the process interactions which are entered on the diagram using the symbols shown in table 2. Clearly the completing of a first attempt at a DYQUID requires qualitative knowledge about the various mechanisms in the process. This knowledge is often available either in the design group or may be inferred from discussions with operating and laboratory personel.

The complexity of the first attempt DYQUID may often be drastically reduced by using a number of simplifying assumptions, e.g. conserning:

a) Quasistationary quantities. If there is negligible capasity for a quantity it may be considered quasistationary. This assumption must be used with great care, especially if the quantity is involved in very fast mechanisms.

b) Lumping quantity capasities. It is often possible to lump several physically different capacities for the same quantity, e.g. energy.

c) Simplifying mechanisms. It is often possible to use a single rate determining step instead of of a series mechanism or to lump several steps into an effective mechanism.

Using a set of simplifying assumptions a DYQUID results, which is based upon quantities. In practise it may be difficult to measure quantities, such as component holdup. Instead intensive variables which represent the quantities, e.g. concentration may be used. With this change of dependent variable for some of the capacities a modification of some of the DYQUID interactions usually will be required. The DYQUID now contains variables representing the quantities which may be measured either directly or indirectly. Note that all variables need not necessarily be measured. This will be dealt with below.

For processes with only lumped capacities a first attempt DYQUID is complete. The resulting linear continuous time state space model structure may be written in the following form:

$$\frac{dx(t)}{dt} = Ax(t)+A_d x(t-T_x)+Bu(t)+B_d u(t-T_u) \quad (1a)$$

with the measurement equation:

$$y(t) = Cx(t) + C_d x(t-T_y) \quad (1b)$$

The parameter structure matrices: A, A_d, B, B_d, C, C_d are in general time varying for simplicity this time dependence is not shown. The parameter structure matrices contain simply unknown paramteres where the DYQUID indicates that there is a direct coupling between a variable and a holdup in the A-matrices and between an input and a quantity hold-up for the B-matrices.

The principal model form in eq.(1a-b) includes the possibillity of modelling three types of delays: input delay of T_u, measurement delay of T_y and of a state delay of T_x. In the above model the delays are pure, they may include more complicated dynamics. The presence of state delay is a manifestation of a quantity capacity distribution. Such pure state delays may be present, e.g. in processes with recycle. More generally distributed quantity capacities gives rise to partial differential equation models, or ordinary differential equation models of infinite order. The dynamics of these quantities can often be approximated using relatively low order models, if the approximation is based on qualitative process knowledge.

For processes with distributed capacities the states are spatially distributed. Lumped staged processes of high order also exhibit distributed characteristics and may be handled similaly to the distributed processes, as far as procedures for measurement and actuator selection goes (Jørgensen et al. (1984). Four criteria are of importance for selection of measurement positions in these processes: sensitivity, ability to describe internal coupling among variables, detection of effects from input loads, and ability to measure controlled quality (if possible). In the selection of mesurement positions it is important to attempt to describe the propagation of disturbances within the distributed capacities by selection of a low number of suitably located states. How to proceed depends upon the particular phenomena at play within the distributed capacities. For capacities where say axial convective transport dominates a space time relationship may be used to select axial positions which may be used to describe the propagation of disturbances whith the appropriate axial velocity. This type of description thus leads to a model with state delays, where the states are coupled in the direction of flow. In this case the model structure may be formulated directly in discrete time. Knowledge of the range of the residence time is necessary in order to be able to approximate this type of infinte order model with a finite order discrete time model.

For processes where a dispersive phenomenon dominates, a number of sensitive locations may be selected for the states, but with bidirectional couplings in the case of a single dominating spatial direction for the capacity distribution.

To convert the above continuous time model structure to discrete time is straight forward. By solving eq. (1) over a sampling time interval (for the state delay matrix equal to zero), the following discrete time model is obtained:

$$x(t+1) = Fx(t) + Gu(t) + G_d u(t-T_u) \quad (2a)$$

$$y(t) = Cx(t) + C_d x(t-T_y) \quad (2b)$$

where $F=\exp(A*1)$, $G=\int_0^1 \exp(Ap)B\,dp$, $G_d=\int_0^1 \exp(Ap)B_d\,dp$ and time is normalized with the sampling interval.

The structure of the above matrices may be directly determined based upon the structure of the continuous time matrices since the continous time variables interact during a sampling interval and thereby introduces interactions into the discrete time structure. If three variables are selected in the continuous time structure, such that (a) only affects (b) which only affects (c), then (a) will affect (c) in discrete time. This parameter may be indicated by a numeral 2 in the discrete time matrix, to indicate that the interaction strength will be proportional to the sampling interval to the second power. Note that this discrete time interaction may generate a further interaction marked by numeral 3 etc. With this nomenclature direct interactions are marked with numeral 1. Pure integral states must be indicated differently, eg. with an encircled numeral 1.

The discrete time model structure is very informative in that structural observability or more correctly detectability and controllability or more correctly reachability may be judged directly from the relationships between input and output. Similar qualitative information may be obtained by inspection of the DYQUID.

The above discrete time model structure may be applied directly for identification purposes if all states are measured. If all states are not measured then it is necessary either to estimate the model parameters and the states simultaneously using a Kalman filter type approach or to modify the model representation. Since the simultaneous estimation of model parameters and states is quite demanding the second approach is used here. The model representation is changed into an equivalent input-output representation. This transformation is straight forward as the structure of the measurement matrix C is quite simple when some states are directly measured whereas the others are not: Assume that the first m states are measured directly, ie. $C = I_m$ and for simplicity that $C_d = B_d = 0$. Then the discrete time model may be partitioned in a measured set and an unmeasured set:

$$\begin{pmatrix} x_y \\ x_u \end{pmatrix}(t+1) = \begin{pmatrix} F_{yy} & F_{yu} \\ F_{uy} & F_{uu} \end{pmatrix} \begin{pmatrix} x_y \\ x_u \end{pmatrix}(t) + \begin{pmatrix} G_y \\ G_u \end{pmatrix} u(t) \quad (3)$$

Where $y(t) = I_m x_y(t)$, and x_u indicates the $r=n-m$ unmeasured states. The bottom set of equations may be solved for $x_u(t+1)$, as:

$$(I-F_{uu}q^{-1})^{-1} = \frac{\mathrm{Adj}(I-F_{uu}q^{-1})}{\det(I-F_{uu}q^{-1})} = \frac{I + \sum_{i=1}^{r-1} R_i q^{-i}}{1 + \sum_{1}^{r} p_i q^{-i}}$$

The resulting input-output model structure becomes eq.(4), where a noise matrix polynomial is added to account for noise sources:

$$A(q^{-1})\, y(t) = B(q^{-1})\, u(t) + C(q^{-1})\, e(t) \quad (4a)$$

Where the matrix polynomials are:

$$A(q^{-1}) = I + A_1 q^{-1} + A_2 q^{-2} + \ldots + A_{r+1} q^{-(r+1)}$$
$$B(q^{-1}) = B_1 q^{-1} + B_2 q^{-2} + \ldots + B_{r+1} q^{-(r+1)}$$
$$C(q^{-1}) = I + C_1 q^{-1} + C_2 q^{-2} + \ldots$$

and

$$\begin{aligned} A_1 &= p_1 I - F_{yy} \\ A_2 &= p_2 I - p_1 F_{yy} - F_{yu} F_{uy} \\ A_3 &= p_3 I - p_2 F_{yy} - F_{yu} R_1 F_{uy} \\ &\;\vdots \\ A_r &= p_r I - p_{r-1} F_{yy} - F_{yu} R_{r-2} F_{uy} \\ A_{r+1} &= - p_r F_{yy} - F_{yu} R_{r-1} F_{uy} \end{aligned} \quad (4b)$$

$$\begin{aligned} B_1 &= G_y \\ B_2 &= p_1 G_y + F_{yu} G_u \\ B_3 &= p_2 G_y + F_{yu} R_1 G_u \\ &\;\vdots \\ B_{r+1} &= p_r G_y + F_{yu} R_{r-1} G_u \end{aligned} \quad (4c)$$

The above model structure, eq (4) can be directly expanded to include the cases of input and measurement delay. The former case follows from eq.(4), by inclusion additional matrices in the B-polynomial. The latter case may be illustrated by measurement equation (1b), where m states are direcly measured without delay, whereas r measurements are delayed:

$$y(t) = \begin{pmatrix} I_m \\ 0 \end{pmatrix} x(t) + \begin{pmatrix} 0 \\ I_r \end{pmatrix} x(t-T_d) \quad (5a)$$

The delayed states may be modelled by augmenting the number states in the state model eq. (2a) with pure delay states to cover the measurement delay:

$$\begin{pmatrix} x_m(t+1) \\ x_d(t+1) \\ x_d(t) \\ x_d(t-1) \\ \cdot \\ \cdot \\ x_d(t-T_d+1) \end{pmatrix} = \begin{pmatrix} F_{mm} & F_{md} & 0 & . & . & 0 & 0 \\ F_{dm} & F_{dd} & 0 & . & . & 0 & 0 \\ 0 & I_r & 0 & . & . & 0 & 0 \\ . & & & & & & . \\ . & & & & & & . \\ . & & & & & & . \\ 0 & 0 & 0 & . & . & I_r & 0 \end{pmatrix} \begin{pmatrix} x_m(t) \\ x_d(t) \\ x_d(t-1) \\ x_d(t-2) \\ \cdot \\ \cdot \\ x_d(t-T_d) \end{pmatrix} + \begin{pmatrix} G_m \\ G_d \\ 0 \\ 0 \\ \cdot \\ \cdot \\ 0 \end{pmatrix} u(t)$$

In this model only the first x_m states at time t and the last x_d at time $t-T_d$ are measured. The input-output structure for eq.(5a-b) may now be determined directly from eq. (4a-c). The handling of state delays follows a similar line and will be covered by the fixed-bed reactor example.

The result of the model development will often be that a set of candidate models developed based on slightly different assumptions. The next question is how to design an experiment to discriminate between the models and how to validate candidate models.

Experimental design.

In order to fullfill the second basic requirement it is appropriate to design an identification experiment where all inputs are perturbed simultaneously, using statistically independent pertubation sequenses, such as pseudo random binary sequences. The amplitudes of the different inputs should be specified such that the different inputs perturb the outputs reasonably well. The experiment may be carried out either in open or in closed loop. In the latter case it is most important to have one of the perurbations in the closed loop, or to apply a variable gain feedback (Soderstrom et al. 1976, Box and MacGregor 1976). The outcome of the experiment is a set of measurements which may be used for model evaluation.

The model parameters may be estimated either on-line or off-line. For a first attempt it may be desirable to use the latter approach since it most easily provides the flexibility of modifying the model structure according to information present in the data. On-line estimation requires the use of a recursive estimation method. In the present work a recursive extended least squares RELS method is used. This method is of the pseudo linear regression type (Ljung 1987). The model parameters are estimated for one output at the time, this is sufficient with the above mentioned mowing average models. If more complex noise models are used then it may be desirable to use a maximum likelihood method in order to estimate noise parameters reliably. For off-line estimation a batch method may be applied where all the data points are treated simultaneously. However here recursive method may also be used, it can be shown to converge to the same minimum as the batch method by performing mutiple sequential runs through the measurement data, Ljung 1987. The recursive estimation methods minimize the one step prediction error of the model:

$$r(t) = y(t) - \hat{y}(t/t-1)$$

Model evaluation and discrimination.

To discriminate between rival model structures it is necessary to define a measure of the fit or the misfit of a model with estimated parameters to the data. For this purpose we have used the accumulated squared prediction error (Rissanen and Wertz (1985)):

$$V = \frac{1}{N} \sum_{t=1}^{N} r(t)^T r(t) \quad (6)$$

This measure has to be counterbalanced by a measure for model complexity since a large complexity or a large number of parameters may provide a very low value for V, but also may yield a very slow convergence. A simple, however not quite satisfactory measure (Willems 1987) is the number of model parameters NP. A better measure at least for adaptive or on-line estimation purposes is the convergence rate of the misfit when applied in recursive estimation. The convergence rate may be calculated as the change of the accumulated squared prediction error after successive runs through the measurement set.

Evaluation of the ability of the selected model to describe the plant dynamics, may be judged by a number of different tools. A simple somewhat crude tool is to calculate a degree of variance explanation by the model for each output:

$$rho_i = 1 - \frac{variance(r_i)}{variance(y_i)} \quad (7)$$

The calculation of covariance functions between one output and other outputs and the inputs can reveal model deficiences and give clues to which of the underlying assumptions may not be satisfied. Evaluation of the eigenvalues of the covariance matrix may also provide valuable information about variables being constant amd about variables being quasistationary (eigenvalue being zero). Finally evaluation of parameter correlations is a valuable tool for detection of whether or not parameters provide independent information.

3. RESULTS AND DISCUSSION

The methodology presented above is illustrated on two examples. The first a two stage evaporator is a process which may be described as lumped and the second a fixed bed reactor which is a process with a distributed thermal capacity. The first example will be based upon a simulation model whereas the second is based upon experimental data. The fixed bed reactor has also been extensively studied in dynamics and control related aspects (Jørgensen 1986). The first example is used to describe the development of models with a reduced measurement set. The second example illustrates model discrimination among different candidate model structures.

The two stage evaporator innvestigated here was modelled and studied by Bruun and Kummel (1979). The DYQUID in figure 1 is based upon the following assumptions:

1) Steam is saturated.
2) The stage wall and liquid heat capacities are lumped.
3) The second stage pressure is controlled.

Thus five significant capacities result. The selected dependent variables for each capacity and the available inputs are shown on fig.1. The resulting discrete time state space model structure is shown in Table 3a. The degree of variance explanation shown in table 3a is quite satisfactory with a sampling time of 12 s. The parameter estimates reveal that the dynamics of the first stage temperature is very fast (time constant of around 1 min compared to 20-25 min of the concentrations). Since this state is not affected by other states it is attempted to assume it to be quasi stationary. Secondly c_1 is not measured thus only the two levels and the product concentration is measured. With these assumptions the resulting input-output model structure, determined by eq. (4), is shown in table 3b. The resulting degree of explanation is shown in table 4. The main effect of the quasistationarity assumption is seen to be an decrease of the variance explanation for the exit concentration of 2.6%. Increasing the sampling time to 1 min, which with the real eigenvalues of this process is around the maksimal value without getting aliasing effects, decreases this variance explanation with another 4.8%.

The second example is a gas phase fixed bed reactor where Hydrogen in large excess is irreversibly oxidized in a tubular bed of Pt-impregnated alumina pellets. This reactor has been used extensivly in the initial development of the identification methodology presented in this paper (Hallager et al 1984). In the present context the unconverted oxygen are recycled, thus giving a process which is quite sensitive to changes in input conditions. The conserved quantities are total mass, Oxygen and energy. The former two have a thousand times shorter residence times than energy and are therefore assumed quasistationary. The dynamic effects of the product water is assumed to be negligible. The process DYQUID is shown in figure 2. The behaviour of the energy balance is dominated by convective effects. The input variables used are inlet temperature and inlet flow of Oxygen. The thermal residence time is 16 min, with the used nominal total mass flow rate. With a choise of five equidistant temperature sensors, a disturbance in inlet temperature will travel between two sensors in about 3.2 min. Thus with a samplin time of two min the convective dynamics can be described using two A-matrices in an AR-model, and with three A-matrices variations down to 50% of the nominal flow rate can be tolerated by the model structure.

The model structure set contains the four structures shown in table 5. Model structure I describes the convective movement of thermal disturbances and the influence of thermal disturbances upon the inlet concentration measurement. The effect of inlet temperature is obvious, and of inlet oxygen flow is that it rapidly will affect all the measurements exept the temperature measurement at 0.2. The measured inlet concentration is assumed to affect all measurements. Model structure 2 Includes this effect, by B(1,2) and allows for dynamics in the first temperature measurement, by including $B_i(1,1)$ with i=2 and 3. This and structures 3 and 4 all include the noise to be first order coloured, by inclusion of a diagonal C_1 matrix. Structure 3 includes two more effects. The first is the coupling between temperature measurements through the oxygen balance which is assumed to have relative short range, by including only the second subdiagonal in A_1. The second effect is dynamic effect of water which adand desorps from the catalyst pellets. This effect is assumed to be approximated by including higher order dynamics of the effect of inlet oxygen concentration upon all the temperature measurements. In structure 4 some of the parameters are removed from structure 3.

The performance of these model structures in two consequtive runs through data obtained from a 120 min long pertubation experiment with $T_i = 54^oC$, feed flow rate of Oxygen was 1000 SCCM and the total mass flow rate 2.6 mg/(cm^2 s) is shown in table 6. Model strusture 3 shows the lowest value of the accumulated squared prediction error, but also shows the slowest rate of convergence. Therefore the model structure to be preferred is no. 4. The variance explanation of the preferred model was better than 95% for all measurements. thus quite satisfactory. A slight modfication of this model structure was later applied in closed loop adaptive control of the reactor with very satisfactory results (Bortolotto and Jørgensen 1985).

5.CONCLUSIONS

A process knowledge based approach to process identification has been presented, which makes it possible to reach relatively rapidly to modelstructures which may be utilized for model based control design, such as robust design, model predictive control or for adaptive control. The approach has been successfully utilized in modelling both a fixed bed reactor and a destillation column.

REFERENCES

Bortolotto, G. and Jørgensen,S.B.(1985): Adaptive Control of a Recycle Reactor. IFAC Workshop on Adaptive Control of Chemical Processes, Frankfurt october.

Box, G.E.P., MacGregor, J.F.(1976): Parameter estimation with Closed-loop Operating Data.Technometrics 18,p.371-384.

Bruun, N.G. and Kummel, M. (1977): Multiloop, feedforward, modal and optimal control of an evaporator. Automatica 15, p.269

Gevers, M. and Werts, V. (1985):"Techniques for the selection of identifiable Parametrizations for Multivariable linear systems. Contributions to CONTROL AND DYNAMIC SYSTEMS, Volume XXIV.

Hallager,L. and Jørgensen, S.B.(1983): "Multivariable adaptive control of chemical engineering processes." IFAC Workshop on adaptive control and signal processing, San Francisco.

Hallager,L. Goldschmidt,L, and Jørgensen, S.B.(1984): "Multivariable adaptive identification and control of a Distributed Chemical Reactor. Large scale systems 6 p.323

Jørgensen, S.B., Goldschmidt,L., and Clement, K.(1984)."A Sensor Location Procedure for Chemical Processes". Comp. Chem Eng.8 p.195.

Jørgensen, S.B., Goldschmidt,L., and Hallager, L.(1985). "Sparse process modelling for robust multivariable adaptive chemical process control". IFAC symposium on Identification and System Parameter identification, York 1985.

Ljung,L.(1987): System Identification: Theory for the User. Prentice Hall, New Jersey

Söderström,T. Ljung,L. and Gustavsson,I. (1976):Identifiability conditions for Linear Multivariable Systems operating under feedback. IEEE Trans. Act. Aut Control,AC-21, p837-840

Willems,J.C.(1986a-b, and 1987)."From Time Series to Linear systems. Part I: Finite Dimensional Linear Time Invariant Systems", "Part II. Exact Modelling", and "Part III Approximate modelling" Automatica 22p561-580, p675-694, and 23,p87-115.

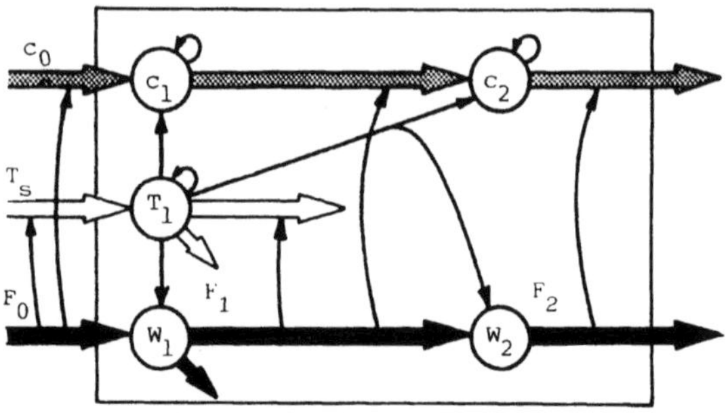

Figure 1: Dyquid for two stage evaporator. Dependent variables are shown in the quantity capacitance symbols. Inputs are shown next to the quantity fluxes. Symbols: c g/kg solution, T °C, W kg solution, and F kg/s.

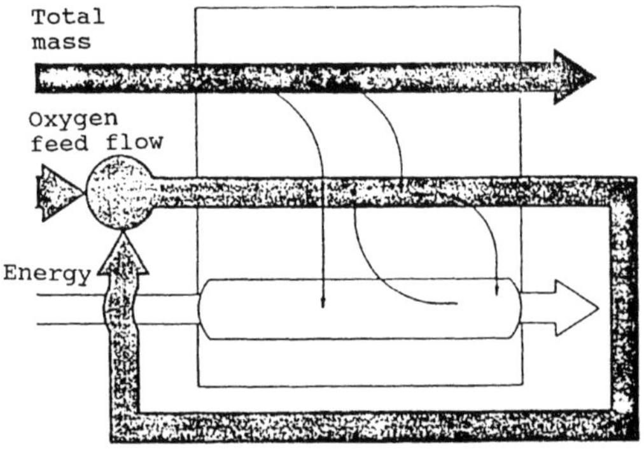

Figure 2: DYQUID for recycle fixed bed reactor. Oxygen is hydrogenated in a large excess of hydrogen. Unconverted Oxygen is recycled.

Table 1: DEFINITIONS FOR DYNAMIC QUANTITY INTERACATION DIAGRAMS.

Phase	: A collection of homogeneous parts of the process with "similar" intensive properties.
Quantity	: An extensive thermodynamic property. (e.g. mass (total or component), momemtum, energy, electric charge)
Process state:	The state of a process is described by the values of a necessary and sufficient number of quantities for each phase of the process.

A quantity may be transported in a phase, transferred between phases, converted into another quantity chemically or physically. The rates of these elementary processes may be influenced by (other) quantities or their intensive properties, or by external conditions. These terms are defined next:

Quantity flow	: Bulk transport of a quantity within a phase by e.g. convection, dispersion, diffusion.

Quantity transfer	: Transfer of a quantity over a phase boundary, e.g. momentum, heat or mass transfer.
Quantity conversion	: Conversion of one or more types of quantity into (an-)other type(-s) by chemical reaction or physically ,e.g. evaporation, power generation.
Rate influence	: Rates of quantity transport, transfer or conversion may be influenced by (other) quantities, their intensive properties, or by external conditions.

Table 2: DYQUID SYMBOLS

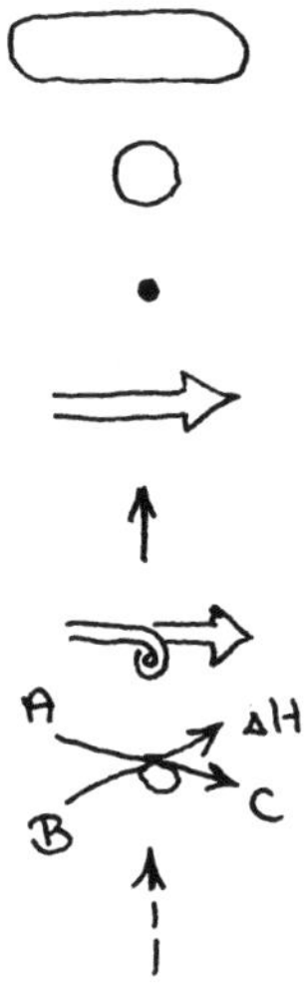

Distributed quantity.

Lumped quantity.

Lumped quasi-stationary quantity.

Quantity flow.

Quantity transfer.

Physical quantity conversion

Quantity reaction.

Rate influence.

Table 3: a) Discrete time input-output model structure for two stage evaporator with measurement of all five states.
b) Input-output model structure for three measurements.

a)

$$A_1 = \begin{matrix} 1 & 0 & 1 & 0 & 0 \\ 0 & 1 & 1 & 0 & 0 \\ 0 & 0 & X & 0 & 0 \\ 0 & 0 & 1 & 1 & 0 \\ 0 & 1 & 1 & 0 & 1 \end{matrix} \qquad B_1 = \begin{matrix} 1 & 0 & 2 & 0 \\ 0 & 0 & 2 & 1 \\ 0 & 0 & 1 & 0 \\ 1 & 1 & 2 & 0 \\ 1 & 0 & 2 & 2 \end{matrix}$$

$$y = (W_1, C_1, T_1, W_2, C_2)^T, \quad u = (F_1, F_2, T_i, C_i)^T$$

b)

$$A_1 = A_2 = \begin{matrix} X & 0 & 0 \\ 0 & X & 0 \\ 0 & 0 & X \end{matrix} \qquad B_1 = B_2 = \begin{matrix} X & 0 & X & 0 \\ X & X & X & 0 \\ X & 0 & X & X \end{matrix}$$

$$y = (W_1, W_2, C_2)^T, \quad u = (F_1, F_2, T_i, C_i)^T$$

Table 4: Fractional variance explanation for two stage evaporator. a) five measurement model structure and sampling time of 12 s. b1) Three measurements and 12 s. b2) Three measurements and sampling time of 60 s.

	a:	b1:	b2:
	rho_i	rho_i	rho_i
W_1	.998	.998	.997
C_1	.993		
T_1	.998		
W_2	.989	.986	.993
C_2	.998	.972	.926

Table 5: Discrete time input-output model structures for fixed bed reactor.

$$y = (T_{0.2}, T_{0.4}, T_{0.6}, T_{0.8}, T_{1.0}, C_{0.0})^T$$
$$u = (T_i, F_{O2})^T$$

MODEL 1

A1	A2	A3	B1
OOOOOX	OOOOOO	OOOOOO	XO
XOOOOX	XOOOOO	XOOOOO	OX
OXOOOX	OXOOOO	OXOOOO	OX
OOXOOX	OOXOOO	OOXOOO	OX
OOOXOX	OOOXOO	OOOXOO	OX
XXXXXX	OOOOOO	OOOOOO	OX

MODEL 2

A1	A2	A3	B1	B2	B3	C1
OOOOOX	OOOOOO	OOOOOO	XX	XO	XO	XOOOOO
XOOOOX	XOOOOO	XOOOOO	OX	OO	OO	OXOOOO
OXOOOX	OXOOOO	OXOOOO	OX	OO	OO	OOXOOO
OOXOOX	OOXOOO	OOXOOO	OX	OO	OO	OOOXOO
OOOXOX	OOOXOO	OOOXOO	OX	OO	OO	OOOOXO
XXXXXX	OOOOOO	OOOOOO	OX	OO	OO	OOOOOX

MODEL 3

A1	A2	A3	B1	B2	B3	C1
XOOOOX	OOOOOX	OOOOOX	XX	XO	XO	XOOOOO
XXOOOX	XOOOOX	XOOOOX	OX	OO	OO	OXOOOO
OXXOOX	OXOOOX	OXOOOX	OX	OO	OO	OOXOOO
OOXXOX	OOXOOX	OOXOOX	OX	OO	OO	OOOXOO
OOOXXX	OOOXOX	OOOXOX	OX	OO	OO	OOOOXO
XXXXXX	OOOOOO	OOOOOO	OX	OO	OO	OOOOOX

MODEL 4

A1	A2	A3	B1	B2	B3	C1
OOOOOO	OOOOOO	OOOOOO	XO	XO	XO	XOOOOO
XOOOOO	XOOOOO	XOOOOO	OO	OO	OO	OXOOOO
OXOOOX	OXOOOX	OXOOOO	OX	OO	OO	OOXOOO
OOXXOX	OOXOOX	OOXOOO	OX	OO	OO	OOOXOO
OOXXXX	OOOXOX	OOOXOO	OX	OO	OO	OOOOXO
OOXXXO	OOOOOO	OOOOOO	OX	OO	OO	OOOOOX

Table 6: Estimation results for fixed bed reactor with recycle. Subscript i of V indicates the i'th successive run through the data.

Model:	1	2	3	4
V_1	5.55	4.44	3.70	3.44
V_2	5.39	3.80	2.81	2.97
number of parameters	29	38	53	37
$\frac{V_1 - V_2}{V_1} 100$	3	14	24	14

AUTHOR INDEX

Printed and bound by CPI Group (UK) Ltd, Croydon, CR0 4YY
13/05/2025
01869555-0003